Glacial Processes Past and Present

Edited by

David M. Mickelson
Department of Geology and Geophysics
University of Wisconsin
1215 W. Dayton Street
Madison, Wisconsin 53706
United States

and

John W. Attig
Wisconsin Geological and Natural History Survey
3817 Mineral Point Road
Madison, Wisconsin 53705
United States

SPECIAL PAPER
337
1999

Copyright © 1999, The Geological Society of America, Inc. (GSA). All rights reserved. GSA grants permission to individual scientists to make unlimited photocopies of one or more items from this volume for noncommercial purposes advancing science or education, including classroom use. Permission is granted to individuals to make photocopies of any item in this volume for other noncommercial, nonprofit purposes provided that the appropriate fee ($0.25 per page) is paid directly to the Copyright Clearance Center, 222 Rosewood Drive, Danvers, MA 01923, USA, phone (978) 750-8400, http://www.copyright.com (include title and ISBN when paying). Written permission is required from GSA for all other forms of capture or reproduction of any item in the volume including, but not limited to, all types of electronic or digital scanning or other digital or manual transformation of articles or any portion thereof, such as abstracts, into computer-readable and/or transmittable form for personal or corporate use, either noncommercial or commercial, for-profit or otherwise. Send permission requests to GSA Copyrights.

Copyright is not claimed on any material prepared wholly by government employees within the scope of their employment.

Published by The Geological Society of America, Inc.
3300 Penrose Place, P.O. Box 9140, Boulder, Colorado 80301

Printed in U.S.A.

GSA Books Science Editor Abhijit Basu

Library of Congress Cataloging-in-Publication Data
Glacial processes past and present / edited by David M. Mickelson and John W. Attig.
    p. cm.
    Outgrowth of a symposium held at the North Central Geological Society of America and Midwest Glaciology meetings in Madison, Wis., Spring 1997.
    Includes bibliographical references.
    ISBN 0-8137-2337-X
      1. Glaciology--Congresses. I. Mickelson, David M. II. Attig, John W. III. Geological Society of America. Meeting. (1997 : Madison, Wis.) IV. Midwest Glaciology Meeting (1997 : Madison, Wis.)
    GB2401.2 .G53 1999
    551.31--dc21   99-050399

Cover: (Top) Modern ice tunnel in Burroughs Glacier, Alaska. Photo by David M. Mickelson. Studying modern glacier processes is one way of understanding the glaciology of past glaciers. (Bottom) Esker in Wisconsin that formed in a similar ice tunnel during the Wisconsin Glaciation. Photo by Kent M. Syverson.

To our friend and colleague, W. Hilton Johnson.

February 14, 1935 – November 30, 1997

Contents

Preface vii

1. *Glaciological and geological implications of basal-ice accretion in overdeepenings* 1
   R. B. Alley, J. C. Strasser, D. E. Lawson, E. B. Evenson, and G. J. Larson

2. *Example of the dependence of ice motion on subglacial drainage system evolution: Matanuska Glacier, Alaska, United States* 11
   S. L. Ensminger, E. B. Evenson, R. B. Alley, G. J. Larson, D. E. Lawson, and J. C. Strasser

3. *Field evidence for the recognition of glaciohydrologic supercooling* 23
   E. B. Evenson, D. E. Lawson, J. C. Strasser, G. J. Larson, R. B. Alley, S. L. Ensminger, and W. E. Stevenson

4. *Isotopic composition of vent discharge from the Matanuska Glacier, Alaska: Implications for the origin of basal ice* 37
   D. D. Titus, G. J. Larson, J. C. Strasser, D. E. Lawson, E. B. Evenson, and R. B. Alley

5. *Microstructures of glacigenic sediment-flow deposits, Matanuska Glacier, Alaska* 45
   M. S. Lachniet, G. J. Larson, J. C. Strasser, D. E. Lawson, E. B. Evenson, and R. B. Alley

6. *Need for three-dimensional analysis of structural elements in glacial deposits for determination of direction of glacier movement* 59
   A. Dreimanis

7. *Tunnel channels formed in Wisconsin during the last glaciation* 69
   L. Clayton, J. W. Attig, and D. M. Mickelson

8. *Spooner Hills, northwest Wisconsin: High-relief hills carved by subglacial meltwater of the Superior Lobe* 83
   M. D. Johnson

9. *Origin of the Driftless Area by subglacial drainage—a new hypothesis* 93
   H. Hobbs

10. *Paleoglacier reconstruction and late Pleistocene equilibrium-line altitudes, southern Sawatch Range, Colorado* 103
    K. A. Brugger and B. S. Goldstein

***11. Pre-Illinoian glacial geomorphology and dynamics in the central United States,
west of the Mississippi*** 113
    J. S. Aber

***12. Wisconsin Episode glacial landscape of central Illinois: A product of subglacial
deformation processes?*** 121
    W. H. Johnson and A. K. Hansel

***13. Reconstruction of the Green Bay Lobe, Wisconsin, United States,
from 26,000 to 13,000 radiocarbon years B.P.*** 137
    P. M. Colgan

***14. Ice-surface profiles and bed conditions of the Green Bay Lobe from 13,000 to
11,000 ^{14}C-years B.P.*** 151
    B. J. Socha, P. M. Colgan, and D. M. Mickelson

***15. Ice sliding over weak, fine-grained tills: Dependence of ice-till interactions
on till granulometry*** 159
    S. Tulaczyk

***16. Quaternary glacial deposits and landforms of the north Timan region,
Russia—a possible center of local glaciation*** 179
    A. V. Matoshko

***17. Discussion of the observed asymmetrical distribution of landforms of the southeastern
sector of the Scandinavian Ice Sheet*** 187
    R. Karukäpp

***18. Role of climate oscillations in determining ice-margin position:
Hypothesis, examples, and implications*** 193
    T. V. Lowell, R. K. Hayward, and G. H. Denton

Preface

The papers in this volume are an outgrowth of a symposium on paleoglaciology that was held in Madison, Wisconsin in the Spring of 1997 at the North-Central Geological Society of America and Midwest Glaciology Meetings. This well-attended symposium was also sponsored by the Commission on Glaciation of the International Quaternary Union (INQUA). It attracted participants whose research on processes at modern glaciers, and the interpretation of landscapes glaciated during the Pleistocene, covers many parts of North America and parts of Europe. The purpose of the symposium was to bring together scientists who are interested in going beyond traditional mapping of glacial deposits and landforms, and interpretations of glacial chronology, to attempting to understand the processes and dynamics of both existing and former glaciers. Although there have been some attempts to quantitatively tie glacial deposits and landforms to the glaciological parameters that controlled their formation, glacial geologists and glaciologists have not commonly interacted in a way allowing truly cooperative research that took full advantage of each other's expertise. In the last 20 years, few attempts been made to incorporate glacial geologic constraints into glaciological models or to place physical constraints on hypotheses expounded by glacial geologists. We hope this volume will help foster communication and cooperative work between glacial geologists and glaciologists.

In general, two approaches have been used in integrating the work of glaciologists and glacial geologists. One approach is to incorporate the results of process studies conducted at existing modern glaciers into the interpretation of the deposits and landforms of areas glaciated in the past. This approach has been used for over 100 years, but often without significant exchange of understanding or information between glacial geologists and glaciologists. A second approach has been the semiquantitative reconstructions of ice surface profiles, basal shear stress, and other parameters of former glaciers, or lobes of glaciers, and relating these glaciological parameters to the distribution of sediments and landforms. This approach to glacial geology and glaciology has the potential for leading to numerical models of the dynamics of former glaciers that are constrained by what is known of landform and sediment distribution, climate, and chronology.

The first series of papers in this volume focus on studies at a modern glacier in Alaska. Several of these papers focus on subglacial water and the refreezing process at the base of the Matanuska Glacier in Alaska. They report evidence of supercooling of meltwater as it rises from overdeepenings in the glacier bed. Alley *et al.* examine the process of basal-ice accretion and broaden the analysis to evaluate erosion processes in similar settings elsewhere. They argue that basal accretion processes are important in determining sediment yield from beneath glaciers where overdeepenings exist. Ensminger et al. examine the relationship between ice-flow velocity and the subglacial drainage system. They report short-term increases in velocity in response to periods when more water is input to the subglacial system as well as an overall increase in velocity during the melt season as the subglacial water system develops. Late in the melt season, the drainage system is partly blocked by ice flow and water storage in the glacier increases. In a related paper, Evenson et al. present field evidence for the process of freezing of subglacial water that has risen from overdeepenings behind the margin. Frazil ice grows unattached and anchor ice grows in cavities, conduits, and at conduit mouths. The geochemical signature of the frazil and anchor ice is similar to that in the basal ice and Titus *et al.* argue for a genetic connection between the two.

The final paper in the series of papers from Matanuska Glacier Lachniet *et al.* discusses microstructures in sediment-flow deposits along the glacier margin. Dry flows tend to have fabrics in thin section that are bi- or polymodal, exhibit greater textural variability, and more signs of microdeformation than wetter, more plastic flows. Presumably these characteristics will aid in recognition of flow types in Pleistocene deposits. Dreimanis proposes using secondary deformation features and a three-dimensional view of clast orientations and other signs for deformation when reconstructing previous ice-flow conditions. He presents examples from drumlin areas in North America and Latvia.

Subglacial water played an important role in landform development during the late Pleistocene. Tunnel channels along the west side of the Green Bay Lobe in Wisconsin may have been produced by catastrophic flows of water from beneath the ice margin as described by Clayton et al. Johnson argues that subglacial water may also have been responsible for erosion of the high-relief hills that formed beneath the Superior Lobe in northwest Wisconsin. Hobbs also evaluates the effects of subglacial on the landscape and suggests that the loss of water through a permeable bed is responsible for the Pleistocene glaciers not reaching the Driftless Area of southeastern Minnesota and southwestern Wisconsin.

Equilibrium-line altitudes and ice-surface slopes are reconstructed for six late Pleistocene glaciers in south-central Colorado by Brugger and Goldstein. They estimate temperatures during the last glacial maximum were 7 to 9° cooler than today. Aber use GIS methods to compile the geomorphology of a broad area west of the Mississippi River outside the Illinoian glacial limit. This work was part of a project for the Work Group on Geospatial Analysis of the INQUA Commission on Glaciation. Based on meltwater routes and moraine distribution, he documents the shapes of glacier lobes and argues for low ice-surface gradients similar to those that have been postulated for the late-Wisconsin Des Moines Lobe.

The question of deposition of till by a soft deforming bed beneath the Lake Michigan Lobe is examined in the paper by Johnson and Hansel. Their model suggests wet bed conditions that allowed development of broad, low-relief hummocky moraines with little supraglacial sediment. They recognize different patterns of moraines representing different conditions of ice margin stability and weigh the evidence of soft bed deformation as an explanation of the landform and sediment distribution. Colgan also suggests differences in ice sheet dynamics in Wisconsin. His reconstruction of the Green Bay Lobe demonstrates the some advances had significantly steeper ice margins than others. Here the ice margin seems to have had a frozen bed in a narrow zone, probably about 5 km wide, at least by the time of retreat from the glacial maximum. After this ice surface, profiles are more gentle, implying less resistance at the bed. Younger ice margins in the same lobe had even more gentle slope as reported by Socha *et al*. This may be because of a soft deforming bed or abundant water causing increased sliding.

Tulaczyk has analyzed coarse and fine tills from beneath Ice Stream B in Antarctica, glaciers in the Yukon and Iceland, and Pleistocene till from Ohio, and compares the properties of these materials with a theoretical analysis of basal sliding and deformation on these materials.

The response of glaciers to outside climatic or other forcing is discussed in several papers. Matoshko argues that the Timan Plateau in northern Russia was the site of valley glaciers before an ice cap as the last Eurasian Ice Sheet expanded and overrode it. Along the southern margin of the same ice sheet, ice lobes were deflected westward and Karukäpp evaluates possible causes. Finally, Lowell and Denton suggest that fluctuations of the margin of the Laurentide Ice Sheet in the North American Midwest appear to have been caused chiefly by changes in ablation in the marginal zone.

Glaciological and geological implications of basal-ice accretion in overdeepenings

Richard B. Alley
Earth System Science Center and Department of Geosciences, Pennsylvania State University, 204A Deike Building, University Park, Pennsylvania 16802
J. C. Strasser
Department of Geology, Augustana College, Rock Island, Illinois 61201
D. E. Lawson
U.S. Army Cold Regions Research and Engineering Laboratory, Anchorage, Alaska 99505
E. B. Evenson
Department of Earth and Environmental Sciences, Lehigh University, Bethlehem, Pennsylvania 18015
G. J. Larson
Department of Geological Sciences, Michigan State University, East Lansing, Michigan 48824

ABSTRACT

Glaciers commonly erode rock basins, also called overdeepenings. Hooke suggested that erosion steepens the adverse slopes on the downglacier sides of overdeepenings until supercooling of subglacial water flow causes ice growth that plugs water channels and reduces the sediment transport needed for further steepening. Erosion then would be localized on the upglacier, headwall sides of overdeepenings, causing overdeepenings to migrate upglacier over time. We hypothesize that an increasing ice-surface slope would reduce or eliminate supercooling of water flowing from an overdeepening, allowing erosional steepening of the adverse slope. Subsequent decrease in ice-air surface slope would favor supercooling and accretion of debris-rich basal ice accompanied by sediment deposition in the overdeepening from disruption of subglacial streams. We further hypothesize that glacial erosion rates and sediment yield depend sensitively on the interactions of subglacial water with bedrock-floored and sediment-floored overdeepenings.

INTRODUCTION

Glaciated valleys commonly exhibit regions in which the local bed rises in the direction of ice flow. Such features may be erosional or depositional, and often produce lakes following deglaciation. This topography has variously been termed basin-and-riegel, basin-and-bar, or overdeepened (e.g., Sugden and John, 1976, ch. 9). We follow recent glaciological usage in calling such a basin an overdeepening, with the upglacier side called the headwall and the downglacier side called the adverse slope.

Subglacial water flowing up the adverse slope of a glacial overdeepening can supercool and cause ice accretion to the glacier (e.g., Röthlisberger, 1968; Röthlisberger and Lang, 1987; Hooke et al., 1988; Lawson et al., 1996). This possibility has figured prominently in interpretation of the hydrology of overdeepened glaciers (e.g., Röthlisberger, 1968, 1972; Lliboutry, 1983; Hantz and Lliboutry, 1983; Hooke et al., 1988; Hooke and Pohjola, 1994). Observations at the Matanuska Glacier, Alaska (Lawson and Kulla, 1978; Lawson et al., 1996; Strasser et al., 1996), have shown that the supercooling of waters flowing from

Alley, R. B., Strasser, J. C., Lawson, D. E., Evenson, E. B., and Larson, G. J., 1999, Glaciological and geological implications of basal-ice accretion in overdeepenings, *in* Mickelson, D. M., and Attig, J. W., eds., Glacial Processes Past and Present: Boulder, Colorado, Geological Society of America Special Paper 337.

overdeepenings causes rapid growth of debris-rich ice on the base of the glacier.

Hooke (1991) identified feedbacks that are likely to produce overdeepenings, and control their shapes, beneath glaciers through which surface melt drains to the bed. Here, we attempt to combine Hooke's model with insights gained from our observations on the Matanuska Glacier (e.g., Lawson et al., 1996) to generate hypotheses for the interactions between glaciers and the overdeepenings they produce.

OVERVIEW

As described by several authors (e.g., Röthlisberger and Lang, 1987; Hooke, 1998, ch. 8), surface meltwater can be routed to the glacier bed through moulins and then along the glacier bed toward the toe from high-pressure regions beneath thick ice. This routing is typical for alpine glaciers with surface melting. The pressure drop along channels can cause supercooling and freezing because a reduction in pressure raises the freezing temperature of water, and water in contact with ice remains close to the freezing point. This tendency is countered by the viscous dissipation ("friction") of the flowing water, as its potential energy is converted to heat.

For glacier beds that slope in the same direction as the ice surface, and those with gradual slopes opposed to the ice surface, the heat liberated by viscous dissipation is sufficient to maintain the water at the pressure-melting point. (Here we assume that the potential energy lost by water flowing through a glacier is converted to heat rather than to kinetic energy, which typically is quite accurate—the kinetic energy in water flowing in a channel at a representative velocity of 1 m/s equals the potential energy released by a decrease in head of only 0.05 m.) The heat from viscous dissipation just equals that required to warm the water to the pressure-melting temperature if the bed slope is 1.2 to 1.7 times that of the surface and in the opposite direction, with the uncertainty related to the air-saturation state of the water (e.g., Röthlisberger and Lang, 1987; Hooke, 1998, ch. 8).

Artesian-type basal water flow will occur provided the magnitude of the adverse bed slope is less than about 11 times the magnitude of the surface slope, assuming that the water-pressure gradient equals the ice-pressure gradient along water flow. (The increase in potential energy for water moving up the adverse slope of an overdeepening is nearly offset by the coupled decrease in ice pressure, causing the ice-surface slope to be about 11 times more important than the bed slope in controlling water flow; e.g., Röthlisberger and Lang, 1987; Hooke, 1998, ch. 8). Thus, for adverse bed slopes of overdeepenings with magnitudes between about 1.2 and 11 times the magnitude of the surface slope, supercooling of basal water is possible.

Other heat sources including the geothermal flux may be significant and prevent freeze-on if the water flux is small, but these additional heat sources are calculated to be insignificant for glaciers discharging much water subglacially (Alley et al., 1997a). Water flowing from appropriately steep overdeepenings then would be denied access to efficient basal channels because those channels are calculated to freeze closed. This water may be diverted to englacial channels that rise less steeply than the bed, to basal channels that flow around the overdeepening, or to a distributed "overland-flow" system, such as has been documented for Storglaciären, Sweden (Hooke et al., 1988; Hock and Hooke, 1993; Hooke and Pohjola, 1994). Supercooling of water in a distributed system could produce widespread ice accretion to the glacier base, releasing latent heat to balance the supercooling. This accretion would not permanently clog the drainage system. As described by Iken et al. (1983) and others, if a glacier is supplied more water than its channels can carry, basal water pressures will rise until that water is forced across the bed between the ice and its substrate.

Numerous data sets from the Matanuska Glacier, Alaska (e.g., Lawson and Kulla, 1978; Lawson et al., 1996; Strasser et al., 1996), have convinced us that such a "hydraulic freeze-on" process is active there. The glacier grows on the bottom, probably on the downglacier side of prominent overdeepenings, incorporating tens of percent by volume of diverse but silt-dominated sediment. Dye added to moulins has average flow velocity through the glacier much slower than expected for well-developed channels (Lawson et al., 1996). Tritium from atmospheric testing of nuclear bombs occurs in high concentrations in the accreted ice, confirming its young age (Strasser et al., 1996). A wide range of observational and theoretical evidence (reviewed by Lawson et al., 1998; Alley et al., 1998) leads us to believe that this mechanism is active beneath other glaciers, possibly many other glaciers, and that it is geomorphologically important.

BALANCE IMPLICATIONS OF HYDRAULIC FREEZE-ON

Hydraulic freeze-on adds debris-laden basal ice to the glacier while removing silt-dominated sediment from the subglacial environment. It thus has the potential to affect the balance of ice, water, and sediment beneath a glacier.

Freeze-on is calculated to have only a minor effect on the water budget of a glacier. As an example, water ascending 100 m from an overdeepening must warm by roughly 0.1 °C to remain at the pressure melting temperature. Because the heat of fusion of water is almost 100 times as large as its specific heat, freezing of only 1/1,000 of the water flow will supply all of the necessary heat. Note that such a small freezing rate is more than sufficient to plug channels, however. For example, a typical water-flow velocity in a channel of 0.3 m/s = 10,000,000 m/yr is about 100,000 times larger than a typical ice-flow velocity of 100 m/yr. If a channel segment could be transported by ice flow across an overdeepening without being plugged, enough water would pass through the segment to fill it 100,000 times. But freezing of 1/1,000 of this water would produce enough ice to fill the channel segment 100 times. Hence, we expect such channels to be plugged.

Freeze-on may be important in the ice balance of a glacier. Basal freeze-on rates of the Matanuska Glacier appear to be in the range of tenths of a meter per year, with surface ablation rates in the range of meters per year (Lawson et al., 1996). The basal freeze-on thus may be significant in some regions, although the total area experiencing ablation is probably much larger than the area experiencing freeze-on. Very near the terminus, ablation from above may leave much or all of the total ice thickness as frozen-on basal ice.

Freeze-on is probably more important in sediment budgets. Net ice accretion appears to have been ~0.1 m/yr or faster for some regions of the Matanuska Glacier, with debris concentration typically 15–60% by volume, equivalent to net entrainment of order 10–100 mm/yr of rock (Lawson et al., 1996). Rates of debris entrainment comparable to glacial erosion rates (0.1–10 mm/yr) are obtainable with a wide range of possible overdeepening models (Alley et al., 1997a). Notice, however, that active subglacial streams fed by abundant surface meltwater typically can transport much more sediment than is moved by debris-rich basal ice or deforming subglacial sediment (reviewed by Lawson, 1993; Alley et al., 1997b).

Freeze-on may be important or dominant in the basal-ice debris budget for many glaciers. Numerous processes serve to add debris to the basal ice of glaciers, but the efficiency of these processes is widely variable (reviewed most recently by Alley et al., 1997b). At the Matanuska Glacier, the meters-thick layer of basal ice containing tens of percent debris by volume flows towards the terminus at tens of meters per year (Ensminger et al., 1997), and this debris appears to have been added to the ice largely by the freeze-on mechanism discussed above. Features possibly associated with other mechanisms of debris acquisition by ice, such as the clast-supported texture expected for regelation into the bed (Iverson, 1993), occur but are not common in the basal ice of the Matanuska Glacier (Lawson et al., 1996). Basal-crevasse fills occur locally but are related to the basal-ice formation in the sense that silt-bearing, pressurized basal water flowing up from the bed apparently transported the sediment into the ice and may have contributed to its subsequent freezing (Evenson et al., 1998). Hence, at least for the Matanuska Glacier and possibly for other glaciers with important freeze-on, the supply of sediment to moraines from the ice is dominated by sediment entrained by freeze-on.

It is likely that freeze-on entrainment also affects the deformation of the ice to some degree, and hence the ice flux in the vicinity of overdeepenings. Most studies indicate that impurities and second-phase particles affect ice rheology. The results of Echelmeyer and Zhongxiang (1987) suggest that included debris softens ice, although the size and even the sign of this effect are likely to depend on the temperature and debris loading, among other factors (e.g., Hooke et al., 1972; Budd and Jacka, 1989). Creep tests with silt-dominated accreted ice would be of interest in this regard.

OVERDEEPENINGS BENEATH STEADY GLACIERS: QUALITATIVE ASPECTS

Although overdeepenings might be produced by differential erodibility of the glacier bed, Hooke (1991) argued that they typically are normal products of glacier erosion and can develop on homogeneous bedrock. To explain the origin of overdeepenings, Hooke (1991) suggested that flow over a convex part of the bed produces surface crevasses that aid development of moulins, funneling water to the bed just downglacier of the convex part of the bed (cf. Shreve, 1972). Water-pressure fluctuations are especially large where moulins reach the bed. The primary mechanism of glacier erosion appears to be quarrying of blocks from the glacier bed, enabled by fracturing driven by water-pressure fluctuations (Hooke, 1991; Iverson, 1991; Hallet, 1996), so erosion is localized where water reaches the bed. This creates a positive feedback: a convex portion of the bed speeds erosion just downglacier, which amplifies the convexity. Overdeepenings result (Hooke, 1991).

Building on the work of Iverson (1991), Hallet (1996) modeled the erosion of a glacier bed limited by growth of fractures through bedrock ledges with water-filled leeside cavities (Walder, 1986). In this model, an increase from low basal water pressure increases the quarrying rate because expansion of water-filled cavities between ice and rock concentrates stresses on remaining regions of ice-rock contact and so drives fractures more rapidly through that rock. With further water-pressure increase, the ice-rock contact stress plateaus at the failure strength of ice, and the quarrying rate decreases as further cavitation reduces the size of the ice-rock contact areas, which in turn reduces the size of the quarried blocks. Hallet's model suggests that the rate of glacial quarrying may vary by orders of magnitude for water-pressure fluctuations within the range exhibited by typical glaciers. The water pressure that maximizes the rate of quarrying depends on several characteristics of the glacier and the rock. These characteristics almost certainly vary over space and time. Because most water-pressure conditions will produce quarrying rates far below the maximum quarrying rate for a glacier, increasing water-pressure variability will increase the quarrying rate by allowing the quarrying rate to reach its maximum value at times. Large water-pressure variability also may cause fatigue and other processes that favor erosion (Hallet, 1996; Creyts and Alley, 1997).

The theory of quarrying is not yet complete, and further insights are likely. Based on available results, the rate of erosion is modeled as being a strongly increasing function of total water-pressure variability (Hooke, 1991; Iverson, 1991; Hallet, 1996; Creyts and Alley, 1997). Because of the strong sensitivity, even a small downglacier decrease in water-pressure fluctuations may reduce erosion sufficiently to allow formation of an overdeepening. Because fluctuations typically are driven by the input and damped by various subglacial processes, such a decrease in fluctuations away from the site of input over a convex region of the bed is expected (Hooke, 1991).

Hooke (1991) also argued that this process of erosional overdeepening must be self-limiting. In Hooke's (1991) model, if erosion of an overdeepening causes its adverse slope to become too steep, supercooling and ice growth will clog channels flowing up that slope. Water flow will be forced to avoid the bed through the overdeepening, or to cross the overdeepening in a distributed basal system. Compared to channels, a distributed drainage system is likely to have reduced sediment-transport capacity, and reduced water-pressure fluctuations causing reduced erosional capacity. Channelized water flow allows low water pressures in two ways: transiently, because channels can drain more rapidly than ice can creep into them during times of decreasing water input; and in steady state, by the heat from viscous dissipation melting channel walls as rapidly as ice creeps into channels (e.g., Röthlisberger and Lang, 1987). If basal channels tend to freeze closed in an overdeepening, as argued by Hooke (1991), then neither mechanism will be active, basal water pressure will remain close to the ice pressure, and erosion caused by fluctuating water pressure will be slow.

Hooke (1991) thus suggested that erosional steepening of the adverse slope of an overdeepening continues until supercooling clogs channels and suppresses erosion, and that thereafter this slope is maintained at an approximately steady value. Water-pressure variability, hence erosion, then would be concentrated on the headwall of the overdeepening, and would not extend to and through the bottom of the overdeepening to steepen its adverse slope further. Sediment produced by headwall erosion or supplied to the overdeepening in other ways but not removed by streams would accumulate, allowing formation of a deforming till layer along the adverse slope of the overdeepening. Sediment continuity would be achieved by a balance between sediment supply to the overdeepening and sediment export in the deforming till. Additional sediment supply by erosion beneath such a deforming till layer is likely to be quite slow or zero (Hooke, 1991; Cuffey and Alley, 1996).

Bedrock erosion would occur primarily by the upglacier migration of successive overdeepenings through headwall erosion, with downglacier sides of overdeepenings nearly passive until removed by the headwall of the next overdeepening (Figs. 1 and 2). This is identical to the step-erosion models envisioned for quarrying (Iverson, 1991; Hallet, 1996), but with overdeepening headwalls of perhaps tens of meters height rather than with ledge cavities of order 0.1 m height. The question of generation of new overdeepenings is considered briefly below.

Hydraulic freeze-on as well as deforming till can remove debris from overdeepenings. Steepening of the adverse slope of the overdeepening can occur until sediment supply from upglacier and local erosion are balanced by till deformation plus export of frozen-on debris. This in no way changes the concept of Hooke's model, but the additional sediment export by freeze-on allows slightly steeper overdeepenings with slightly more-disrupted stream transport than modeled previously.

From these modeling studies, one can expect overdeepenings to become deeper and steeper until stream transport of sediment is significantly slowed by the supercooling of waters flowing upslope, and a balance is achieved between sediment supply (erosion and transport from upglacier, especially by channels) and sediment removal (bed deformation, hydraulic freeze-on, and any continuing channel transport; Hooke, 1991). Hence, hydraulic supercooling and freeze-on should be widespread processes subglacially, occurring wherever glaciers have been able to modify their beds significantly.

OVERDEEPENINGS BENEATH STEADY GLACIERS: MODEL

Consider the following simple one-dimensional implementation of the Hooke (1991) model for development of overdeepenings, referenced to Figure 1. Overdeepenings numbered 1, 2, 3, . . . , i, . . . occur along a glacier with water assumed to enter surface crevasses and reach the bed at distances $x_1, x_2, x_3, \ldots, x_i$, . . . from the upglacier limit of runoff at $x = 0$. (Water in general will not descend vertically from crevasses to the bed, Shreve, 1972; but if the distance along ice flow remains constant from where the water enters the glacier to where it reaches the bed, our equations are accurate without introducing further complexity.) For simplicity, the specific rate of runoff (m^3/s/m^2) from surface melt averaged over a few days during the melt season is taken to increase linearly with distance as $K_w x$ with K_w a constant, and the crevasses at the headwall of each overdeepening collect all runoff supplied to them from upglacier. Then runoff per unit width supplied to the ith overdeepening from the surface, averaged over a few days, is

$$dQ_{wi} = \int_{x_{i-1}}^{x_i} K_w x \, dx = K_w \left(x_i^2 - x_{i-1}^2 \right) / 2. \quad (1)$$

Treating inwash and basal melt or freeze-on as insignificant, the total water flux per unit width at an overdeepening is

$$Q_{wi} = \sum dQ_{wi} = K_w x_i^2 / 2. \quad (2)$$

Water input is strongly diurnal, so the water flux is assumed to vary from zero in the early morning to its maximum in the afternoon. Water pressure near overdeepening headwalls is assumed to track this input, with pressure fluctuations increasing with the water input. When the adverse slope of an overdeepening is steep enough to cause supercooling, the overdeepening is assumed to act as a filter removing pressure fluctuations in water passing through. Pressure fluctuations in an overdeepening then are driven by the water input to that overdeepening from the surface, causing erosion to be sensitive to dQ_{wi} but not to Q_{wi}. Were supercooling not acting, then water-pressure fluctuations and erosion would depend on Q_{wi} as fluctuations from upglacier were added to the effects of input fluctuations (cf. Creyts and Alley, 1997). Larger water channels downglacier can drain more quickly during transient decreases in water flux, and can reach lower steady water pressures than can smaller channels upglacier, so water-pressure fluctuations are expected to be larger downglacier.

This assumption, that an overdeepening filters out pressure fluctuations from upglacier, merits further comment. Both Hantz and Lliboutry (1983) and Hooke and Pohjola (1994) have discussed the tendency for water pressures in overdeepenings to remain closer to the overburden pressure of the glacier than in other regions. This is not true of channels that flow around

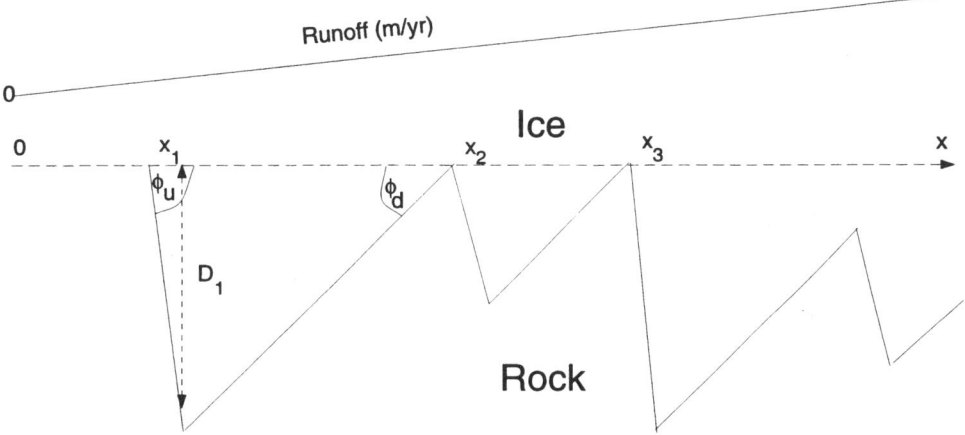

Figure 1. Model geometry. Overdeepenings with constant or nearly constant angles ϕ_u on the upglacier side and ϕ_d on the downglacier side relative to the mean bed, migrate in the negative-x direction. Specific runoff, in water thickness per time, increases with distance x. Overdeepenings are located at positions $x_1, x_2, \ldots$ and have depths $D_1, D_2, \ldots$.

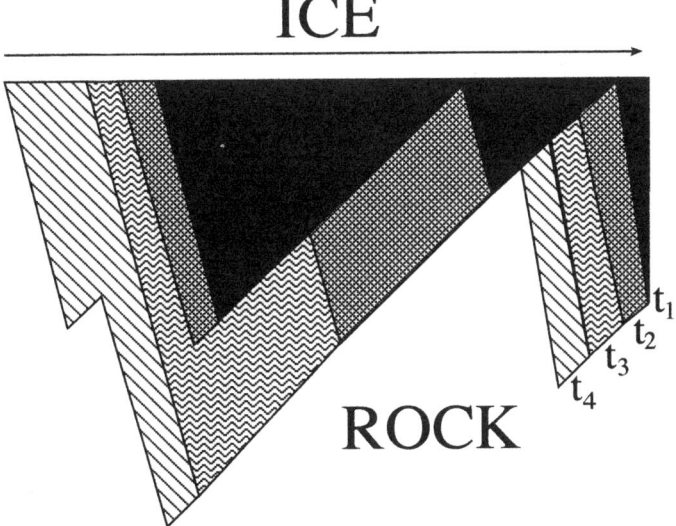

Figure 2. Cartoon showing migration of overdeepenings by headwall erosion. The ice-flow direction is shown by the arrow. Headwalls of initial overdeepenings (time t_1, shown darkest) are eroded, as shown by successively lighter fill (times t_2, t_3, and t_4). Merger of two overdeepenings is shown to have occurred between t_2 and t_3, and headwall splitting to create a new overdeepening is shown to have occurred between t_3 and t_4.

overdeepenings, and probably is less true of englacial water flow across overdeepenings than of subglacial flow up adverse slopes of overdeepenings. Following Hooke (1991), this behavior is fully consistent with the adverse slope of the overdeepenings causing basal channels to freeze closed. Thus, at least for basal flow, one can reasonably assume that water pressure approaches overburden pressure in overdeepenings, and thus that any excursions to lower pressure from upglacier will be filtered out by the overdeepening and not transmitted farther downglacier. In what follows, we assume that overdeepenings do filter out pressure variations, except as noted, although we recognize that englacial and ice-marginal water flow may transmit pressure variations past overdeepenings.

Sediment generation dQ_{si} in the ith overdeepening during times of supercooling, in response to pressure fluctuations driven by the local water input dQ_{wi}, is parameterized as:

$$dQ_{si} = K_s dQ_{wi}^a = K_s \left(x_i^2 - x_{i-1}^2 \right)^a \quad (3)$$

with K_s a constant; for heuristic purposes we set $a = 1$, although it is possible or likely that it is a stronger function of water supply (Hallet, 1996). The total sediment flux is then

$$Q_s = \sum dQ_{si} = K_s x_i^2. \quad (4)$$

Assuming some (uncharacterized) mechanism to maintain a constant slope ϕ_u for the upglacier side of an overdeepening of depth D_i, the overdeepening migrates at

$$v = dQ_{si}/D_i = K_s \left(x_i^2 - x_{i-1}^2 \right)/D_i \quad (5)$$

when supercooling is active. Were supercooling not active, and the basal system capable of removing all sediment supplied to it, then erosion would be sensitive to water-pressure fluctuations driven by the integrated water flux along the entire glacier, yielding

$$v = K_s x_i^2/D_i. \quad (6)$$

If this set of equations describes aspects of glacier behavior, and overdeepenings have been eroded sufficiently to cause supercooling of basal water flow, then several results follow.

1. Shallow overdeepenings migrate more rapidly than deep ones.

2. However, overdeepenings never merge, because an upglacier overdeepening creates a "runoff shadow" that reduces erosion just downglacier and slows any approaching fast-moving overdeepenings.

3. With increasing distance downglacier, x, overdeepenings are likely to be either more-closely spaced or more-rapidly migrating. This is seen in equation (5), if we write $x_i = x_{i-1} + \Delta x$ to obtain $v = K_s(2\ x_{i-1}\Delta x + \Delta x^2)$, which increases with x. This increase results from the increased ablation rate with increasing distance along the glacier, which causes water supply, pressure fluctuations, and erosion rate to increase downglacier in our model. Note that we assume that overdeepenings are generated efficiently somewhere downglacier; we discuss this assumption below.

4. For long glaciers, sediment supply from upglacier will be important or dominant compared to local sediment production. For local sediment to exceed that from upglacier along a glacier with many overdeepenings, the erosion rate in any overdeepening must be at least as large as the square of that in the previous overdeepening; for local sediment to dominate, erosion rates must increase downglacier more rapidly than this.

Two additional inferences that can be drawn from these results follow.

5. The degree to which the adverse bed slope of an overdeepening causes supercooling may decrease downglacier for a steady glacier (less-steep adverse slopes of overdeepenings with increasing distance downglacier, assuming constant surface slope), to allow increasing stream transport to balance the rapid downglacier increase in sediment flux. This assumes that the freeze-on entrainment of debris is less efficient than subglacial streams in normal flow, and that some stream transport is maintained across some of the downglacier overdeepenings.

6. The bedrock slopes leading out of overdeepenings may be related to ice velocity and overdeepening size. High ice velocity may advect a channel across a shallow and short but steep overdeepening before the channel can freeze closed and so lose most of its sediment-transport capacity, allowing smaller overdeepenings to have steeper rising slopes.

These last two inferences may suggest a general downglacier decrease in the importance of glaciohydraulic supercooling; however, the considerations on nonsteady glaciers, next, lead us to expect supercooling to be prominent even well downglacier during portions of a climatic cycle.

OVERDEEPENINGS BENEATH NONSTEADY GLACIERS

The preceding discussion assumed that the typical ice-air surface slope of a glacier is steady for a time that is long compared to that for significant modification of bedrock by erosion. This almost certainly is not realized in most cases; ice geometry typically varies more rapidly than bedrock geometry (although sediment geometry may vary at rates similar to those for ice; Nolan et al., 1995). Changes in surface slope may be related to ice advance or retreat, ice stagnation, surges or kinematic waves, and perhaps other glacial changes (e.g., Paterson, 1994, ch. 13). Response to climatic forcing may be especially important.

We recognize that supercooling is not directly linked to surface slope, but to the balance between heat generation and pressure-melting-point rise along water flow. However, because these typically can be closely estimated based on surface and bed slopes in the case for which overdeepenings cause supercooling (Hooke and Pohjola, 1994), we refer only to surface and bed slopes for convenience.

If a glacier and its bed reach steady state, Hooke (1991) has postulated that typically a little supercooling of the water rising along the adverse slope of the overdeepening will occur. A logical consequence of Hooke's model is that any increase in ice surface slope will initially suppress supercooling, leading to enhanced channelized drainage, enhanced sediment transport in channels, erosion of any sediment fill (e.g., Nolan et al., 1995), and then further bed erosion to produce deeper overdeepenings with steeper adverse slopes. A decrease in surface slope will lead to enhanced supercooling, decrease in channelized transport of sediment, and sedimentation to produce a lower effective adverse slope. Again, we are assuming that active channelized transport, not till deformation or sediment entrainment by freeze-on, dominates sediment budgets in the absence of supercooling (cf. Alley et al., 1997b).

If an increase in surface slope suppresses supercooling so that water-pressure fluctuations are not damped by overdeepenings, then our interpretation of the Hooke (1991) model leads to the following hypotheses.

1. Water-pressure fluctuations increase downglacier with the water flux, as discussed above.

2. Channelized sediment transport increases downglacier more rapidly than does water flux, so that the basal system retains the capacity to remove sediment supplied to it.

3. Erosion rates increase downglacier.

4. Small overdeepenings migrating upglacier can overtake and merge with larger ones.

5. Steepening of bedrock along adverse slopes continues until a little supercooling occurs under the steepest ice surface slope commonly maintained during a climatic cycle.

6. This bedrock slope would cause supercooling during most of a climatic cycle if not modified by sediment fill.

7. Subsequent decrease in ice-surface slope causes supercooling and freeze-on debris entrainment to be especially prominent, accompanied by sedimentation to partially fill overdeepenings if sediment transport by freeze-on and bed deformation cannot balance sediment supply from upglacier by channelized water and other processes.

OVERDEEPENINGS WITHOUT BEDROCK: MORAINE SHOALS

Similar continuity arguments can be advanced for moraines and moraine shoals. Subglacial water flux increases along the glacier length, and the sediment-transport capacity in channels increases along flow more rapidly than does channelized water flux (reviewed by Alley et al., 1997b), so subglacial sedimentation requires some disruption of the channelized transport system such as is produced by overdeepenings. At the ice terminus, a decrease in hydraulic gradient and the possibility for splitting of streams into distributaries may cause rapid sedimentation. Deforming-bed transport also ends where the ice ends, as does transport in ice unless debris-bearing icebergs are produced. The result is a moraine or moraine shoal. Some subglacial streams may cut through shoals, but we consider the situation in which the shoal is not pierced by streams and water rises along the bed to exit at the ice-sediment interface.

We postulate that the angle of the upglacier side of such a moraine shoal evolves to produce a little supercooling of streams rising along it. Were the slope steeper than this, basal streams would lose transport capacity by supercooling and ice growth, as well as due to the steep slope. This would cause deposition of sediments at the foot of and along the shoal, reducing its slope (still assuming that channelized sediment transport exceeds the potential transport by deforming beds and freeze-on debris entrainment).

If the upglacier slope of a shoal were too gradual, we speculate that channelized sediment transport would be unaffected by the shoal. Deforming glacier beds can be 0.5 m thick or possibly more, but continuity may cause thinner till to overlie bedrock; if so, the thickness of any deforming subglacial till would increase onto a moraine shoal as bedrock limitation on thickness disappeared (cf. Humphrey et al., 1993; Iverson et al., 1995; Jenson et al., 1995). Hence, the transport across the shoal would exceed supply to the shoal, and the upglacier end of the shoal would be eroded and steepened. We thus hypothesize that moraine shoals behave in the same way as overdeepenings in the Hooke (1991) model, reaching a steepness just sufficient to cause supercooling of channelized water drainage.

For tidewater glaciers advancing their moraine shoals rapidly, erosion exceeds sediment supply on the upglacier side of the shoal; thus, stream transport may be occurring across the shoal, and hydraulic freeze-on may be absent or slow. Freeze-on is more likely along stationary or slowly advancing moraine shoals. During rapid retreat of marine-ending tidewater glaciers from their moraine shoals, the rise of basal waters will occur between ice and seawater rather than between ice and sediment. Any ice accretion or frazil formation then is unlikely to trap much sediment.

GENERATION OF OVERDEEPENINGS

At least for tidewater glaciers, strong longitudinal extension associated with approach to flotation and loss of ice to calving may cause crevasses that allow enhanced access of surface waters to the bed, aided by the strong downglacier increase in ablation rates. Close juxtaposition of no-erosion zones under shoals and enhanced-erosion zones under extending ice may generate overdeepenings that then migrate upglacier over time.

During times of decreasing surface slope, overdeepenings with high headwalls might split. Erosion of the upper part of a headwall may occur while supercooling is suppressing water-pressure fluctuations and erosion along the lower part. Separation of the two by a small overdeepening would reduce water-pressure fluctuations downglacier, and separation would increase. Thus, while increase of surface slope may allow merger of overdeepenings, decrease of surface slope may allow splitting of overdeepenings. This last is highly speculative; we have built some simple models of ledges in overdeepenings that seem to point this way, but more work is clearly needed.

If a mechanism of headwall splitting is not active, then a "leading overdeepening"—one with a headwall that is not eroding the adverse slope of an overdeepening farther upglacier—may tend to become deeper and deeper with time. This will slow down its migration. A cirque is a clear example of this (Hooke, 1991). It is possible that overdeepenings in fjords will behave similarly, creating spectacular headwalls with icefalls.

Assuming that glacier erosion ends near the toe of a glacier where moraine deposition begins, and that erosion rates increase downglacier in response to downglacier increase in water flux if overdeepenings are absent, there will always be a tendency for a glacier to erode bedrock most rapidly slightly upglacier of its moraine. Subsequent advance will create a subglacial overdeepening.

DISCUSSION

Based on simple thermodynamic arguments and direct observation, it is known that subglacial water flowing up an adverse slope tends to supercool, grow ice that clogs channels, and find other flow paths that include high-pressure, distributed systems (e.g., Röthlisberger and Lang, 1987; Hooke, 1998, ch. 8). Observations and theory (Lawson et al., 1996; Strasser et al., 1996) show that supercooling in distributed systems is capable of growing ice subglacially, and that this ice entrains significant loads of sediment.

Hooke (1991) has argued that glacier flow over a convex portion of the bed causes crevassing that allows water access to the bed just downglacier. Fluctuations in water pressure are maximized where water reaches the bed, and modeling shows that glacial quarrying, which is believed to dominate glacial erosion, is maximized where water-pressure fluctuations are largest. Generation of overdeepenings thus is expected to be a general property of glaciers with abundant surface meltwater that reaches the bed.

As an overdeepening is eroded, basal water must flow upward toward the glacier's toe along the adverse slope of the overdeepening. When the adverse slope becomes sufficiently steep, models indicate that supercooling will cause freezing that constricts channels and reduces their transport of sediment. Deposition of sediment is then expected to armor the adverse slope and reduce erosion, consistent with observation and theory (Hooke, 1991; Cuffey and Alley, 1996). Deformation in this till can remove some sediment, as can sediment entrainment by freeze-on from distributed basal water forced out of the constricted channels.

Based on this model, the time-evolution of the surface slope of the ice will change the bed slope required to achieve sediment balance. During times of increasing ice-surface slope, bedrock erosion may deepen overdeepenings and steepen their adverse slope. Assuming that changes in ice geometry are fast compared to those in bedrock geometry, supercooling and freeze-on can be reduced or eliminated during times of increasing ice-surface slope, causing subglacial channels to be efficient, sediment flux to be large, erosion to be rapid, and englacial transport of sediment entrained by freeze-on to be slight. Thus, times of increasing ice-surface slope will cause erosion to dominate beneath a glacier.

We further hypothesize that decreasing ice-surface slope over an overdeepening will favor supercooling, basal accretion, reduction in channelized water and sediment transport, sedimentation to reduce the adverse slope of the overdeepening resulting in subglacial sediment storage, and increased sediment transport by till deformation and basal freeze-on of sediment that subsequently can build moraines. Times of decreasing ice-surface slope thus will be dominated by deposition.

Models suggest that during times of decreasing ice-surface slope, disruption of the efficient channelized drainage system should produce sedimentation and high basal water pressures. Together, these will allow efficient ice motion by sliding and deformation of the sediment (e.g., Iverson et al., 1995), which will speed ice export and promote further reduction in surface slope. During times of increasing ice-surface slope, channelized water will tend to lower basal pressures and thus decrease sliding speeds (e.g., Iken and Bindschadler, 1986) and bed deformation (e.g., Jenson et al., 1995), and will tend to remove the till bed and so suppress bed deformation. This will tend to slow ice motion and cause further increase in surface slope.

Sedimentation at the end of a glacier, especially if it is a tidewater glacier, can create an overdeepening without bedrock expression. We hypothesize that the adverse slopes of such overdeepenings similarly should tend towards a value with some supercooling to achieve sediment continuity. Supercooling is more likely for a steady terminus position than for one advancing rapidly behind a moraine shoal being recycled in conveyor-belt fashion.

The common occurrence of overdeepenings beneath large ice sheets (e.g., the Laurentian Great Lakes) as well as beneath small glaciers suggests a possible role of basal accretion in generation of the sediments of these ice sheets. For example, the prominent overdeepening of the sill of Hudson Strait may have played some role in the Heinrich events of the north Atlantic. However, we have calculated that some water source other than the basal melting produced by thermal dissipation of a MacAyeal (1993a, b)-type surge would be required to allow sufficient freeze-on to explain most of the debris in Heinrich layers (Alley and MacAyeal, 1994). For the tills deposited by ice flowing out of Lake Michigan, reconstructed freeze-on is sensitive to the details of the assumed surface slope and of the water flux, but a role remains possible.

We note that in our models of freeze-on, ice accretion on the adverse slope of one overdeepening should alternate with melting on the next downward slope. However, the tendency of water to become channelized when freeze-on is not occurring may allow survival of some accreted debris. A channel on a downward slope may melt through accreted ice and dissipate much of its energy melting clean ice above, while adjacent accretion ice is melted only slowly by the geothermal heat and heat of sliding. The preservation potential of accreted ice thus may be larger than first appears likely.

The view of glacier-bed erosion by migrating overdeepenings driven by water-pressure fluctuations (Hooke, 1991) has implications for parameterizations of glacier erosion in models of mountain-belt evolution (Creyts and Alley, 1997). One can imagine many ways to parameterize erosion, based on ice flux, ice velocity, or basal shear stress, for example. Compared to these, the overdeepening-dominated, water-pressure-driven erosion model leads to dependence on melt rate or total water flux, as discussed above. In either case, erosion will be maximized farther downglacier than in other likely parameterizations, with implications for the long-term evolution of slopes, isostatic response, and other behavior (Creyts and Alley, 1997). This tendency may provide a means of testing various possible parameterizations.

We echo Hooke's (1991) contention that overdeepenings are a typical product of glacier erosion, and that understanding their effects on hydrology, ice motion, and sediment budget is key to understanding the glacial system, its deposits and erosional products. The discussions above only begin to explore these effects.

ACKNOWLEDGMENTS

We thank the National Science Foundation Office of Polar Programs for funding, and T. Creyts, T. Dupont, R. Hooke, N. Iverson, B. Parizek, and an anonymous reviewer for helpful suggestions.

REFERENCES CITED

Alley, R. B., and MacAyeal, D. R., 1994, Ice-rafted debris associated with binge/purge oscillations of the Laurentide Ice Sheet: Paleoceanography, v. 9, p. 503–511.

Alley, R. B., Evenson, E. B., Lawson, D. E., Strasser, J. C., and Larson, G. J., 1997a, Conditions for Matanuska-type glaciohydraulic supercooling beneath other glaciers: Geological Society of America Abstracts with Programs, v. 29, p. 1.

Alley, R. B., Cuffey, K. M., Evenson, E. B., Strasser, J. C., Lawson, D. E., and Larson, G. J., 1997b, How glaciers entrain and transport sediment at their beds: Physical constraints: Quaternary Science Reviews, v. 16, p. 1017–1038.

Alley, R. B., Lawson, D. E., Evenson, E. B., Strasser, J. C., and Larson, G. J., 1998, Glaciohydraulic supercooling: a freeze-on mechanism to create stratified debris-rich basal ice: II. Theory: Journal of Glaciology, v. 44, no. 148, p. 562–568.

Budd, W. F., and Jacka, T. H., 1989, A review of ice rheology for ice sheet modelling: Cold Regions Science and Technology, v. 16, p. 107–144.

Creyts, T. T., and Alley, R. B., 1997, Possible Glacier-Erosion Parameterizations for Landscape-Evolution Models [abs.]: Eos (Transactions, American Geophysical Union), v. 78, p. F248.

Cuffey, K. M., and Alley, R. B., 1996, Erosion by deforming subglacial sediments: Is it significant? (Toward till continuity): Annals of Glaciology, v. 22, p. 126–133.

Echelmeyer, K., and Zhongxiang, W., 1987, Direct observation of basal sliding and deformation of basal drift at sub-freezing temperatures: Journal of Glaciology, v. 33, p. 83–98.

Ensminger, S. L., Evenson, E. B., Lawson, D. E., Alley, R. B., Larson, G. J., and Strasser, J. C., 1997, Velocity Measurements Demonstrate Dependence on Evolution of Subglacial Drainage System Configuration: Geological Society of America Abstracts with Programs, v. 29, no. 4, p. 14.

Evenson, E. B., Ensminger, S., Larson, G., Lawson, D., Alley, R. B., and Strasser, J., 1998, "Shear planes" and "the dirt machine" revisited: the debris bands at the Matanuska Glacier are not "shear planes": Geological Society of America Abstracts with Programs, v. 30, no. 1, p. 16–17.

Hallet, B., 1996, Glacial quarrying: a simple theoretical model: Annals of Glaciology, v. 22, p. 1–9.

Hantz, D., and Lliboutry, L., 1983, Waterways, ice permeability at depth, and water pressures at Glacier D'Argentière, French Alps: Journal of Glaciology, v. 29, p. 227–239.

Hock, R., and Hooke, R. LeB., 1993, Evolution of the internal drainage system in the lower part of the ablation area of Storglaciären, Sweden: Geological Society of America Bulletin, v. 105, p. 537–546.

Hooke, R. LeB, 1991, Positive feedbacks associated with erosion of glacial cirques and overdeepenings: Geological Society of America Bulletin, v. 103, p. 1104–1108.

Hooke, R. LeB., 1998, Principles of glacier mechanics: Upper Saddle River, New Jersey, Prentice Hall, 248 p.

Hooke, R. LeB., and Pohjola, V. A., 1994, Hydrology of a segment of a glacier situated in an overdeepening, Storglaciären, Sweden: Journal of Glaciology, v. 40, p. 140–148.

Hooke, R. LeB., Dahlin, B. H., and Kauper, M. T., 1972, Creep of ice containing dispersed fine sand: Journal of Glaciology, v. 11, p. 327–336.

Hooke, R. LeB., Miller, S. B., and Kohler, J., 1988, Character of the englacial and subglacial drainage system in the upper part of the ablation area of Storglaciären, Sweden: Journal of Glaciology, v. 34, p. 228–231.

Humphrey, N., Kamb, B., Fahnestock, M., and Engelhardt, H., 1993, Characteristics of the bed of the lower Columbia Glacier, Alaska: Journal of Geophysical Research, v. 98, p. 837–846.

Iken, A., and Bindschadler, R. A., 1986, Combined measurements of subglacial water pressure and surface velocity of Findelengletscher, Switzerland: conclusions about drainage system and sliding mechanism: Journal of Glaciology, v. 32, p. 101–119.

Iken, A., Röthlisberger, H., Flotron, A., and Haeberli, W., 1983, The uplift of Unteraargletscher at the beginning of the melt season—a consequence of water storage at the bed?: Journal of Glaciology, v. 29, p. 28–47.

Iverson, N. R., 1991, Potential effects of subglacial water-pressure fluctuations on quarrying: Journal of Glaciology, v. 37, p. 27–36.

Iverson, N. R., 1993, Regelation of ice through debris at glacier beds: implications for sediment transport: Geology, v. 21, p. 559–562.

Iverson, N. R., Hanson, B., Hooke, R. LeB., and Jansson, P., 1995, Flow mechanism of glaciers on soft beds: Science, v. 267, p. 80–81.

Jenson, J. W., Clark, P. U., MacAyeal, D. R., Ho, D., and Vela, J. C., 1995, Numerical modeling of advective transport of saturated deforming sediment beneath the Lake Michigan Lobe, Laurentide Ice Sheet: Geomorphology, v. 14, p. 157–166.

Lawson, D. E., 1993, Glaciohydrologic and glaciohydraulic effects on runoff and sediment yield in glacierized basins: U.S. Army Cold Regions Research and Engineering Laboratory Monograph 93-2, 108 p.

Lawson, D. E., and Kulla, J. B., 1978, An oxygen isotope investigation of the origin of the basal zone of the Matanuska Glacier, Alaska: Journal of Geology, v. 86, p. 673–685.

Lawson, D. E., Evenson, E. B., Strasser, J. C., Alley, R. B., and Larson, G. J., 1996, Subglacial supercooling, ice accretion, and sediment entrainment at the Matanuska Glacier, Alaska: Geological Society of America Abstracts with Programs, v. 28, p. 75.

Lawson, D. E., Strasser, J. C., Evenson, E. B., Alley, R. B., Larson, G. J., and Areone, S. A., 1998, Glaciohydraulic Supercooling: a freeze-on mechanism to create stratified, debris-rich basal ice: I. Field evidence: Journal of Galciology, v. 44. no. 148, p. 547–561.

Lliboutry, L., 1983, Modifications to the theory of intraglacial waterways for the case of subglacial ones: Journal of Glaciology, v. 29, p. 216–226.

MacAyeal, D. R., 1993a, A low-order model of growth/purge oscillations of the Laurentide Ice Sheet: Paleoceanography, v. 8, p. 767–773.

MacAyeal, D. R., 1993b, Binge/purge oscillations of the Laurentide Ice Sheet as a cause of the North Atlantic's Heinrich events: Paleoceanography, v. 8, p. 775–784.

Nolan, M., Motkya, R. J., Echelmeyer, K., and Trabant, D. C., 1995, Ice-thickness measurements of Taku Glacier, Alaska, U.S.A., and their relevance to its recent behavior: Journal of Glaciology, v. 41, p. 541–553.

Paterson, W. S. B., 1994, The physics of glaciers (third edition): Oxford, United Kingdom, Pergamon Press, 480 p.

Röthlisberger, H., 1968, Erosive processes which are likely to accentuate or reduce the bottom relief of valley glaciers: International Association of Hydrological Sciences Publication Number 79, p. 87–97.

Röthlisberger, H., 1972, Water pressure in intra- and subglacial channels: Journal of Glaciology, v. 11, p. 177–203.

Röthlisberger, H., and Lang, H., 1987, Glacial hydrology, in Gurnell, A. M., and Clark, M. J., eds., Glaciofluvial sediment transfer—an alpine perspective: New York, Wiley, p. 207–284.

Shreve, R. L., 1972, Movement of water in glaciers: Journal of Glaciology, v. 11, p. 205–214.

Strasser, J. C., Lawson, D. E., Larson, G. J., Evenson, E. B., and Alley, R. B., 1996, Preliminary results of tritium analyses in basal ice, Matanuska Glacier, Alaska, U.S.A.: evidence for subglacial ice accretion: Annals of Glaciology, v. 22, p. 126–133.

Sugden, D. E., and John, B. S., 1976, Glaciers and landscape: London, Edward Arnold Publishers, 376 p.

Walder, J. S., 1986, Hydraulics of subglacial cavities: Journal of Glaciology, v 32, p. 439–445.

MANUSCRIPT ACCEPTED BY THE SOCIETY OCTOBER 8, 1998

Example of the dependence of ice motion on subglacial drainage system evolution: Matanuska Glacier, Alaska, United States

Staci L. Ensminger
Department of Earth and Environmental Sciences, Lehigh University, Bethlehem, Pennsylvania 18015
Edward B. Evenson
Department of Earth and Environmental Sciences, Lehigh University, Bethlehem, Pennsylvania 18015
Richard B. Alley
Earth System Science Center and Department of Geosciences, The Pennsylvania State University, University Park, Pennsylvania 16802
Grahame J. Larson
Department of Geological Sciences, Michigan State University, East Lansing, Michigan 48824
Daniel E. Lawson
U.S. Army Corps of Engineers Cold Regions Research Engineering Laboratory, Anchorage, Alaska 99505
Jeffery C. Strasser
Department of Geology, Augustana College, Rock Island, Illinois 61201

ABSTRACT

The horizontal ice motion of the Matanuska Glacier, Alaska, was monitored along the western portion of the terminus during late May to the end of August of 1996 and 1997. The daily positions of five (1997) to six (1996) stations anchored into the ice, were measured using total station surveying equipment. The velocity curves for each station were stacked and smoothed to generate one velocity profile for each of the two study periods. The velocity curves for each summer show a similar seasonal trend of velocity increasing abruptly and significantly in early June, attaining a seasonal high in late June and early July, then generally decreasing through mid-August. Data from both field seasons show that velocity increases after mid-August through the end of the field season. In addition, short-term, small-scale fluctuations in the two records indicate the glacier responds to meteorological events, such as sunny periods and rainfall, as illustrated by abrupt changes in velocity. The velocity records are compared to input and output proxies in an attempt to develop a conceptual model for the evolution of the subglacial drainage system throughout the melt season. The current hypothesis is that Matanuska Glacier's subglacial drainage system consists of low, broad canals in subglacial sediment. The velocity records have a hysteretic relationship with the discharge of a stream that is sourced by glacial discharge vents. At the start of the melt season, there are large increases in ice velocity with very little change in stream discharge, indicating the subglacial drainage system has not fully developed to accommodate the increasing influx of meltwater generated by the onset of summer temperatures. It is hypothesized that meltwater influx increases through late June and early July, and subglacial water storage causes increased basal water pressure. Drainage system evolution and increased basal water pressure are suggested to control the horizontal ice velocity. Late melt season ice flow reduces drainage system capacity faster than the meltwater inputs decrease. Some drainage areas may cease discharging water because subglacial pathways become disconnected, which causes storage to again increase.

Ensminger, S. L., Evenson, E. B., Alley, R. B., Larson, G. J., Lawson, D. E., and Strasser, J. C., 1999, Example of the dependence of ice motion on subglacial drainage system evolution: Matanuska Glacier, Alaska, United States, *in* Mickelson, D. M., and Attig, J. W., eds., Glacial Processes Past and Present: Boulder, Colorado, Geological Society of America Special Paper 337.

INTRODUCTION

The dominant subglacial drainage configuration(s) is (are) important to the theories of the growth and accretion of basal ice. Subglacial drainage systems are generally modeled as flow-through channels or as thin distributed flow (Röthlisberger, 1972; Weertman, 1961). Channels can be conduits and tunnels in ice (Röthlisberger, 1972) or canals in till (Hock and Hooke, 1993; Walder and Fowler, 1994). Distributed flow systems have been further modeled as having linked-cavity configurations (Kamb, 1987). Ice accretion may take place in basal cavities that open during overpressurization of the water system due to large meltwater influxes (Iken et al., 1983; Alley et al., 1998). The accretion of discontinuous ice layers onto the sole of the glacier, or basal freeze-on, may also occur in low, broad canals (Alley et al., 1998; Lawson et al., 1998). Because water pressure increases with water flux in soft-bedded, Walder-Fowler canals as well as in linked-cavity or other distributed-system models (reviewed by Alley, 1996), we use the heuristic device of lumping the canals with the distributed models.

Water flow up the adverse slopes of over-deepened basins can cause the pressure-melting temperature to rise more rapidly than the water warms, causing supercooling (Hooke and Pohjola, 1994; Lawson et al., 1996). Supercooled water has been shown to lead to ice nucleation and growth (Forest, 1994; Strasser et al., 1996). The western portion of Matanuska Glacier's terminus has active ice that is flowing out of several over-deepened basins, one of which is adjacent to the margin (Arcone et al., 1995). Terraces of frazil ice have been observed growing during the melt season at the terminus of Matanuska Glacier (Strasser et al., 1992; Evenson et al., this volume, Chapter 3). Processes similar to those known to produce ice terraces are hypothesized to cause accretion of frazil ice to the base of the glacier, trapping sediment in a basal ice facies (Lawson, 1979a; Lawson et al., 1996, 1998; Strasser et al., 1996; Alley et al., 1998). Basal freeze-on is an important mechanism of debris entrainment near the terminus (Lawson et al., 1996; Knight, 1997; Alley et al., 1997, 1998; Alley et al., this volume, Chapter 1).

The stage of evolution and the drainage system configuration (i.e., the amount of pressure in the subglacial drainage system) are integral to understanding mechanisms of debris entrainment in the basal ice facies. Based on preliminary dye-tracer tests (Lawson et al., 1998) and ground-penetrating radar (Arcone et al., 1995), the current hypothesis is that Matanuska Glacier's subglacial drainage system consists of distributed flow through low, broad canals in subglacial sediment (Alley et al., 1998; Lawson et al., 1998). An ice-motion survey was conducted over the 1996 and 1997 melt seasons to develop a conceptual model of the seasonal evolution of the subglacial drainage system.

BACKGROUND

The Matanuska Glacier is a large valley glacier that is approximately 48 km long and reaches widths of 5 km at the terminal lobe. It flows north from the ice fields of the Chugach Mountains in south-central Alaska, into the upper Matanuska River Valley (Fig. 1), and terminates about 135 km northeast of Anchorage (Lawson, 1979b). The glacier drains 647 km^2 of the Chugach Mountains (Williams and Ferrians, 1961). Elevation decreases from 3,500 m in the accumulation zone to approximately 500 m at the terminus. The terminus has been located within 4 km of its present position for the past 8,000 years and at its present location for about 200 years (Williams and Ferrians, 1961). The southwestern portion of the terminus has been the focus of previous and current research on the growth and accretion of basal ice (Lawson, 1979a, b; Strasser et al., 1992, 1996; Lawson et al., 1996, 1998; Alley et al., 1998).

Subglacial water discharges from several vents (triangle symbols, Fig. 2) along the southwestern ice margin into a proglacial lake. A stream-gauging station is located approximately 400-m downstream from the ice margin (diamond shape, Fig. 2). The station monitors the combined discharge from the ice margin vents into the South Fork of Matanuska River. Winter discharge is estimated at 10 m^{3}s^{-1} or less and is derived from subglacial ground water and basal melting (Lawson et al., 1998).

During late May and early June, 1996, there were few discharge vents along the ice margin. There were 3 large discharge vents proximal to the study area, which were as large as 2–3 m in diameter. These large discharge vents sometimes fountained to approximately 1 m in height. Associated with the large discharge vents were several much smaller discharge vents. The small discharge vents had orifice diameters of 2–3 cm at the beginning of the melt season in May. By midsummer discharge vents had grown to nearly 1 m wide and in many areas, had seemingly merged with several other vents along the anchor ice terraces (see also Evenson et al., this volume, Chapter 3). As the smaller discharge vents grew in size and in number through the summer, the fountains from the larger discharge vents decreased in height until they upwelled gently. One feature common to all of the fountaining vents was the growth of frazil

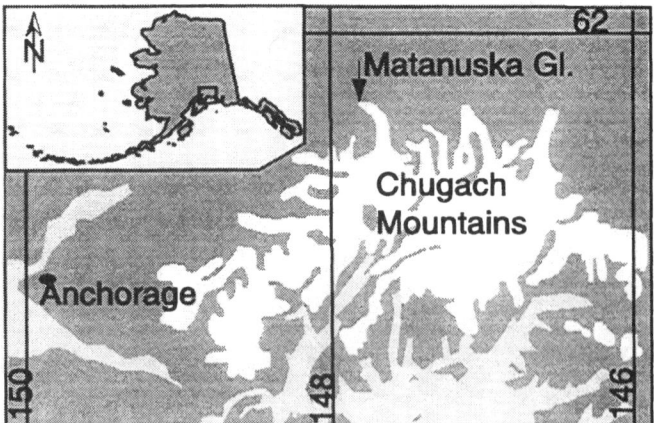

Figure 1. Matanuska Glacier flows north out of the Chugach Mountains and terminates in the Matanuska River Valley. The terminus is indicated by the arrow located at approximately 62° N, 148° W.

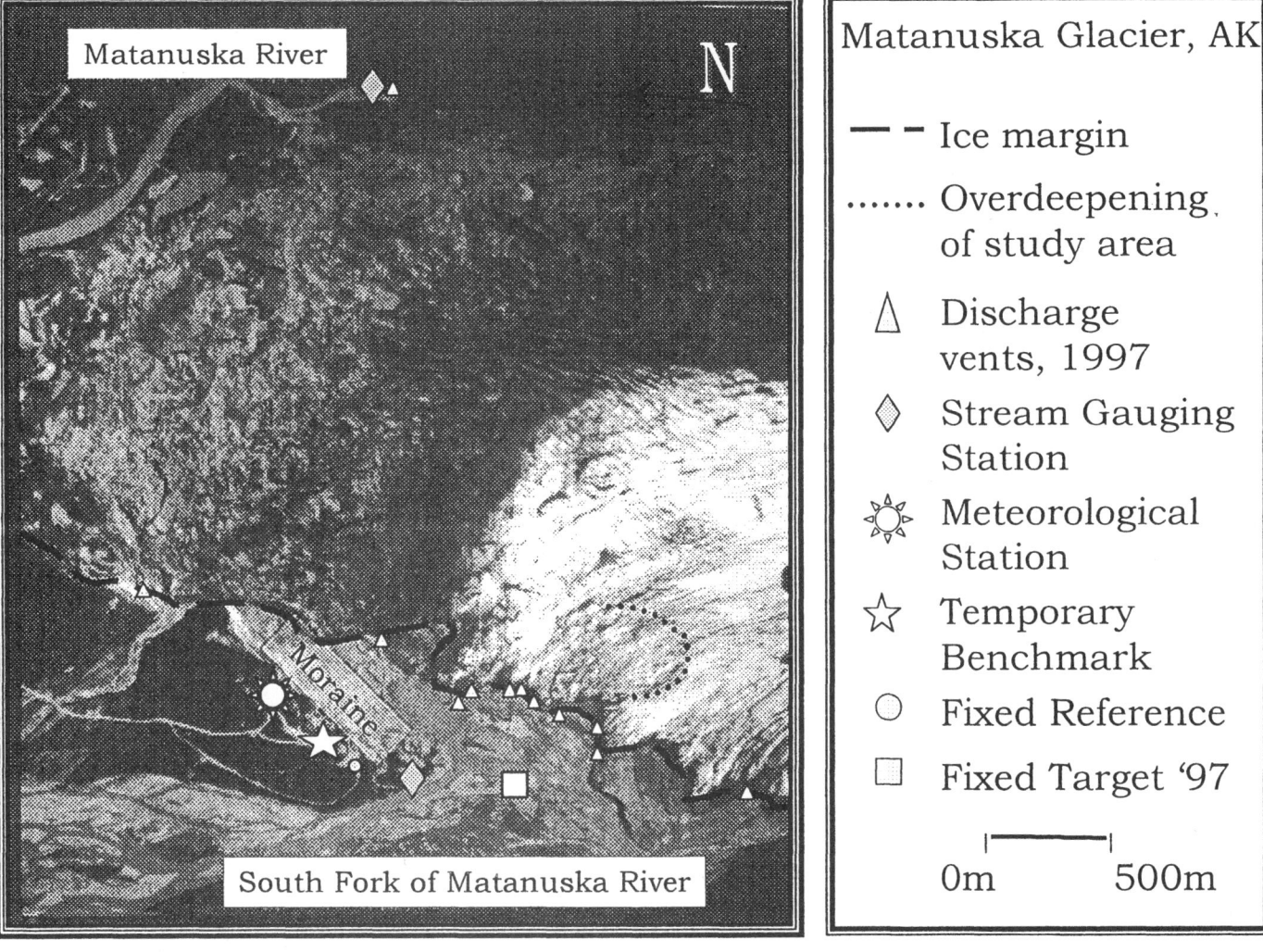

Figure 2. Annotated aerial photograph of the western portion of Matanuska Glacier's terminus.

ice, which in many localities grew into large terraces (Evenson et al., this volume, Chapter 3). Frazil ice growth was common during periods of warm summer air temperatures and high vent discharge. Similar observations were made along the ice margin in 1997.

The glacier ice became highly fractured during late July, 1996, where the terminus is flowing out of an over-deepened basin (dotted outline, Fig. 2). Very small upwellings of dirty water were observed within some of the fractures (see Evenson et al., this volume, Chapter 3). The fracturing was accentuated in the 1997 melt season. Many moulins were opened, and several formerly active, englacial conduits became exposed. Falling water level marks were observed within one such conduit. Water levels were marked with mud that had adhered to the conduit walls, suggesting the conduit had been a passageway for silty, subglacial waters. Hooke et al. (1988) demonstrated that basal water will flow through englacial conduits for ice situated in an over-deepened basin in order to remain at the pressure melting point.

METHODOLOGY

Target stations and procedure

During mid-May, 1996 and 1997, 6 stakes (1996) and 5 stakes (1997) were positioned vertically in holes drilled in the western terminus region of the Matanuska Glacier. The stakes were white PVC (Østrem and Brugman, 1991), 3.05 m length. Stakes were placed in hand-augured holes in the ice initially 2.5–3 m deep. Stake positions were selected so that ice motion would be almost exactly along the range distance. Figure 3 shows plots of each station's daily motion so that almost all of the ice motion recorded was along the measured range distance. When necessary, stakes had to be relocated because auguring was prohibited by rocks, dirty ice, or the area became unsafe for a rodperson.

Stake positions were measured at least daily over each of the three-month melt seasons. Three schedules of surveying were followed in 1996 to monitor events of high, low, and intermediate frequency. A short-term "high-intensity" schedule consisted of

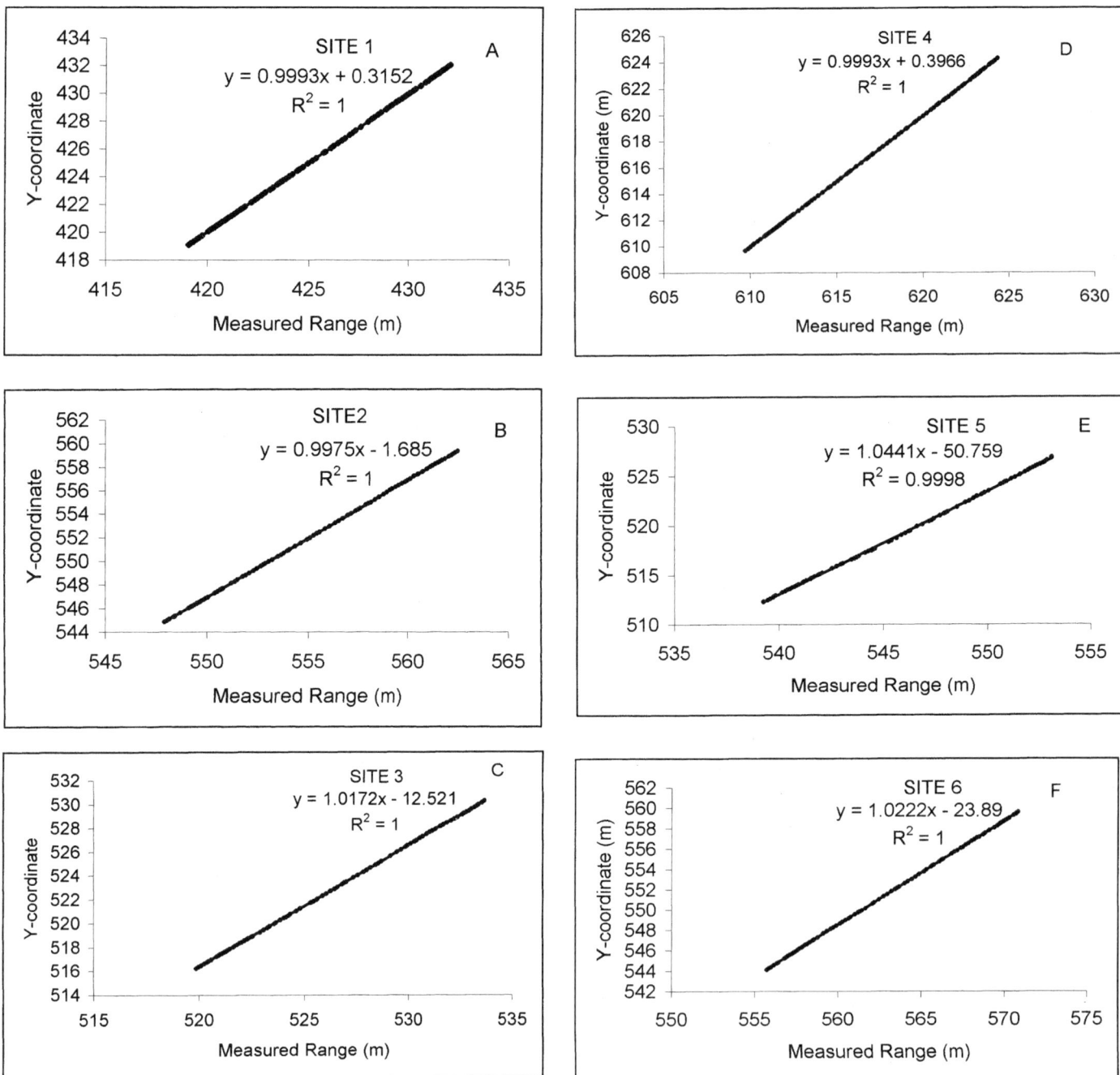

Figure 3. Plots of measured range distance against the calculated y-coordinate of all six target stations in the 1996 survey (shown as A to F). Also shown are the equations of a line fit through the data, demonstrating that daily changes in the range are reflected in the y-coordinate.

measuring all six target positions on the ice and the stationary target at 15-minute intervals (Krimmel and Vaughn, 1987) for periods of 6–8 hours at least once a week. A long-term schedule, consisting of measurement of all station positions at a particular time on the same day each week, was planned in case velocity or uplift "events" were frequent enough to obscure the seasonal signal. An intermediate-term schedule consisted of station measurements at times of daily low- and high-vent discharges. Daily low-vent discharge was, as a summer average, approximately 10 AM, and daily high-vent discharge was approximately 10 PM (Lawson, personal communication, 1996). In 1997, station positions were recorded only at times of daily low-vent discharge.

The stakes also served as references against which daily ablation measurements were made. A daily record of ablation was obtained by measuring from the top of the stake down to the ice surface. A piece of plywood 1.3 cm thick and 15 cm by 15 cm was used as a removable "foot" that fit around the stakes to average the ice surface for ablation readings. Ablation was measured to the nearest centimeter. The removable "foot" was also used for precise placement of the electronic distance measurement (EDM) reflecting prism adjacent to the stakes during surveys. The stakes became prone to tilting as surface ablation proceeded. To minimize tilting, the stakes were shortened and placement holes in the ice were periodically redrilled. During mid- to late July, when ablation was the most rapid, stakes were cut every 3–4 days and holes were redrilled every week.

Survey station and procedure

The reference station from which surveying was conducted (a temporary benchmark) was located on the end moraine, just west of the terminus (star symbol, Fig. 2). A fixed reference for turning angles and a stationary anchored target for determining surveying errors were also located on the moraine. The fixed reference was a marked point on a large stable rock located 40 m due south of the benchmark (circle symbol, Fig. 2). The stationary target was an EDM reflecting prism located on the moraine in 1996, due north of the benchmark (near the sun symbol, Fig. 2). In 1997, the stationary target was relocated to a position on outwash south of the benchmark because of land development near the previous year's location (square symbol, Fig. 2). There was also a meteorological (met) station recording various weather data in the vicinity of the 1996 stationary target on the moraine (sun symbol, Fig. 2). The met data were used for atmospheric corrections applied to the survey data, as well as for the development of the conceptual model.

The elevation of the temporary benchmark surveyed in 1997, was 559.30 ± 0.02 m. The distance from the benchmark to the stationary target was chosen to be approximately an equal distance as from the benchmark to the closest stake on the glacier. The survey instrumentation was a Top-Con total-station theodolite. Standard survey procedures were observed (Davis and Foote, 1953). Each survey consisted of measuring the distance to each of the target stakes and the stationary target.

Atmospheric corrections, applied during data reduction following the field season, were based on meteorological data collected at the same time as the survey data. Average calculated atmospheric corrections for the station farthest from the temporary benchmark were ±3.5 mm. An unquantified amount of error may be associated with the properties of the air mass through which the survey was conducted. Most of the travel path of the theodolite's infrared beam was over the same moraine and debris flows occupied by the met station, so we expect the met station data to characterize this air mass accurately. However, part of the beam's path was over the glacier and the properties of the air are undoubtedly different a few meters above the ice surface (Paterson, 1994). Because the beam did not have far to travel over ice (generally much less than 30% of the beam's path), it is assumed that these errors are small relative to the distances measured (Harper et al., 1996).

Over the 96 days of surveying in 1996, the repeated measurements of the stationary target, after atmospheric corrections were made, yielded a range distance uncertainty of ±0.2 cm (1 standard deviation). Horizontal angle and vertical angle uncertainties were determined to be ±28 seconds and ±29 seconds, respectively. The 1997 data have the same range uncertainty as the 1996 data, but have horizontal angle uncertainties of ±13 seconds and vertical angle uncertainties of ±8.5 seconds or approximately ±1 cm at typical range. However, errors at stations may exceed those at the stationary target because of stake tilting and EDM prism placement on the uneven ice surface.

The survey data on slope distance, vertical angle, and horizontal angles were reduced to x-y-z coordinates (Fig. 4). The survey station is at the origin of the coordinate system (corresponding to the star symbol, Fig. 2) and the fixed reference is at approximately 40 m along the across-glacier axis (corresponding to the circle symbol, Fig. 2).

Horizonal velocity record construction

Examination of the 1996 "high-intensity" data suggests that distances obtained with a 15-minute sampling interval differ by less than the location error (±0.2 cm) and an ice-motion record could not be constructed from that data. A record of horizontal velocity from the intermediate schedule is somewhat noisy, so for clarity only the data collected during the daily low discharge were used to construct each melt season's velocity profile. The recorded ablation was added to the surveyed ice-surface elevation to provide an ablation-free, ice-motion record.

Due to the high precision in the y direction and larger errors in the 1996 x and z directions, daily changes in the y direction were used to construct the horizontal velocity record. This required a fundamental assumption that the daily change in the y-coordinate is approximately equal to the daily change in the range distance measured for each of the stations (i.e., most of the motion was in the y direction rather than the x direction). Because each of the target stations was at a horizontal angle of approximately 90° to the fixed reference, this assumption was successfully tested for each station (Fig. 3). The daily position of each station was plotted and an average direction of motion relative to the temporary benchmark was determined (inset, Fig. 4).

The horizontal velocity, v_y, was determined for each station by:

$$v_y = (\Delta y / \Delta \text{time})/\cos ¥, \qquad (1)$$

where ¥ is the average angle the target station's direction of motion made relative to the temporary benchmark throughout the summer. The velocity data of all of the target stations were averaged to generate one velocity profile for each melt season (Fig. 5). A 7-point moving average was applied to the velocity profile to accentuate the seasonal signal. The uncertainty of the 1996 velocity with this method of record construction is ±0.33 cm/d for an average velocity of 16.7 cm/d. Conversely, for an average 1996 velocity of 20.1 cm/d, calculated as:

$$v_{xyz} = [(\Delta x^2 + \Delta y^2 + \Delta z^2)^{1/2}]/\Delta t, \qquad (2)$$

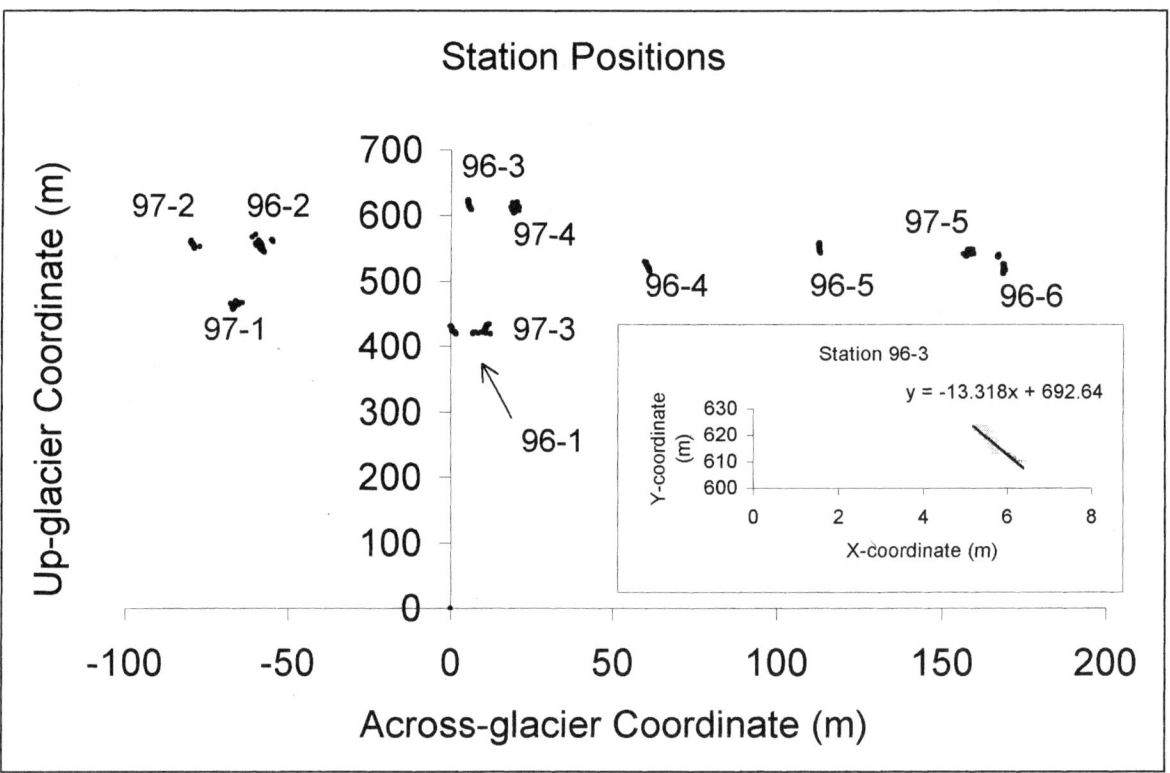

Figure 4. Plot of daily x- and y-coordinates of all stations in survey. Stations are labeled with the year, then their site number corresponding to plots in Figure 3. Note that some stations (e.g., 97-3) reflect relocation of the station because augering was inhibited. Inset illustrates one target station's motion, which is mostly in the y direction.

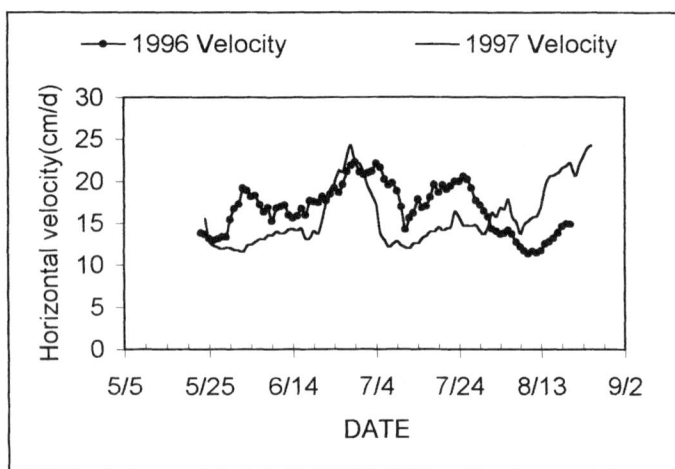

Figure 5. Horizontal velocity for each of the study periods. Note they peak at approximately the same date.

the uncertainty is ±8 cm/d. The 1997 data yield velocities similar to 1996 data, but because the 1997 data had smaller errors than the 1996 data using either method of calculation, the 1997 data were reduced using equation 1 for the sake of consistency and comparison.

Daily motion in the z direction was much less than the error limits in 1996 and was, therefore, considered for neither velocity record construction, nor for an uplift record. While the error in the 1997 record is considerably less, the daily changes in elevation are still occasionally smaller than one standard deviation (±0.01 m; Fig. 6). Over periods of several days, broad changes are observable (Fig. 6). The elevation of all stations increased throughout the summer. Elevation records were averaged together to generate one profile (Fig. 6).

Water samples were collected from four glacial discharge vents and the stream using ISCO water samplers on either a one-hour or a two-hour schedule. Samples were filtered with a vacuum pump to capture silt and larger grain sizes, then baked to remove water, and weighed to determine sediment concentration. Seabird Model 19 Conductivity-Temperature-Depth (CTD) profilers were suspended in the discharge vents to characterize the water flowing through the drainage system (see Evenson, this volume, Chapter 3).

SURVEY ANALYSIS

Horizontal velocity

The velocity (v_h) curves for both summers are similar (Fig. 5). Surveying in mid- to late May records the increasing ice surface velocity associated with the onset of the melt season. The

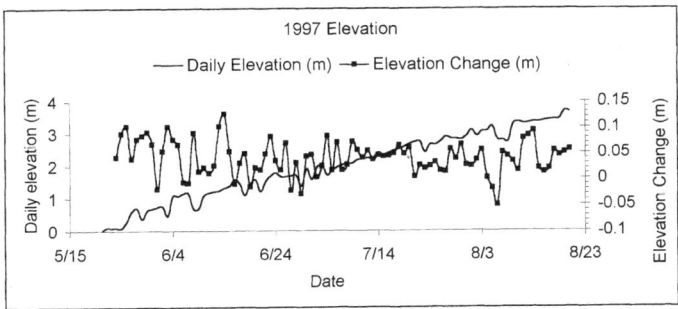

Figure 6. Seasonal and daily changes in elevation for 1997. The seasonal record shows ice flowing out of an over-deepened basin.

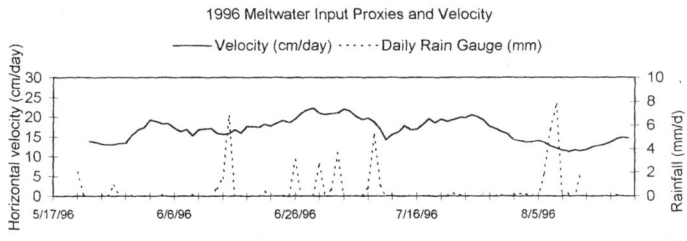

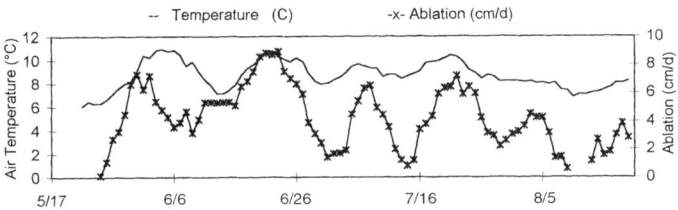

Figure 7. 1996 horizontal velocity and meltwater input proxies.

velocity reaches its seasonal maximum on June 29, 1996, and June 28, 1997. Velocity decreased through the middle of August, after which time both records show increasing velocities. The velocity in late August, 1997, was nearly equal to the melt season maximum in June. The velocity records of both years have higher-frequency signals superimposed on the seasonal trend. The "spikes" and "dips" are hypothesized to reflect the system's response to weather events.

Stenborg (1970) compared the total discharge of the stream at the terminus with the meltwater calculated from measured meteorological parameters. The velocity curves are compared to meltwater-input proxies and drainage system output parameters collected from the discharge vents and stream. The met station collected weather data that include rainfall, air temperature, and incoming and outgoing long- and short-wave radiation. Combined with the ablation data, the weather data serve as meltwater-input proxies. The drainage system output parameters include a stream discharge record from the South Fork (diamond symbol, Fig. 2), total suspended sediment (TSS) concentration of stream water collected at the stream-gauging station, TSS concentration of glacial vent discharge, and vent temperature records.

Subglacial drainage inputs. Ablation is assumed here to be a function of air temperature and net radiation (Pollard, 1980), with air temperature variations accounting for more of the variation in the ablation rate (Braithwaite, 1981). Rainfall is also assumed to pass through the subglacial drainage system. Figures 7 and 8 summarize the variations of the meltwater input proxies throughout each of the melt seasons. The 1996 data (Fig. 7) show that most of the rainfall occurred prior to or during maximum melt season velocities. Ablation rates and air temperatures were also greatest just prior to maximum velocities. Early in our 1997 survey (Fig. 8), moderate air temperatures and a large rainfall event followed a very warm first week. Both the warm air temperatures and rainfall may have contributed to the initial increase in the velocity. Unfortunately, the rain gauge of the met station was inoperable at the start of the study, so the magnitude of the early rainfall event is unknown. The maximum ablation occurs with warm air temperatures following the peak velocity. Rainfall also becomes an important influx of the subglacial drainage system after peak velocities.

Subglacial drainage outputs. When the velocity and stream discharge from both years are compared (Fig. 9), it is apparent that there is a relationship between the velocity and stream discharge, although it is not a linear one. The likeness of the velocity curves to each other and their respective discharge curves suggests that processes controlling the velocity (i.e., hypothesized here to be the evolution and configuration of the subglacial drainage system) evolve throughout the melt season similarly from year to year. Hooke et al. (1985) also suggest consistency in the evolution of the drainage system at Bondhusbreen, Norway. Theakstone and Knudsen (1979) attribute the details of the glacier river hydrograph at Austre Okstindbreen to the development of the glacier's internal drainage system, and the general shape of the hydrograph to climatic conditions. They also hypothesize that observed discharge patterns are related to water storage within the glacier.

The drainage system output records of 1997 are the focus of

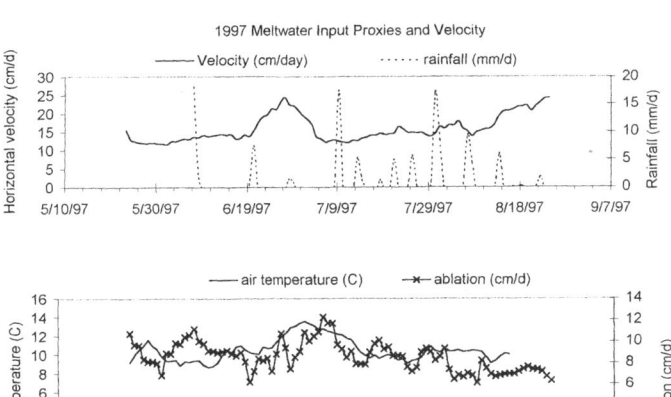

Figure 8. 1997 horizontal velocity and meltwater input proxies.

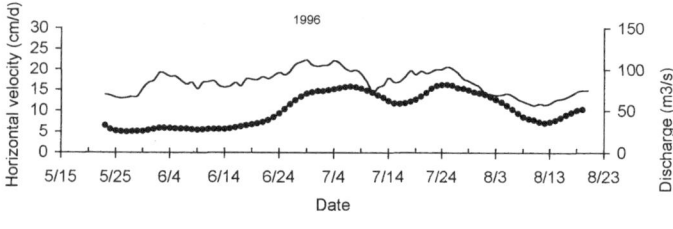

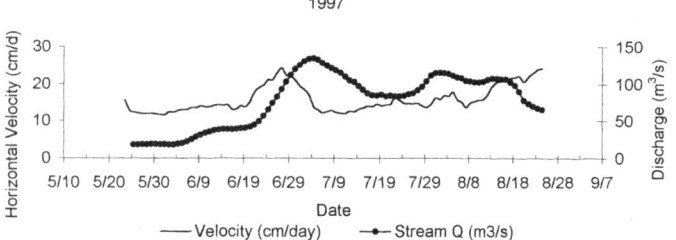

Figure 9. A comparison of the 1996 velocity and discharge to the 1997 velocity and discharge shows how similar the velocities are to one another, as well as their respective discharges. The discharge lags behind the velocity both years. Also note the discharge for 1997 is much greater than for 1996. Discharge in 1997 reached the bank-full stage.

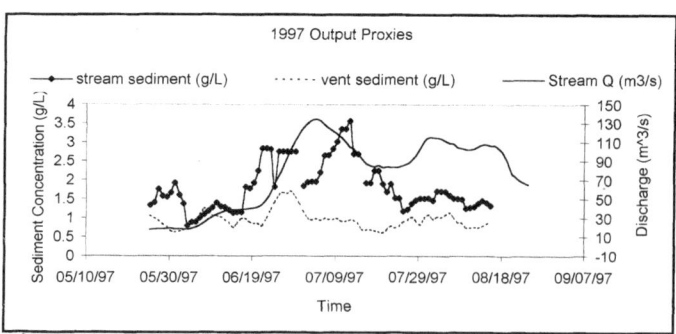

Figure 10. Three parameters of the subglacial drainage system output. Values for vent total suspended sediment (TSS) are at a maximum value on June 28, while discharge is increasing and velocity attains its maximum. Vent TSS values decrease while stream discharge and stream TSS maintain high values.

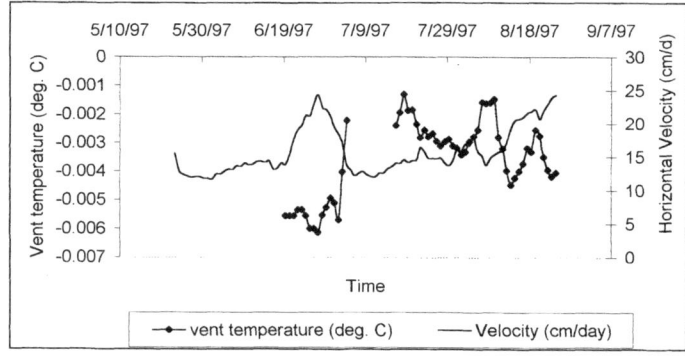

Figure 11. Supercooled temperatures were measured throughout the 1997 melt season.

this section because they present a much more complete package than the 1996 data, as no stream TSS data were collected in 1996. Average vent TSS concentrations in the 1997 melt season were approximately 1.0 g/L during ordinary discharges, and reached 2.5 g/L during increasing discharge (Fig. 10). Vent TSS attains seasonal maximum on June 28, which is the same date as the velocity maximum. Stream TSS concentrations were 1.8 g/L on average, but reached concentrations as high as 3.7 g/L. A complex relationship resides among and between the stream discharge, stream TSS, and vent TSS data. Figure 10 illustrates that the 3 outputs have the same seasonal nature, but the durations of maximal values are quite different. The vent TSS seasonal maximum may have a shorter duration than the stream's seasonal TSS maximum because of sediment availability at the glacier's bed. For example, while stream discharge is still increasing, the vent's sediment load begins decreasing because the subglacial hydrologic system is no longer eroding sediments at the bed, either because the supply was depleted, or because the drainage system is no longer growing by erosion (Hooke et al., 1985; Lawson, 1993).

Temperatures of vent discharge remained at or below freezing throughout the summer while air temperatures are well above freezing (Fig. 11). It should be noted that the depth at which the instruments were suspended in the vents is unknown at times during the summer because some data were lost due to errors during downloading of CTD records. Variations of temperature may be reflecting variations in depth, not changes in the drainage system. However, supercooled water was measured throughout the melt season. Evenson et al. (this volume, Chapter 3) present field evidence for the phenomenon of supercooled vent discharge at the Matanuska Glacier, which includes frazil ice flocs, anchor ice terraces, and frazil ice growth on instruments suspended in discharge vents. Alley et al. (this volume, Chapter 1) present the theory of the growth and accretion of the basal ice facies from processes like those that form frazil ice in subglacial conduits.

Fleisher et al. (1994, 1997) have recognized the association between supercooled waters with related frazil ice features and the ice surface behavior of a surge-type glacier. As basal water pressure increases with increasing meltwater inputs, the subglacial drainage system becomes pressurized such that the freezing point is depressed, thereby permitting the formation of supercooled conditions (Hooke et al., 1988; Alley et al., 1998, and this volume, Chapter 1) and associated increases in ice velocity with reduced effective pressure (Iken, 1981; Iken et al., 1983; Hooke et al., 1997).

CONCEPTUAL MODEL OF DRAINAGE SYSTEM EVOLUTION

Field observations at the ice margin are used to develop a conceptual model from which inferences may be drawn about the evolution of the subglacial drainage system. Meltwater input proxy data and subglacial drainage system output data obtained throughout the melt season provide "limits" for data interpretation. Figure 12 illustrates a conceptual model of the subglacial drainage system evolution based on ice motion and glacially sourced stream discharge. We assume changes in the velocity to be a function of the current water pressure and the water-pressure history of the subglacial drainage system (Bahr and Rundle, 1996). We also anticipate a hysteretic relationship between basal water pressure and velocity. Arrow directions indicate chronology along the hysterysis loop of discharge versus velocity. Steeply increasing trends on the loop are hypothesized periods of reduced effective pressure (increased basal water pressure) due to water storage at the bed. Decreasing trends are hypothesized periods of increased effective pressure (decreased basal water pressure) due to drainage of water in storage or reduced meltwater influx.

Early melt season system configuration (Fig. 12A) is an immaturely developed system of canals in till that cannot accommodate large meltwater influxes resulting in meltwater storage at the bed. The system evolves with increased erosion of the bed so that most meltwater passes through the system, but water storage is still increasing (Fig. 12B). The drainage system is fully developed by the mid- to late melt season so that all meltwater influx passes through as stored water is draining (Fig. 12C). Once the drainage system has reached a fairly stable late-season configuration, further adjustments to the configuration are on the order of days (Röthlisberger and Lang, 1987). Cooler autumn weather reduces meltwater production and increases the effective pressure, making the canal configuration unstable (Fig. 12D). Drainage areas that are higher than the glacier's decreasing water table level will cease to discharge water. A canal in till may also cease to discharge water by closing due to the creep of the overlying ice and inward creep of till (Patterson, 1994).

Data from both melt seasons show a hysteretic relationship between the velocity and stream discharge with a chronology like that of the conceptual model (Fig. 13). Increases in meltwater proxies and low values of output parameters are generally observed

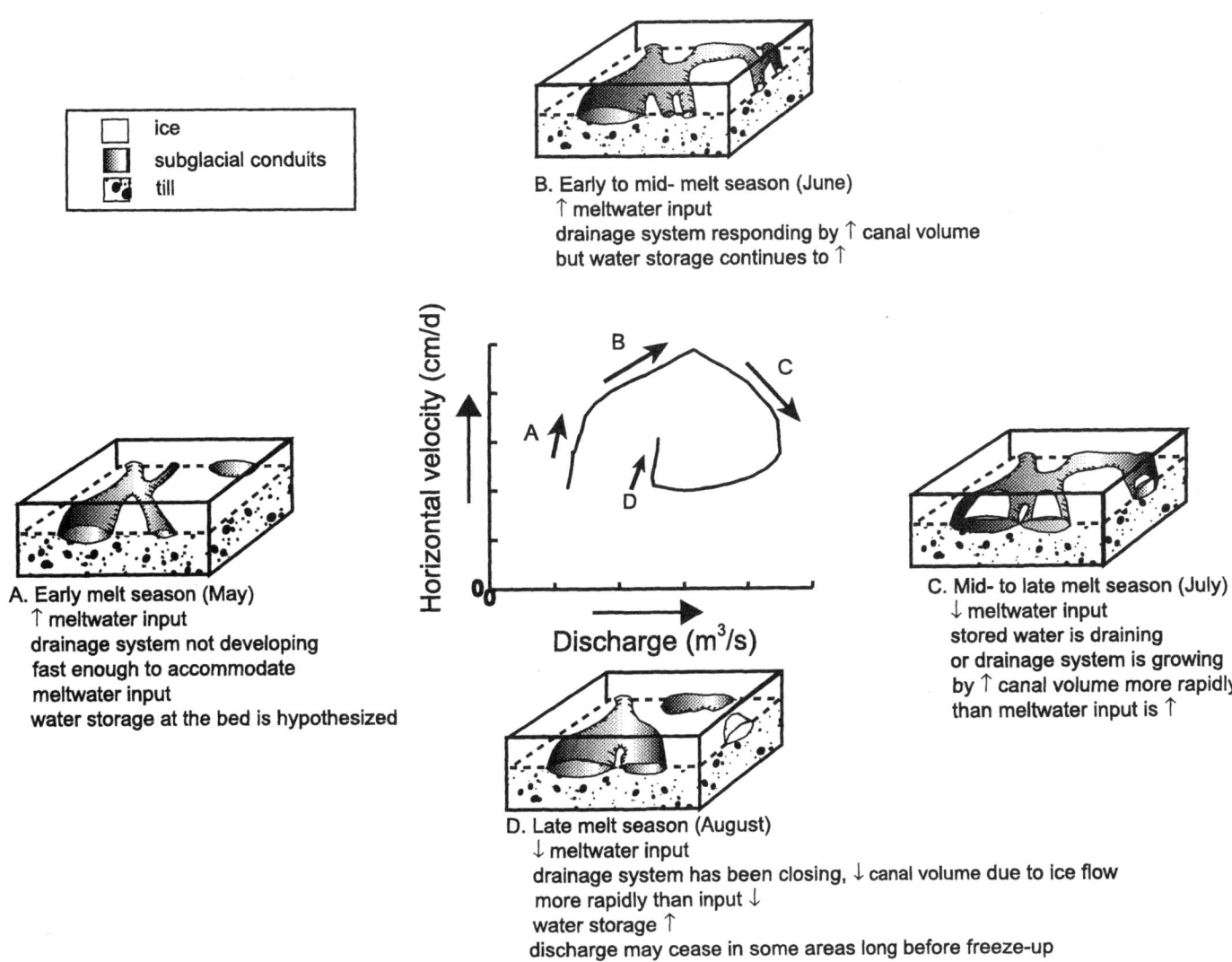

Figure 12. A conceptual model of the evolution of the subglacial drainage system. Drainage system conduits are assumed to have a capacity that transmits water. Filled conduits are represented by the gradient pattern. Arrows and letters indicate chronology of the hysteretic relationship.

over the initial increasing interval of the hysteris loop during the early melt season (Figs. 7, 8, and 13). Discharge vents are few in number and most have diameters on the scale of centimeters.

1997 output parameters increase mid-melt season with maximum subglacial sediment erosion occurring during maximum velocity (Figs. 8, 13B). Discharge vents have grown in size and number and have seemingly merged with one another forming terraces, which suggests an increase in braiding of the system. 1996 meltwater proxies generally decrease in the mid- to late melt season with the falling limb of the hysteresis loop (Figs. 7, 13B).

1997 meltwater proxies remain at high values into the late melt season, suggesting the drainage system has evolved to accommodate large influxes (Figs. 8, 13B). Output parameter values decline through the late melt season (Fig. 10) also suggesting the drainage system is no longer growing, but has the capacity to transmit influxes it receives. Late melt season increases in velocity unaccompanied by increases in meltwater proxies or output parameters suggest water storage is increasing (Figs. 7, 8, 10, 13). The cessation of discharge from some areas indicates abandonment or closure is occurring in the subglacial drainage system. The data do not unequivocally support the model, but the model does seem reasonable.

The smaller loop occurring within the 1996 hysteresis loop (Fig. 13A) is hypothesized to reflect short-term changes of meltwater input rather than changes in system configuration because it occurs over approximately a 10-day period. The 1997 hysteresis loop (Fig. 13B) also shows short-term changes during the late season, which are hypothesized to reflect changes of meltwater input.

DISCUSSION

Variations of ice-surface velocities have been shown to reflect changes in meltwater storage within the subglacial drainage system, as well as changes in the system configuration itself (Iken, 1977, 1981; Iken et al., 1983; Iken and Bindschadler, 1986; Kamb, 1987; Iverson et al., 1995; Iken and Truffer, 1997). Increased surface velocities suggest storage is occurring at the glacier bed, while decreasing velocities suggest the release of stored water as the capacity of an evolving drainage system accommodates large influxes of meltwater (Iken et al., 1983). Subglacial water storage might decouple the glacier from its bed and increase the sliding rates (Llibutry, 1968; Kamb, 1970). Force balance calculations by Hooke et al. (1989) suggest that the basal drag at Storglaciären was reduced by 16%–40% during high-velocity events.

The ability of the subglacial drainage system to remain open throughout the winter will affect how the system handles a large influx of meltwater at the start of the melt season. Water pressure increases associated with rising water storage may result in accelerated sliding rates (Iken, 1981; Lawson, 1993) and possibly even "uplift" events (Iken and Bindschadler, 1986). The basal water pressure is determined by the water supply and the resistance of the subglacial drainage channels to ice flow (Hooke et al., 1985;

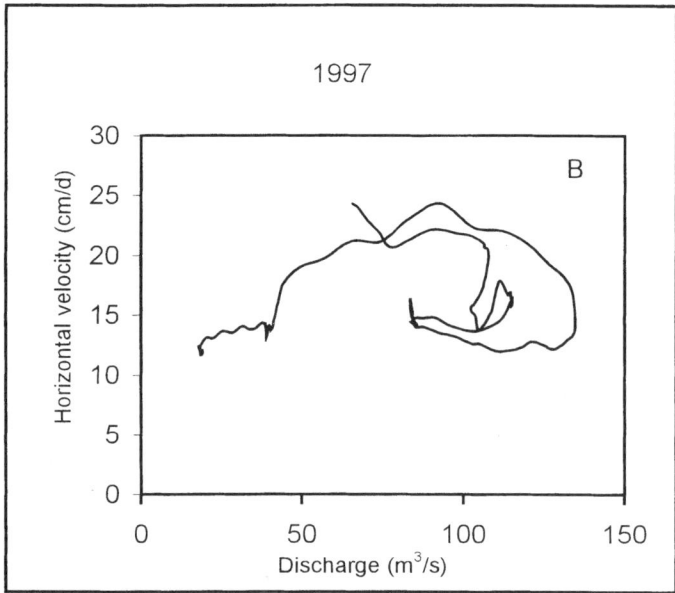

Figure 13. A, Hysteresis loop defined by the 1996 velocity and stream discharge. Chronology around the loop is the same order as in the model. B, Hysteresis loop defined by the 1997 velocity and stream discharge with a chronology similar to that in A.

Lawson, 1993). Diurnal variations of meltwater production or large melt and rain events may increase the water supply faster than channels can enlarge, building up the water pressure.

Basal water pressure plays an important role in basal sliding as it counteracts the weight of the glacier (Paterson, 1994) thus reducing the effective pressure. Increased basal water pressure may also reduce the shear strength of underlying permeable sediment, perhaps allowing it to deform more rapidly under the shear stress of the ice and increasing ice motion (Blake et al., 1994; Iverson et al., 1995). The amount of motion due to deforming subglacial sediments is unknown at this time, but the glacier does overlie sediments in at least some localities (Fig. 14), so it might be anticipated that the velocity includes some bed deformation (Iverson et al., 1995).

The percentage of water content in the basal ice layers may also strongly influence the deformation rate of the ice. Creep rate and recrystallization may be enhanced in the basal ice layers due to the presence of water at the ice-grain boundaries (Glen, 1955; Mellor and Testa, 1969; Baker and Gerberich, 1979). It is also unknown what effect the thick section of dirty basal ice that is present in the study area has on ice motion. While Baker and Gerberich (1979) found that secondary creep is enhanced in ice >-10 °C with debris inclusions, their results show that inclusion volume fractions greater than 0.05, secondary creep is not substantially enhanced and recrystallization is inhibited.

CONCLUSIONS

The ice motion survey at Matanuska Glacier during the 1996 and 1997 melt seasons records ice flowing out of an over-deepened basin, as evidenced by the seasonal changes of elevation.

Figure 14. Wintertime photograph of subglacial sediment frozen on to the base of the glacier and exposed by ramping of the ice. Single arrow indicates sediment contact with basal ice. Double arrow points to slickensided surface. A metal ice chipper near center of photo is resting on a large flap of frozen sediment that is curling downward under its own weight.

The velocity peaks in late June of both studied melt seasons, which follows increases of the meltwater proxies. The meltwater input is hypothesized to exceed the capacity of the drainage system early in the melt season. Glacially sourced tream discharge lags behind the ice velocity, suggesting that there was water storage at the glacier's bed. Frazil ice growth around discharge vent orifices indicate the subglacial drainage system is pressurized so that the subglacial water becomes supercooled. The vent TSS data show increased erosion at the bed during maximum melt season velocities. The drainage system is hypothesized to evolve to accommodate the large fluxes of seasonal meltwater by eroding canals in till. Field observations and dye-tracer tests (Lawson et al., 1998) suggest the canals are increasingly braided as the drainage system increases its capacity to transmit water throughout the melt season. During the late stages of the melt season, the drainage system is hypothesized to adjust to a reduced meltwater influx by abandonment and possibly even closure of some drainage paths.

The occurrence of supercooled water and a dirty basal ice facies at Matanuska Glacier make it an important location for studying the seasonal evolution of the subglacial drainage system. The subglacial drainage system configuration is hypothesized to be braided canals in till throughout the melt season. The stage of evolution of the subglacial drainage system configuration did control the effective pressure and therefore, the horizontal sliding velocity during the 1996 and 1997 ice-motion surveys. There are alternative hypotheses to the conceptual model of drainage system evolution presented here, but we believe this model best represents the observations made along the ice margin during current (Alley et al., this volume, Chapter 1, Evenson et al., this volume, Chapter 3) and past research (Strasser et al., 1996; Alley et al., 1998; Lawson et al., 1998) at Matanuska Glacier.

ACKNOWLEDGMENTS

The authors would like to thank the National Science Foundation and U.S. Army Corps of Engineers Cold Regions Research Engineering Laboratory in Anchorage, Alaska, (CRREL) for financial support, as well as the people at CRREL who have proven to be invaluable in data collection and analysis. We would also like to extend our thanks to the community of Glacier View, Alaska, especially Bill and Kelly Stevenson of Glacier Park Resort, whose help and hospitality have made the project successful and fun every year. Also, we thank the students from Lehigh University, Michigan State University, Augustana College, and the University of Kansas that have spent their summers collecting these data. The manuscript was significantly improved by the helpful comments of P. J. Fleisher and an anonymous reviewer.

REFERENCES CITED

Alley, R. B., 1996, Towards a hydrologic model for computerized ice-sheet simulations: Hydrological Processes, v. 10 p. 649–660.

Alley, R. B., Cuffey, K. M., Evenson, E. B., Lawson, D. E., and Strasser, J. C., 1997, How glaciers entrain and transport basal sediment: Physical constraints: Quaternary Science Reviews, v. 16, p. 1017–1038.

Alley, R. B., Lawson, D. E., Evenson, E. B., Strasser, J. C., and Larson G. J., 1998, Glaciohydraulic supercooling: a freeze-on mechanism to create stratified, debris-rich basal ice. 2. Theory: Journal of Glaciology, v. 44, p. 562–568.

Arcone, S. A., Lawson, D. E., and Delaney, A. J., 1995, Short-pulse radar wavelet recovery and resolution of dielectric contrasts within englacial and basal ice of Matanuska Glacier, Alaska, U.S.A.: Journal of Glaciology, v. 41, p. 68–86.

Bahr, D. B., and Rundle, J. B., 1996, Hysteresis in the relationship between glacier sliding velocities and basal water pressure *in* Colbeck, S. C., ed., Glaciers, ice sheets and volcanoes: A tribute to Mark F. Meier: U.S. Army Corps of Engineers Cold Regions Research Engineering Laboratory, Special Report 96–27, p. 10.

Baker, R. W., and Gerberich, W. W., 1979, The effect of crystal size and dispersed-solid inclusions on the activation energy for creep of ice: Journal of Glaciology, v. 24, p. 179–194.

Blake, E. W., Fischer, U. H., and Clarke, G. K. C., 1994, Direct measurement of sliding at the glacier bed: Journal of Glaciology, v. 40, p. 595–599.

Braithwaite, R. J., 1981, On glacier energy balance, ablation, and air temperature: Journal of Glaciology, v. 27, p. 381–391.

Davis, R. E., and Foote, F. S., 1953, Surveying: Theory and practice: New York, McGraw-Hill Book Company, 1021 p.

Fleisher, P. J., Muller, E. H., Cadwell, D. H., Rosenfeld, C. L., Thatcher, A., and Bailey, P. K., 1994, Measured ice-front advance and other surge related changes, Bering Glacier, Alaska [abs.]: Eos (Transactions, American Geophysical Union), v. 75, p. 63.

Fleisher, P. J., Cadwell, D. H., Tormey, B. B., Lissitschenko, P., and Dell, J., 1997, Post-surge changes and abundant frazil ice, eastern sector, Bering Glacier, AK: Geological Society of America Abstracts with Programs, v. 29, no. 6, p. 216.

Forest, T., 1994, Physics of frazil ice, *in* Daly, S. F., ed., International association for hydraulic research working group on thermal regimes: Report on frazil ice: U.S. Army Corps of Engineers Cold Regions Research Engineering Laboratory, Special Report 94–23, p. 1–4.

Glen, J. W., 1955, The creep of polycrystalline ice: Proceedings of the Royal Society of London, v. 128, p. 519–538.

Harper, J. T., Humphrey, N. F., Pfeiffer, W. T., and Welch., B. C., 1996, Short wavelength variations in the horizontal velocity field of a valley glacier, *in*

Colbeck, S. C., ed., Glaciers, ice sheets and volcanoes: A tribute to Mark F. Meier: U.S. Army Corps of Engineers Cold Regions Research Engineering Laboratory, Special Report 96–27.

Hock, R., and Hooke, R. L., 1993, Evolution of the internal drainage system in the lower part of the ablation area of Storglaciären, Sweden: Geological Society of America Bulletin, v. 105, p. 537–546.

Hooke, R. L., and Pohjola, V. A., 1994, Hydrology of a segment of a glacier situated in an over deepened basin, Storglaciären, Sweden: Journal of Glaciology, v. 40, p. 140–148.

Hooke, R. L., Wold, B., and Ove Hagen, J., 1985, Subglacial hydrology and sediment transport at Bondhusbreen, southwest Norway: Geological Society of America Bulletin, v. 96, p. 388–397.

Hooke, R .L., Miller, S. B., and Kohler, J., 1988, Character of the englacial and subglacial drainage system in the upper part of the ablation area of Storglaciären, Sweden: Journal of Glaciology, v. 34, p. 228–231.

Hooke, R. L., Calla, P., Holmlund, P., Nilsson, M., and Stroeven, A., 1989, A 3 year record of seasonal variations in surface velocity, Storglaciären, Sweden: Journal of Glaciology, v. 35, p. 235–247.

Hooke, R. L., Hanson, B., Iverson, N. R., Jansson, P., and Fischer, U. H., 1997, Rheology of till beneath Stroglaciären, Sweden: Journal of Glaciology, v. 43, p. 172–179.

Iken, A., 1977, Variations of surface velocities of some alpine glaciers measured at intervals of a few hours. Comparison with Arctic glaciers: Zeitschrift für Gletscherkunde und Glazialgeologie, v. 13, p. 23–35.

Iken, A., 1981, The effect of the subglacial water pressure on the sliding velocity of a glacier in an idealized numerical model: Journal of Glaciology, v. 27, p. 407–421.

Iken, A., and Bindschadler, R. A., 1986, Combined measurements of subglacial water pressure and surface velocity of the Findelngletscher, Switzerland. Conclusions about drainage system and sliding mechanism: Journal of Glaciology, v. 32, p. 101–119.

Iken, A., and Truffer, M., 1997, The relationship between subglacial water pressure and velocity of Findelengletscher, Switzerland, during its advance and retreat: Journal of Glaciology, v. 43, p. 328–338.

Iken, A., Röthlisberger, H., Flotron, A., and Haeberli, W., 1983, The uplift of Unteraargletscher at the beginning of the melt season. A consequence of water storage at the bed?: Journal of Glaciology, v. 29, p. 28–47.

Iverson, N. R., Hanson, R. B., Hooke, R. L., and Jansson, P., 1995, Flow mechanism of glaciers on soft beds: Science, v. 267, p. 80–81.

Kamb, B., 1970, Sliding motion of glaciers: Theory and observation: Reviews of Geophysics and Space Physics, v. 8, no. 4, p. 673–729.

Kamb, B., 1987, Glacier surge mechanism based on linked-cavity configuration of the basal water conduit system: Journal of Geophysical Research, v. 92, no. B9, p. 9083–9100.

Krimmel, R. M., and Vaughn, B. H., 1987, Columbia Glacier, Alaska: Changes in velocity 1977–1986: Journal of Geophysical Research, v. 92, no. B9, p. 8961–8968.

Knight, P. G., 1997, The basal ice layer of glaciers and ice sheets: Quaternary Science Reviews, v. 16, p. 975–993.

Lawson, D. E., 1979a, Characteristics and origins of the debris and ice, Matanuska Glacier, Alaska: Journal of Glaciology, v. 23, p. 437–438.

Lawson, D. E., 1979b, Sedimentological analysis of the western terminus region of the Matanuska Glacier, Alaska: U.S. Army Corps of Engineers Cold Regions Research Engineering Laboratory, Monograph 79–9, 112 p.

Lawson, D. E., 1993, Glaciohydrologic and glaciohydraulic effects on runoff and sediment yield in glacierized basins: U.S. Army Corps of Engineers Cold Regions Research Engineering Laboratory, Monograph 93–2, 108 p.

Lawson, D. E., Evenson, E. B., Strasser, J. C., Alley, R. B., and Larson, G. J., 1996, Subglacial supercooling, ice accretion, and sediment entrainment at the Matanuska Glacier, Alaska: Geological Society of America, Abstracts with Programs, v. 28, no. 3, p. 75.

Lawson, D. E., Strasser, J. C., Evenson, E. B., Alley, R. B., Larson, G. J., and Arcone, S. A., 1998, Glaciohydraulic supercooling: a freeze-on mechanism to create stratified, debris-rich basal ice. 1. Field evidence: Journal of Glaciology, v. 44, no. 14.

Lliboutry, L., 1968, General theory of subglacial cavitation and sliding of temperate glaciers: Journal of Glaciology, v. 7, p. 21.

Mellor, M., and Testa, R., 1969, Effects of temperature on the creep of ice: Journal of Glaciology, v. 8, p. 131–145.

Østrem, G., and Brugman, M., 1991, Glacier mass-balance measurements: a manual for field and office work: National Hydrology Research Institute Report no. 4.

Paterson, W. S. B., 1994, The physics of glaciers: New York, Pergamon Press, 480 p.

Pollard, D.,1980, A simple parameterization for ice sheet ablation rate: Tellus, v. 32, p. 384–388.

Röthlisberger, H., 1972, Water pressure in intra- and subglacial channels: Journal of Glaciology, v. 11, p. 177–203.

Röthlisberger, H., and Lang, H., 1987, Glacial hydrology, in Gurnell, A. M., and Clark, M. J., eds., Glacio-fluvial sediment transfer: New York, John Wiley and Sons, p. 207–284.

Stenborg, T., 1970, Delay of run-off of a glacier basin: Geografiska Annaler, v. A52, p. 1–30.

Strasser, J. C., Lawson, D. E., Evenson, E. B., Gosse, J. C., and Alley, R. B., 1992, Frazil ice growth at the terminus of the Matanuska Glacier, Alaska, and its implications for sediment entrainment in glaciers and ice sheets: Geological Society of America, Abstracts with Programs, v. 24, p. 78.

Strasser, J. C., Lawson, D. E., Larson, G. J., Evenson, E. B., and Alley, R. B., 1996, Preliminary results of tritium analyses in basal ice, Matanuska Glacier, Alaska, USA: Evidence for subglacial ice accretion: Annals of Glaciology, v. 22, p. 126–133.

Theakstone, W. H., and Tvis Knudsen, N., 1979, Englacial and subglacial hydrology, Austre Okstindbreen, Norway: Journal of Glaciology, v. 23, p. 434.

Walder, J. S., and Fowler, A., 1994, Channelized subglacial drainage over a deformable bed: Journal of Glaciology, v. 40, p. 3–15.

Weertman, J., 1961, Mechanism for the formation of inner moraines found near the edge of cold ice caps and ice sheets, in Goldthwait, R. P., ed., Glacial deposits: Stroudsburg, Pennsylvania: Dowden, Hutchinson and Ross, Inc., p. 208–221.

Williams, J. R. and Ferrins, O. J., Jr., 1961, Late Wisconsin and recent history of the Matanuska Glacier, Alaska: Arctic, v. 14, p. 82–90.

MANUSCRIPT ACCEPTED BY THE SOCIETY OCTOBER 8, 1998

Field evidence for the recognition of glaciohydrologic supercooling

Edward B. Evenson
Department of Earth and Environmental Sciences, Lehigh University, Bethlehem, Pennsylvania 18015
Daniel E. Lawson
U.S. Army Cold Regions Research and Engineering Laboratory, Anchorage, Alaska 99505
Jeffery C. Strasser
Department of Geology, Augustana College, Rock Island, Illinois 61201
Grahame J. Larson
Department of Geological Sciences, Michigan State University, East Lansing, Michigan 48824
Richard B. Alley
Earth System Science Center and Department of Geosciences, The Pennsylvania State University, University Park, Pennsylvania 16802
Staci L. Ensminger
Department of Earth and Environmental Sciences, Lehigh University, Bethlehem, Pennsylvania 18015
William E. Stevenson
Glacier Park Resort, HCO3 Box 8449, Palmer, Alaska 99645

ABSTRACT

Glaciohydrologic supercooling at Matanuska Glacier results in abundant and conspicuous summer ice growth even when air temperatures are constantly and significantly above freezing. Ice grows as frazil ice and anchor ice. Frazil ice grows unattached in high-velocity discharge water. Anchor ice grows at vent orifices, in subglacial conduits and canals, in moulins, and in fractures. Anchor ice can grow as large, debris-free platy crystals or, more commonly, as fine-grained, debris-rich laminated ice. To date, these types of summer ice growth have been reported at few glaciers and studied in detail at the Matanuska Glacier but we anticipate that glaciohydrologic supercooling and the associated ice growth are operating at numerous other glaciers with the appropriate geometries. The objective of this paper is to provide a photographic atlas, and brief explanation, of the ice growth forms and features observed at the Matanuska Glacier. We hope that it will enable other researchers to recognize the evidences of glaciohydrologic supercooling at additional glaciers and eventually establish glaciohydrologic supercooling, and the associated basal freeze-on, as a major debris entrainment and transportation mechanism.

INTRODUCTION

In a series of recent papers we have demonstrated the occurrence of glaciohydrologic supercooling and basal ice accretion (freeze-on) at the Matanuska Glacier in Alaska (Strasser et al., 1992, 1994a, b, 1996; Lawson et al., 1998) and developed the theoretical basis for the process in general (Alley et al., 1998, this volume). In these papers we demonstrated, using tritium and isotopic analysis, that the debris-laden basal ice at the Matanuska Glacier is young and accreted in an open hydrologic system. The interested reader is referred to these papers for details of the glaciohydrologic supercooling process and our interpretation of the origin of stratified basal ice.

Assuming our interpretations at Matanuska Glacier are

Evenson, E. B., Lawson, D. E., Strasser, J. C., Larson, G. J., Alley, R. B., Ensminger, S. L., and Stevenson, W. E., 1999, Field evidence for the recognition of glaciohydrologic supercooling, *in* Mickelson, D. M., and Attig, J. W., eds., Glacial Processes Past and Present: Boulder, Colorado, Geological Society of America Special Paper 337.

essentially correct the critical remaining question is, "does this process occur at other modern glaciers?" To answer this question it is necessary to have a broad spectrum of Quaternary scientists looking for, and able to recognize, the field evidence for glaciohydrologic supercooling and basal freeze-on, which are separate, but related, processes.

Glaciohydrologic supercooling can be documented directly by water temperature measurement in the field (Priscott and Fleisher, 1993; Fleisher et al., 1994; Ensminger et al., this volume). However, collecting accurate measurements of supercooling of hundredths of a degree is difficult and expensive and few field geologists have the equipment necessary to make sufficiently accurate temperature measurements. Fortunately, supercooling also manifests itself in the field in a variety of easily recognized, unique, ways. Fleisher et al. (1993, 1995, 1997), Natel and Fleisher (1994), and Molnia et al. (1994) have observed frazil ice and frazil ice rafts in supercooled meltwater discharging into ice-marginal lakes at the Bering Glacier, Alaska. Roosa and Fleisher (1997) suggest the rapid rates of sedimentation in conditions that are not conducive to chemical clay flocculation may be due, in part, to the occurrence of supercooled water. The purpose of this paper is to present and explain the field evidences for the occurrence of the supercooling process at the Matanuska Glacier.

At the Matanuska Glacier, evidence for supercooling and ice growth exists at the very margin of the glacier where, in the summer, numerous subglacial conduits discharge into shallow standing water. Evidence of supercooling and ice growth also occurs upglacier in debris bands and moulins of the white glacial ice.

EVIDENCE OF SUPERCOOLED WATER

Free frazil and frazil aggregates

At the Matanuska Glacier, frazil crystals grow even during the summer in high-velocity subglacial discharge waters. Initially, frazil crystals are microscopic but rapidly evolve and grow into lozenge-shaped crystals several millimeters in size (Daly, 1984). These larger crystals sinter into "frazil aggregates" (Fig. 1) that float in the discharge water at vent orifices. The formation and growth of frazil crystals and aggregates is most prolific when air temperatures remain well above freezing day and night and meltwater production is at a maximum. Frazil crystals, growing in turbulent silt-laden subglacial water, contain no debris. It is excluded from the crystals in the freezing process (Fig. 1). Sintered frazil aggregates, however, form a complex network of interlocking crystals that efficiently trap suspended silt (rock flour) in the interstitial pore space. As a result the interiors of frazil aggregates quickly become packed with silt (Fig. 2). The presence of frazil plates and aggregates in summer meltwater is possible only if supercooling conditions exist in the subglacial system (Ashton, 1979, 1986).

Frazil growth on submerged instruments

To measure temperature, conductivity, discharge, and suspended sediment concentration we regularly suspend instruments in subglacial conduits and vents (Fig. 3). Under conditions of high discharge it is common for ice crystals to grow as "rosettes" on instruments and ropes (Fig. 4) used to suspend them in the vents. Fleisher et al. (1996) have similarly reported ice growth on instruments at the Bering Glacier. Again, the growth of crystals, underwater, when air temperatures are continuously above freezing necessitates glaciohydrologic supercooling.

GROWTH OF ANCHOR ICE TERRACES

One of the most striking features at the Matanuska is the development of anchor ice (Daly, 1994) terraces which grow around subglacial discharge vents in the shallow "lake" that forms in front of the glacier every summer. The terraces are completely covered by water and inaccessible at high discharge. At lower discharge rates, water level drops and the growth process and structure of the terraces can be studied.

High discharge

"High discharge" occurs diurnally on sunny, warm days and for several days following heavy precipitation. At high discharge the terraces are rapidly growing, but completely covered with vent discharge water (Fig. 5). At this stage vents have a large hydrostatic head and appear as fountains or upwellings. They most often have a large, flat (Fig. 5) or slightly domed central upwelling area within a "tread" of anchor and frazil ice, surrounded by drop off, or "riser." At large discharge vents the riser may be as much as 1.5 m in height and the tread may be 10–15 m across (Fig. 6). We have measured rates of vertical growth, using spikes driven into the terraces near the tread/riser contact, of 7 cm in two weeks. Jon Boothroyd (personal communication, 1997) reports having seen terraces of the above description along the Malispina Glacier's margin in the 1980s.

Intermediate discharge

At intermediate discharge water flow is low enough that parts of the anchor ice terraces are exposed. The clean, freshly formed anchor ice can be seen growing upward and outward extending the breadth and height of the terrace. The uppermost, newest, terrace surface anchor ice (Fig. 7) resembles clean frazil-floc ice. It is weakly sintered together when new, although it will support a person's weight. After a short period of subaerial exposure much of the clean frazil-floc ice melts off and begins to expose older terrace ice (Fig. 8), which is better investigated at low discharge.

Low discharge

At low discharge, many of the anchor ice terraces are partially exposed (Figs. 3, 4, and 8) by the falling water levels. Com-

Figure 1. An aggregate of frazil ice collected from a summertime discharge vent orifice. Note especially how clean the newly grown crystals are compared to the remainder of the interlocking network of frazil crystals that have trapped silt. Green tobacco can is 6.35 cm in diameter.

Figure 3. Instrumentation that measures the conductivity-temperature-depth (CTD) of subglacially discharged waters is suspended by nylon rope into the center of a vent orifice at the ice margin. Roughly circular terraces and anchor ice terraces are well exposed during periods of low discharge. Stakes are used to measure height of upwelling during high discharge and to measure rate of ice growth on terraces.

Figure 2. The silt-laden interior of a frazil ice aggregate. The complex network created by the platy crystals traps sediment carried in suspension in subglacial discharge waters.

Figure 4. Frazil rosettes on CTD (conductivity-temperature-depth) ropes and cage. Frazil ice growth proceeds rapidly on instruments and ropes once it begins. Instruments must be removed from the discharge vents at least once a day during peak flow to prevent the freeze-in of the equipment. Anchor ice terraces exposed during intermediate discharge from vents.

plex, roughly circular terraces (Fig. 3) can be seen around upwellings fed by individual vents that discharged under shallow lake water (Fig. 9). Terraces fed by lines of vents at, or above, lake level are elongate and almost completely exposed (Fig. 10) at low discharge. During exposure at low discharge, most of the weakly sintered, clean anchor ice melts off or is removed by flowing water exposing the better indurated, deeper anchor ice on the terraces. This ice has crude primary stratification (Fig. 8) caused by ice grain size differences (Fig. 11) and differing debris content and looks somewhat like "immature" basal stratified ice (Strasser et al., 1996). It is this similarity in appearance that first led the authors to suggest a genetic relationship between the

Figure 5. Large anchor ice terraces submerged by high subglacial discharge. The large hydrostatic head causes these terraces to have a broad, flat tread. Riser height is approximately 60 cm. Person for scale on right.

Figure 7. Newly accreted anchor ice on a terrace surface (center) resembles clean frazil-flock ice. Turbid, silt-laden water discharges at bottom of photo. Silt-packed frazil floc ice visible along top of photo.

Figure 6. A large discharge vent with a prominent riser (about 1.5 m in height), and a large central upwelling. Note turbid discharge water.

Figure 8. Older anchor ice exposed on the "riser" of an anchor ice terrace at low discharge after most of the recently accreted anchor ice has been melted off. The anchor ice is laminated and has a strong resemblance to the stratified basal ice facies. Ice axe is 105 cm long. Upwelling that produced this terrace is visible in left center of photo. See Figure 10 for view of entire terrace.

modern terrace anchor ice and the stratified basal ice facies (Fig. 12) of the Matanuska Glacier.

EXPOSED VENTS

In the late fall, temperatures drop and water level in the "lake" falls completely exposing the vents (described above) that were active during the summer melt season. At this time, detailed examination of the vent orifices and anchor ice terraces is possible. The exposed vents are very complicated because the current year's active vents almost always develop within previously accreted, debris-laden basal freeze-on material. This occurs because basal freeze-on material is at the sole of the glacier and has a higher porosity than the overlying white glacial ice. Therefore, when the spring subglacial drainage begins to develop, basal water takes advantage of the higher permeability of the basal debris zone and develops within it, as well as taking advantage of remnants of the previous year's drainage system.

Interpretation of vent development, stratigraphy, and ice growth is quite complicated because the host material ice, within which vents develop by karstlike melting, look strikingly similar to the anchor and frazil ice. The similar appearance of the host ice and anchor ice is not surprising though, because they have a common origin and are different only in age. Both are weakly laminated and debris-rich and a karstlike system of channels (Fig. 13) permeates the entire deposit. In the winter, sublimation enhances the visible

Figure 9. Large anchor ice terrace exposed during low discharge. Orifice of discharge vent is in the center of a roughly oval pond (2.5 × 8 m) located to the right of the person in the photo. Newly accreted anchor/frazil ice is white and lines the edge of the upwelling pond.

Figure 11. The difference in ice grain sizes causes the deeper anchor ice from ice terraces to have a crude primary stratification.

Figure 10. Prograding anchor ice terrace at intermediate discharge. Water just covers terrace tread and riser. This is the same terrace as shown in Figure 8.

Figure 12. A block of the stratified basal ice facies. Note the strong resemblance in appearance (stratification, debris layers) to the modern terrace anchor ice.

contrast between ice and debris, making it easier to see the structure and debris distribution in the vent ice (Fig. 14). The similarity of vent ice to the basal stratified facies ice, described later in this paper and our previous papers, is striking (Strasser et al., 1994a, b, 1996).

OBSERVABLE FEATURES DUE TO THINNED ENGLACIAL ICE

Unroofed conduits

Investigations along the edge of the glacier demonstrate that at most places debris-laden ice underlies (underplates) the clean white glacier ice we refer to as "englacial ice." Debris-rich ice also underlies the white ice some distance upglacier from the terminus as is demonstrated both by ground-penetrating radar (GPR) profiles (Arcone et al., 1995; Lawson et al., 1998) and the occurrence of "fensters" or windows (Fig. 15) that allow us to look beneath the white ice some distance (10s to 100s of meters) back from the terminus. When the white ice thins enough by ablation that the englacial ice over a subglacial conduit melts off, or collapses, it often exposes black, debris-rich ice (Fig 15) identical, in all respects, to that seen in and around active and recently exposed vents. In these roughly circular windows, frazil-flocs floating in basal meltwater are common and the sharp, usually angular, unconformity (Fig. 15) between the accreted, debris-laden ice and the clean englacial ice is striking. The sharp angular unconformity is due to the fact that the stratification in the englacial ice is most commonly

Figure 13. A summertime close-up view of vent ice and karstlike channels in an uplifted discharge vent. White ruler on right of photo is 6 in. long.

Figure 14. A wintertime close-up view of the uplifted ice. Sublimation has enhanced visibility of the structure and debris distribution in the ice, making its appearance even more strikingly similar to that of the basal ice facies. Ice axe is 105 cm long.

Figure 15. Fenster located approximately 100 m upglacier from the ice margin in the englacial ice exposing a gentle upwelling from a subglacial vent and debris-rich anchor ice. Frazil ice flocs floating in subglacial meltwater are common in these "windows" through the ice. The typically angular unconformity between the debris-rich accreted ice and the clean englacial ice overlying it is visible in shadows below person for scale. A "silt-bar" deposited as the velocity dropped at the vent mouth is visible in the right center of the upwelling pond.

Figure 16. The escape of pressurized, silt-laden subglacial water through a very small hole in the thin ice roof of a conduit.

nearly horizontal (snow stratification) while the lamination in the underplated ice, for reasons as yet unknown, usually dips 15° to 20° upglacier.

Frazil cones

Another feature that develops as a result of thinning of the englacial ice above the active hydrostatically charged subglacial drainage system is the growth of rare, but spectacular, "miniature frazil cones." As the englacial ice "roof" over a subglacial conduit thins, small holes sometimes develop and pressurized, silt-laden subglacial water escapes. This pressurized water will initially spatter out through a small hole (Fig. 16). With time the hole will enlarge and a stream of pressurized, supercooled water, about the size of that from a garden hose, may be emitted. As the pressure on this supercooled water is released, frazil ice will grow as a small cone or terrace surrounding the opening (Fig. 17). A subaerially constructed feature such as this, observed forming when the air temperature is significantly above freezing, could only develop in a supercooled system.

Figure 17. A frazil cone formed by the escape of supercooled subglacial waters from a conduit. As pressure is released upon exiting the conduit, the supercooled water freezes and ice accumulates, forming the cone feature. Note that ice growth excludes all sediment so ice is debris free. These delicate features melt quickly after water flow ceases. Tobacco can for scale.

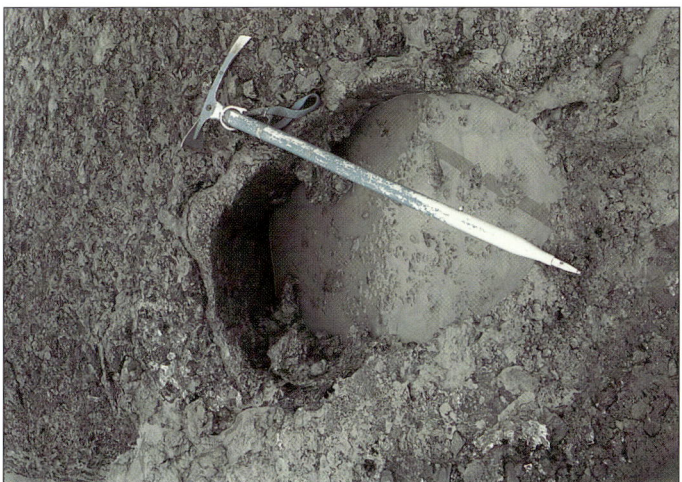

Figure 19. Annular growth of ice in an exposed conduit. Platy frazil crystals floating in silt laden water.

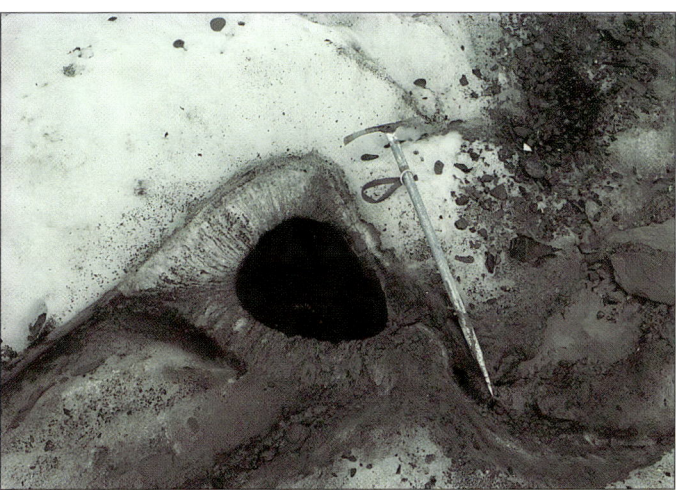

Figure 20. Large, debris-free ice crystals grow orthogonally into a conduit from the conduit walls in all directions. This is one of two common types of ice growth that occur in a conduit that carried subglacial waters. Debris on ice crystals at bottom of photo is surface debris, not debris in conduit ice.

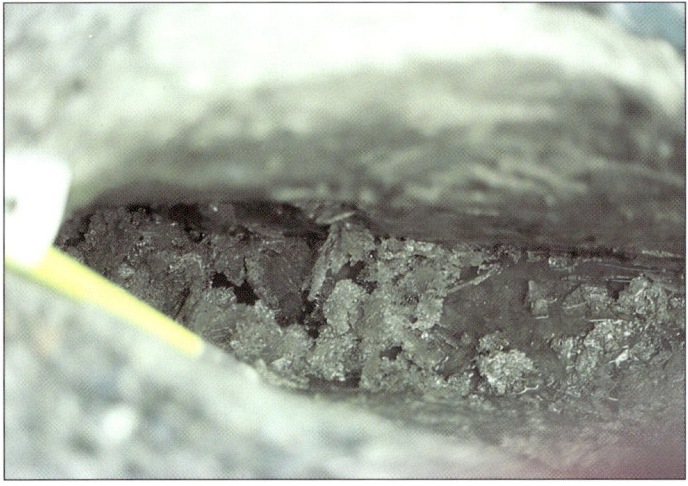

Figure 18. Large platy ice crystals attached to a moulin wall after water level has dropped. The water is subglacially sourced, as evidenced by the presence of debris. Yellow blur to left is the handle of an 80-cm-long ice axe.

OTHER FEATURES DUE TO ICE CRYSTAL GROWTH

In moulins

The Matanuska Glacier has numerous moulins through white englacial ice in its terminal zone. Our previous dye-tracing studies (Lawson et al., 1998) demonstrate that they are connected directly to the subglacial drainage system. Occasionally, subglacial water will "back up" and partially fill moulins from below with pressurized, gently upwelling, silty, subglacial water. As the water level in these moulins drops, large platy ice crystals attached to the moulin walls are exposed (Fig. 18). The platy crystals are also observed floating in the water (Fig. 19). Annular ice growth on the moulin walls (Fig. 19) also occurs during these highstands of subglacial water.

In conduits

Two different, but relatively common, types of ice growth in exposed conduits have been observed. The first type consists of large, debris-free crystals growing into the conduit orthogonally from the conduit walls (Fig. 20). The second type of ice growth in conduits consists of debris-rich anchor ice growing on the con-

Figure 21. Subglacially sourced waters may also generate anchor ice growth into a conduit, trapping debris in interstitial pore space so that the anchor ice on the conduit walls looks identical to the debris-rich anchor ice that grows at vents and terraces. This is the second of two types of ice growth that occur in a conduit transporting subglacial waters.

Figure 23. In the presence of a unidirectional current, ice crystal growth along the sides of the ice-marginal fracture will be in a direction oriented into the current. The ice crystals grown in this case are large and platy, like those that grow in water-filled moulins. Pattern in water in base of fracture demonstrates that the water is gently upwelling and flowing left to right.

Figure 22. Following a 10-cm drop in water level, ice crystals growing inward from opposite sides of an upwelling in a fracture are clearly visible. Ice axe and pack for scale.

Figure 24. Debris bands crisscross the terminus area of the Matanuska Glacier, Alaska, and are a most noticeable feature. Melting and surface debris makes the prominent debris band in the center of the photo look much thicker than its average 0.5 m thickness.

duit walls (Fig. 21). The contact between the debris-rich ice and the clean englacial ice containing the conduit is very sharp and the ice linings are commonly 7.5–15 cm (3–6 in.) in thickness. We interpret them to be anchor ice linings of conduits that have trapped silt in the interstitial pore space. The debris-rich ice is identical in appearance to that formed in active vents and terraces.

In water-filled fractures

Near the edge of the glacier where water stands in fractures at the same level as the ice marginal lake (Fig. 22) and subglacial water gently upwells, anchor ice crystals commonly grow inward on the sides of the fractures. The crystals are identical in size and shape to crystals that grow around active vents. If unidirectional currents rather than upwelling are operating, large platy crystals, oriented into the current (Fig. 23) can develop. Both types of ice growth can eventually cause the fracture to heal completely closed by inward growth from the opposing walls. Again, these types of ice growth, occurring when temperatures remain consistently above freezing, can only occur if the subglacial water is glaciohydrologically supercooled.

Figure 25. A fine-grained silt debris band can be traced directly to the basal freeze-on ice. The thick band on the right side of the photo (arrow) traces toward the center of the photo until it joins the basal ice (double arrow), demonstrating a genetic relationship between this type of debris band and the subglacial hydrologic system.

Figure 27. A bilateral symmetry is commonly preserved by the continued growth of debris-trapping anchor ice inward from the opposing walls of the fracture. Tobacco can rests 3 cm below the "middle" of the debris band.

Figure 26. Anchor ice growing in from both sides of a fracture (center of photo) has closed off most of the fracture with debris-laden anchor ice.

Figure 28. A silt band with several centrally located orifices.

Debris bands

Numerous, conspicuous, black debris bands (Fig. 24), which elsewhere in the literature are commonly referred to as "shear planes," characterize the terminus area of the Matanuska (Evenson et al., 1998). At the Matanuska there are two common, distinct and different types of debris bands: (1) fine-grained silt bands, and (2) coarse-grained slate-chip bands. The slate-chip bands are the result of supraglacial rock debris finding its way into surficial fractures and crevasses—details of their origin is not the concern of this paper. The fine-grained silt bands contain only silt and occasional sand and water rounded pebbles—all clearly of subglacial fluvial origin. Careful study of fine-grained silt debris bands clearly demonstrates that they are the result of anchor ice growth in fractures that were hydraulically connected to the subglacial drainage system. In some cases the fractures can be traced to direct contact with basal freeze-on ice (Fig. 25) but in most cases this connection is at depth and unexposed. The most persuasive argument for the subglacial origin of debris bands comes from seeing them in all stages of development. The earliest recognizable stage of development is that of an open fracture with platy ice crystals lining the fracture walls. At this stage, ice growth in the fractures looks nearly identical to that described in conduits, moulins, and vents. The only difference is the shape of the enclosing ice (i.e., an elongate fracture verses a circular conduit or vent). As the fractures continue to accrete anchor ice on the opposing fracture walls, they begin to grow closed along most of their length and water is forced into a series of small orifices along the fracture's length, partially healing the fracture closed (Fig. 26). If water flow, and hence ice growth, is cut off before the orifices have completely closed, they become "fossilized" and are easily recognizable as previous water passages (Fig. 26). As anchor ice continues to grow in the orifices and trap debris, the debris bands "heal" completely closed but commonly preserve a bilateral symmetry (Fig. 27) demonstrating inward growth from the opposing fracture walls. All variants ranging from concentrically banded fracture fillings with multiple central orifices (Fig. 28) to massive debris bands with no internal structure exist at the Matanuska Glacier. Our preliminary investigation (Evenson et al., 1998) of the tritium and isotopic signatures of debris bands demonstrates that they are the same age and have the same source as basal laminated ice—a conclusion supported by the field relationships.

SEASONAL EXPOSURES

The characteristics of the thick section of laminated basal ice at the Matanuska, which we have demonstrated to be the result of glaciohydrologic supercooling and the resultant basal freeze-on (Strasser et al., 1996; Lawson et al., 1998), are distinctive enough to be used as proof that this process is occurring at the Matanuska Glacier and/or where similar basal ice is encountered. The laminated basal ice manifests itself quite differently during summer (melting) and winter (sublimation), and we will discuss them separately.

Summer exposures of basal ice

In the summer, laminated basal ice is accessible at numerous localities where it is exposed by headward sapping by subglacial streams (Fig. 29) and along icefall cliffs (Fig. 30) where it is not covered by debris flows generated by its own melting. The black basal ice has a very low albedo and melts prolifically through the summer melt season producing the debris flows described so masterfully by Lawson (1979). Surface melting obscures the details of the laminated basal ice but layering, on both the macroscale (Figs. 29 and 30) and microscale (Figs. 31 and 32), is clearly evident and has been described in detail by Lawson (1979). Even in the summertime the characteristic suspension of angular silt clasts in clear glacial ice can be seen if a freshly broken piece is washed free of surface debris and viewed with the sun in the background (Fig. 33).

Winter exposures of vent and basal stratified (laminated) ice

In the winter surface melting stops, many vents and basal sections are elevated by ice flow, and sublimation etches out the details of the recent vent ice and laminated basal ice. At this time the unique characteristics of basal ice are best displayed. In recent papers (Strasser et al., 1994a, 1996; Lawson et al., 1998) we have discussed the physical and isotopic characteristics of laminated basal ice in detail and attempted to explain the origin of many of the features observed. In this paper we will restrict

Figure 29. Laminated basal ice exposed during summer by the headward sapping of a subglacial stream. Basal laminated ice zone is approximately 5 m thick at this site. Large boulder on left is approximately 1 m in diameter.

Figure 30. The laminated basal ice is readily accessible along cliffs not plagued by debris flows. This section is approximately 5 m thick.

our review of basal ice features to those that are always present, field recognizable, and most diagnostic of glaciohydrologic supercooling and basal freeze-on.

Uplifted vents. These are the same vents referred to earlier in this paper as "exposed vents" and we have followed the uplift (ice flow exceeding ablation), draining, and exposure of many from summer activity to winter quiescence. When they are uplifted and drained, it is possible to enter the larger ones and examine their geometry and ice in detail. Most are complicated "karst systems" with many orifices and complex ice growth (Fig. 34). Some are lined with large platy crystals that formed in the late stages of winter freeze-up. Ice from uplifted vents displays exactly the same characteristics (Figs. 13 and 14) as that described for exposed vents because it is the same ice; it has simply been

Figure 31. Summer melting of the basal ice facies obscures some of the fine details within the basal ice, however layering on the macroscale is still readily apparent.

Figure 33. Silt clasts in clear ice are visible during the summer when the surface debris is removed from a sample and it is held up to sunlight.

Figure 32. Layering of basal ice on the microscale is still apparent during the summer melt months. Room key (7.5 cm long) for scale.

Figure 34. Uplifted vents, which have been drained of their pore water, are composed of the same ice as that of the previously discussed exposed vents. Sublimation has enhanced contrast, showing the debris distribution within the ice more clearly. Yellow ice axe is 80 cm long.

elevated, drained of interstitial water, and its visible contrast has been enhanced by sublimation.

Stratified, debris-rich ice. We have already published (Strasser et al., 1996; Lawson et al., 1998) detailed descriptions of the characteristics, and inferred origin, of stratified, debris-rich basal ice and will only briefly review them here. The outstanding characteristic of winter exposed basal ice is its stratified or laminated appearance (Fig. 35). The stratification is subhorizontal when viewed along strike and dips gently upglacier when viewed perpendicular to ice flow (Fig. 35). Stratification is caused by the alternation of lenses of debris-rich, debris-poor, and debris-free ice (Figs. 35 and 36). In some exposures, the stratification resembles cross-bedding (Figs. 35 and 37), but the cross-bedding is apparent rather than real as the silt-sized debris that is responsible for the layering is carried in suspension, not traction. Occasionally, the fine-grained debris will be suspended in crystal clear ice (Fig. 38, lower half of photo), or it will abruptly transition from lacking preferred orientation to having distinct lamination (Fig. 38, center of photo). Another common characteristic of basal ice is the inclusion of large silt clasts suspended in bubble-free clear ice (Fig. 36). In some cases the silt clasts are broken into angular aggregates that could be readily reassembled like a puzzle by the removal of the inter-silt ice (Fig. 39).

Figure 35. A winter exposure of well-laminated basal ice. Sharp contact between the laminated debris-rich basal ice and clean, white englacial ice is clearly visible on upper right of photo. Angular unconformity between laminations in basal ice and overlying white ice is well displayed.

Figure 37. The stratified ice resembles cross-beds in some exposures.

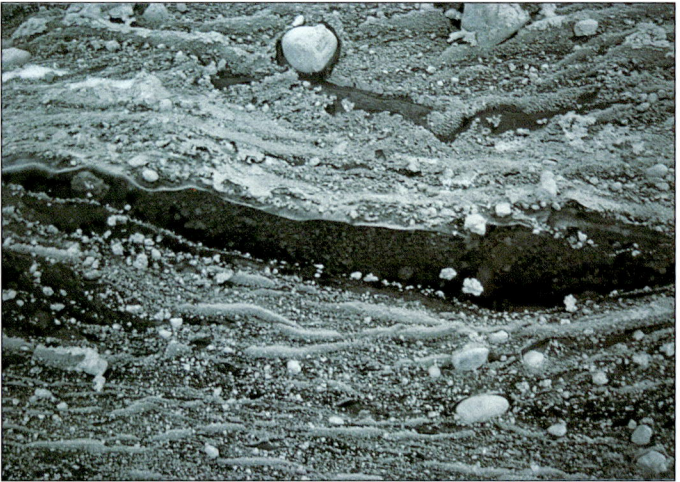

Figure 36. Alternating layers of debris-rich, debris-poor, and debris-free ice cause the observed stratification. Small blebs in crystal clear ice in center of photo are silt "clasts." Large stone surrounded by rim of clear ice (top center) is 8 cm in diameter.

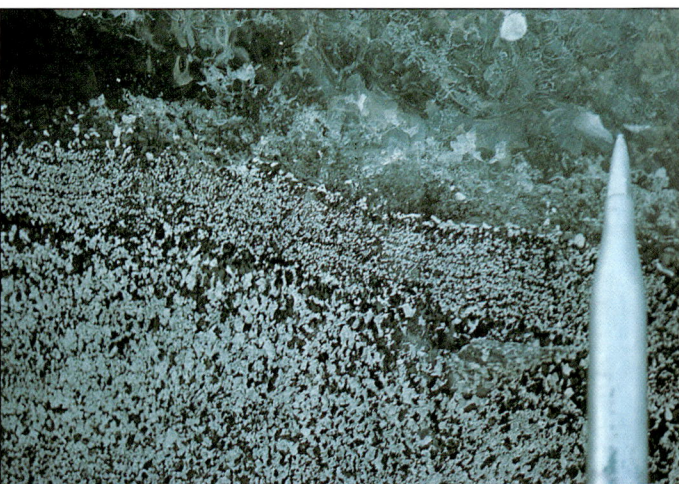

Figure 38. A transition caused by the suspended debris lacking lamination and a preferred orientation (bottom half of photo) to the debris being laminated or having a preferred orientation (center portion of photo) to clear ice (top of photo).

CONCLUSIONS

The ice facies described here and in Lawson et al., 1998, occur in multiple exposures around the margin of the Matanuska Glacier and are distinctive and reoccurring. We have demonstrated that ice of this type must have a unique genesis that is the result of a process that operates when, and only when, conditions are appropriate for glaciohydrologic supercooling and basal freeze-on (Alley et al., 1998, this volume; Lawson et al., 1998; Strasser et al., 1996).

ACKNOWLEDGMENTS

Funding for ongoing research of the glaciohydrologic system at the Matanuska Glacier has been provided through grants from the National Science Foundation and the U.S. Army Cold Regions Research and Engineering Laboratory (CRREL).

REFERENCES CITED

Alley, R. B., Lawson, D. E., Evenson, E. B., Strasser, J. C., and Larson G. J., 1998, Glaciohydrologic supercooling: A freeze-on mechanism to create stratified, debris-rich basal ice. II. Theory: Journal of Glaciology, v. 44, no. 148, p. 563–569.

Arcone, S. A., Lawson, D. E., and Delaney, A. J., 1995, Short-pulse radar wavelet recovery and resolution of dielectric contrasts within englacial and basal ice of Matanuska Glacier, Alaska, and U.S.A.: Journal of Glaciology, v. 41, p. 68–86.

Ashton, G. D., 1979, River ice: American Scientist, v. 67, no. 1, p. 38–45.

Figure 39. Stratified basal ice overlies clear ice with abundant angular silt aggregates (above ice axe head) in the basal ice facies. This section was exposed when winter movement uplifted basal ice forming an overhang and a cave.

Ashton, G. D., 1986, River and lake ice engineering: Littleton, Colorado, Water Resources Publications, p. 261–372.

Daly, S. F., 1984, Frazil ice dynamics: Hanover, New Hampshire, U.S. Army Cold Regions Research Engineering Laboratory Monograph 84-1, 46 p.

Daly, S. F., 1994, Evolution of frazil ice in natural water bodies, in Daly, S. F., ed., International Association for Hydraulic Research Thermal Regimes: Hanover, New Hampshire, U.S. Army Cold Regions Research and Engineering Laboratory Special Report 94-23, p. 11–17.

Evenson, E. B., Ensminger, S. L., Larson, G. J., Lawson, D. E., Alley, R. B., Strasser, J. C., 1998, "Shear planes" and the "dirt machine" revisited: The debris bands at the Matanuska Glacier are not "shear planes": Geological Society of America Abstracts with Programs, v. 30, no. 2, p. 64.

Fleisher, P. J., Franz, J. M., and Gardner, J. A., 1993, Bathymetry and sedimentary environments in proglacial lakes at the eastern Bering Piedmont Glacier of Alaska: Journal of Geological Education, v. 41, p. 267.

Fleisher, P. J., Muller, E. H., Cadwell, D. H., Rosenfeld, C. L., Thatcher, A., and Bailey, P. K., 1994, Measured ice-front advance and other surge related changes, Bering Glacier, Alaska [abs.]: Eos (Transactions, American Geophysical Union), v. 75, p. 63.

Fleisher, P. J., Muller, E. H., Cadwell, D. H., Rosenfeld, C., Gerhard, D., Shaw, L., and Mitteager, W., 1995, Resurgence and ice-front activity, eastern sector, Bering Glacier, Alaska: Geological Society of America, Abstracts with Programs, v. 27, no. 6, p. A–210.

Fleisher, P. J., Muller, E. H., Bailey, P. K., and Cadwell, D. H., 1996, Subglacial discharge and sediment load fluctuations during the 1993–1995 surge of Bering Glacier, Alaska: Geological Society of America Abstracts with Programs, v. 28, no. 7, p. 110–111.

Fleisher, P. J., Cadwell, D. H., Tormey, B. B., Lissitschenko, P., and Dell, J., 1997, Post-surge changes and abundant frazil ice, Eastern Sector, Bering Glacier, AK: Geological Society of America Abstracts with Programs, v. 29, no. 6, p. 216.

Lawson, D. E., 1979, Sedimentological analysis of the western terminus region of the Matanuska Glacier, Alaska: Hanover, New Hampshire, U.S. Army Cold Regions Research and Engineering Laboratory Report 79-9, p. 112.

Lawson, D. E., Strasser, J. C., Evenson, E. B., Alley, R. B., Larson, G. J., and Arcone, S. A., 1998, Glaciohydrologic supercooling: A freeze-on mechanism to create stratified, debris-rich basal ice. I. Field evidence: Journal of Glaciology, v. 44, no. 148, p. 547–562.

Molnia, B. F., Post, A., and Fleisher, P. J., 1994, Unusual hydrological events related to the 1993–1994 surge of the Bering Glacier, Alaska [abs.]: Eos (Transactions, American Geophysical Union), v. 75, p. 63.

Natel, E. and Fleischer, P. J., 1994, Pre-surge meltwater properties in ice-contact lakes, Eastern Peidmont Lobe, Bering Glacier, AK: Geological Society of America Abstracts with Programs, v. 26, no. 3, p. 65.

Priscott, G. W., and Fleisher, P. J., 1993, Subglacial sources of meltwater discharge in an emerging ice-marginal channel, Bering Glacier, Alaska: Geological Society of America Abstracts with Programs, v. 25, no. 2, p. 72.

Roosa, B. S., and Fleisher, P. J., 1997, Chemical and physical factors that influence clay flocculation in ice-contact lakes; Bering Glacier, Alaska: Geological Society of America Abstracts with Programs, v. 29, no. 1, p. 76.

Strasser, J. C., Lawson, D. E., Evenson, E. B., Gosse, J. C., and Alley, R. B., 1992, Frazil ice growth at the terminus of the Matanuska Glacier, Alaska, and its implications for sediment entrainment in glaciers and ice sheets: Geological Society of America, Abstracts with Programs, v. 24, no. 3, p. 78.

Strasser, J. C., Evenson, E. B., Larson, G. J., and Lawson, D. E., 1994a, Sediment entrainment and net freeze-on to the base of the Matanuska Glacier, Alaska, as indicated by bomb-produced tritium: Geological Society of America Abstracts with Programs, v. 26, no. 3., p. 75.

Strasser, J. C., Lawson, D. E., Evenson, E. B., and Larson, G. J., 1994b, Crystallographic and mesoscale analyses of basal zone ice from the terminus of the Matanuska Glacier, Alaska: Evidence for basal freeze-on in an open hydrologic system: Geological Society of America Abstracts with Programs, v. 26, no. 7, p. 177.

Strasser, J. C., Lawson D. E., Larson G. J., Evenson, E. B., and Alley, R. B., 1996, Preliminary results of tritium analyses in basal ice, Matanuska Glacier, Alaska, USA: evidence for subglacial ice accretion: Annals of Glaciology, v. 22, p. 126–133.

MANUSCRIPT ACCEPTED BY THE SOCIETY OCTOBER 8, 1998

ic composition of vent discharge from the Matanuska Glacier, Alaska: Implications for the origin of basal ice

Daniel D. Titus*
Department of Geological Sciences, Michigan State University, East Lansing, Michigan 48824
Grahame J. Larson
Department of Geological Sciences, Michigan State University, East Lansing, Michigan 48824
Jeffrey C. Strasser
Department of Geology, Augustana College, Rock Island, Illinois 61201
Daniel E. Lawson
U.S. Army Cold Regions Research and Engineering Laboratory, Anchorage, Alaska 99505
Edward B. Evenson
Department of Earth and Environmental Sciences, Lehigh University, Bethlehem, Pennsylvania
Richard B. Alley
Earth System Science Center and Department of Geosciences, The Pennsylvania State University,
University Park, Pennsylvania 18015

ABSTRACT

The ^{18}O, D composition, and ^{3}H activity was measured in subglacial discharge samples collected during the summer of 1995 from vents along the margin of the Matanuska Glacier, Alaska. Application of a simple open two-component reservoir model indicates that $^{18}\delta$O, δD, and ^{3}H activity for discharge are within the requisite ranges to form basal ice, and that there is a genetic relationship between the two. Additionally, the temporal variability of δ^{18}O in the basal ice can be attributed to relative amounts of meltwater and rainfall runoff present in the subglacial discharge.

INTRODUCTION

Historically, formation of debris-rich basal ice observed in many temperate and subpolar glaciers has been attributed to the regulation process (Weertman, 1961, 1964; Boulton, 1972; Iverson, 1993). This process occurs as the result of phase changes caused by either variations in temperature at the glacier sole or pressure fluctuations in a glacier near its pressure-melting point. These phase changes may occur in response to (1) isolated cold patches or permafrost at the glacier sole (Robin, 1976), (2) as cold winter air penetrates to the sole (Weertman, 1961; Clarke et al., 1984), or (3) through melting on the upstream side and subsequent refreezing on the downstream side of an obstacle encountered by moving ice at the base of the glacier (Kamb and LaChapelle, 1964; Iverson and Semmens, 1995). Irrespective of the particular cause of the phase change, clean englacial ice is melted and then refrozen thereby incorporating basal debris. Physical constraints on the regelation process dictate, however, that this mechanism cannot incorporate a net thickness of basal ice beyond 1 cm y^{-1} (Alley et al., 1996). Thus, regelation alone cannot explain the >1-m exposures of basal ice observed at the Matanuska Glacier, Alaska (Lawson et al., 1996; Strasser et al., 1996), or at the Greenland ice sheet (Knight, 1994).

Alternatively, it has been shown on theoretical grounds (Alley et al., 1998; Alley et al., this volume, chapter 1) that when water moving through an open linked-cavity drainage system flows out of an overdeepening, at or near the terminus of a glacier, a net accretion of ice through freeze-on can occur at the glacier bed. This freeze-on process happens because subglacial discharge flowing up-gradient out of the overdeepening

*Current address: HRP Associates, Inc., 167 New Britain Avenue, Plainville, Conneticutt 06062

Titus, D. D., Larson, G. J., Strasser, J. C., Lawson, D. E., Evenson, E. B., and Alley, R. B., 1999, Isotopic composition of vent discharge from the Matanuska Glacier, Alaska: Implications for the origin of basal ice, *in* Mickelson, D. M., and Attig, J. W., eds., Glacial Processes Past and Present: Boulder, Colorado, Geological Society of America Special Paper 337.

becomes supercooled due to a rapidly increasing pressure melting point and insufficient influx of heat to compensate for the pressure change (Strasser et al., 1996; Alley et al., 1998; Alley et al., this volume, chapter 1). Supercooling of the discharge results in nucleation of ice crystals and the release of latent heat, allowing the discharge to maintain thermal equilibrium. Ice crystals formed in the supercooled discharge then affix themselves to drainage cavity surfaces resulting in a net accretion of basal ice (Strasser et al., 1996; Alley et al., 1998; Lawson et al., 1998; Alley et al., this volume, chapter 1).

OBJECTIVE

Recently, Strasser et al. (1996), Lawson et al. (1998), and Evenson et al. (this volume, chapter 3) described the characteristics of both debris-rich basal ice exposed along the margin of the Matanuska Glacier, Alaska, and debris-rich frazzil ice terraces forming in and around subglacial discharge vents. They point to similarities between basal and terrace ice, especially in their ^{18}O compositions, and conclude that each formed by the supercooling process described by Alley et al. (1998) and Alley et al. (this volume, chapter 1). In addition, Strasser et al. (1996) and Lawson et al. (1998) suggest that some of the supercooled discharge is of recent meteoric origin and support this argument by the observation of bomb tritium in both types of ice.

Evidence that the ^{18}O, D, and ^3H composition of basal ice reflects that of subglacial discharge (corrected for fractionation) would further support the argument that basal ice at the Matanuska Glacier is created by the freeze-on process. In this paper, we present the results of ^{18}O, D and ^3H analyses of vent discharge samples and attempt to show a genetic relationship between the isotopic composition of vent discharge and basal ice.

FRACTIONATION DUE TO FREEZING

Based on the physical parameters of a simple closed subglacial reservoir, Jouzel and Souchez (1982) derived equations to describe the fractionation path of ^{18}O and D during freezing of subglacial water. In this closed system the physical parameters of the reservoir are defined as follows: input (I) to the reservoir is equal to zero; output (O) from the reservoir is equal to zero; the freezing rate (F) of the reservoir is less than the isotopic diffusion coefficient (e.g., 10^{-5} cms^{-1}) eliminating the possibility of isotopic gradients in the water; and δR_I is the initial $\delta^{18}O$ and δD values of the reservoir. The $\delta^{18}O$ and δD values of basal ice formed in this conceptual closed reservoir are given by the equations (Jouzel and Souchez, 1982):

$$\delta_s^{18}O = 10(1000+\delta_I^{18}O)[(1.1-K)^\alpha-(1-K)^\alpha]-1000, \text{ and} \quad (1)$$

$$\delta_s D = 10(1000+\delta_I D)[(1.1-K)^\beta-(1-K)^\beta]-1000, \quad (2)$$

where $\delta_s^{18}O$ and $\delta_s D$ are isotope values of basal ice formed at each progressive fraction of the initial reservoir frozen (e.g., 0.1–1) denoted by K, and α and β are equilibrium fractionation coefficients for $^{18}O/^{16}O$ (1.00291) and D/H (1.0212), respectively (Lehmann and Siegenthaler, 1991).

Jouzel and Souchez (1982) plotted δ_s^{18} and $\delta_s D$ values from equations (1) and (2) for corresponding values of K and showed that $\delta^{18}O$ and δD values fall on a freezing line with a slope less than that of the Global Meteoric Water Line (e.g., 8). They also concluded that when basal ice sampled from a closed reservoir plots on a freezing line, the initial δR_I composition of the reservoir can be estimated by the intersection of the freezing line and the Global Meteoric Water Line. Thus, the relationship between δR_I and the $\delta^{18}O$ and δD values of basal ice can be determined by the slope of the line and independent of the K value of individual ice samples (Jouzel and Souchez, 1982). According to Jouzel and Souchez (1982) the freezing slope (S) of a closed reservoir is given by the equation:

$$S = [(\alpha-1)/(b-1)][(1000+d_I D)/(1000+d_i^{18}O)] \quad (3)$$

where $\delta_i D$ and $\delta_i^{18}O$ are initial reservoir values and α and β are maximum equilibrium fractionation coefficients for $^{18}O/^{16}O$ (1.00291) and $^2H/^1H$ (1.0212), respectively. An example for a closed reservoir freezing line with a hypothetical δR_I composition is given in Figure 1A.

Souchez and Jouzel (1984) also developed an equation for interpretation of the relationship between δR_I composition and $\delta^{18}O$ and δD values of basal ice formed in an open meltwater subglacial reservoir. In this open meltwater reservoir the values of I, O, and F can vary according to the physical characteristics of the system. The isotopic value of the input (δI) is incorporated to account for changes in the ^{18}O and D content of I during freezing of the reservoir. Souchez and Jouzel (1984) showed that when δI was not significantly different from δR_I (which is reasonable if δR_I and δI are composed primarily of englacial ice melt) the freezing slope (S) is given by the equation:

$$S = (\alpha/\beta)[\alpha-1)/(\beta-1)][(1000+\delta_I D)/(1000+\delta_i^{18}O)]. \quad (4)$$

It is important to note that the equation yields a slope that is nearly identical to that derived from equation (3) (closed reservoir). Thus, for the open and closed reservoirs the only significant difference in the freezing process is that ice produced in an open reservoir is always enriched in ^{18}O and D relative to δR_I. This occurs because there is continuous reconstitution of the reservoir by I, where δI and δR_I are equal and I is not <F. Alternatively, in an open meltwater reservoir ice depleted in ^{18}O and D relative to δR_I may be produced if the value of F is >I. This situation, however, would be limited to reservoirs that receive only small fluxes of meltwater, like in high elevation cirque glaciers. A freezing line for an open meltwater reservoir, where I is not <F, is given in Figure 1B. The $\delta^{18}O$ and δD values used in this simulation are the same as those used for the closed reservoir (Fig. 1A).

Recently, Hubbard and Sharp (1995) pointed out that single

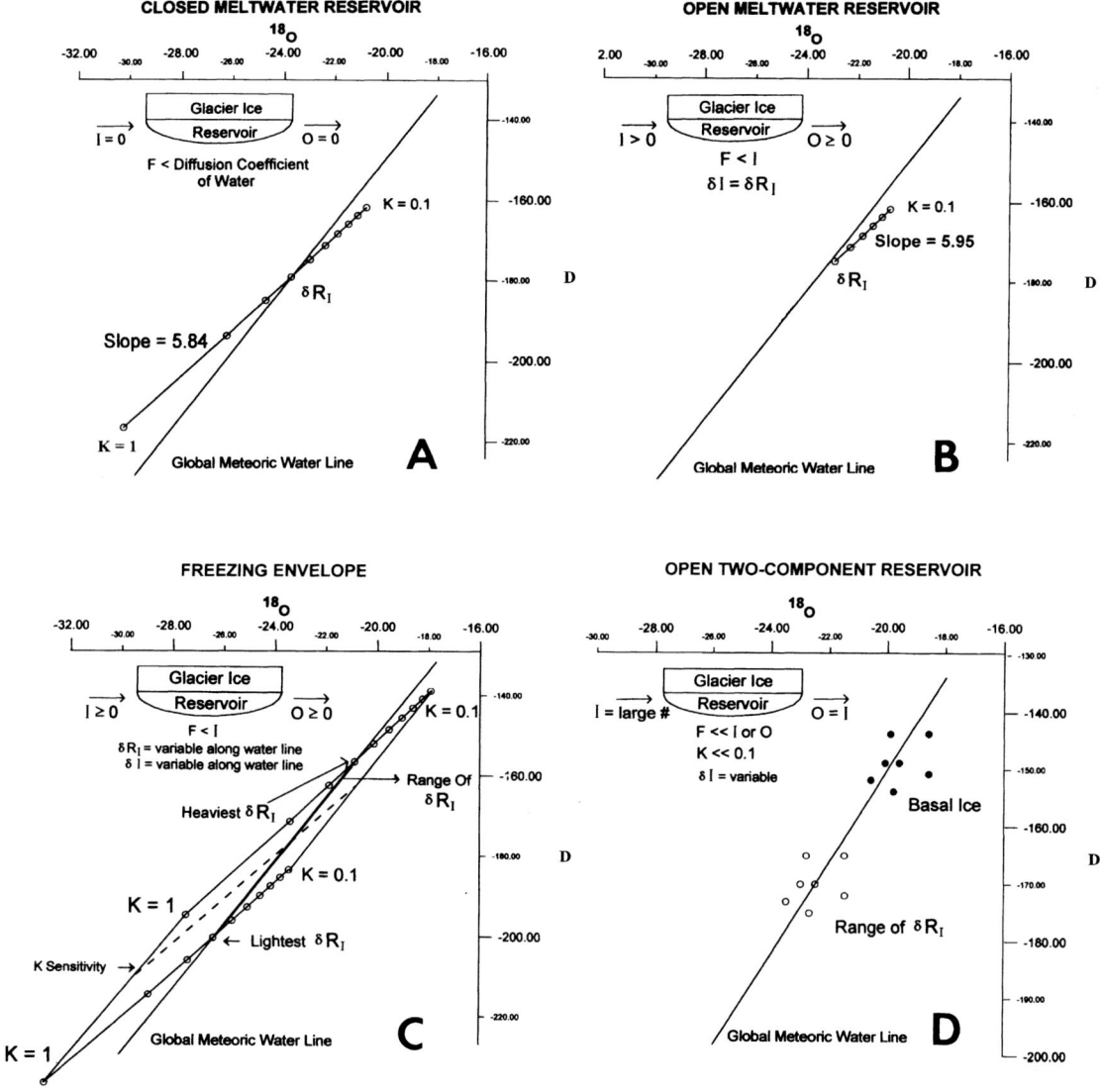

Figure 1. A, Closed meltwater reservoir freezing slope (after Jouzel and Souchez, 1982); B, open meltwater reservoir freezing slope (after Souchez and Jouzel, 1984); C, open meltwater reservoir freezing envelope (after Hubbard and Sharp 1995); D, fractionation due to freezing in an open two-component reservoir.

co-isotopic freezing slopes derived from equations (3) and (4) may not be adequate for interpretations of basal ice $\delta^{18}O$ and δD values when: (1) ice is formed during discrete freezing episodes from reservoirs with different δR_I compositions, or (2) if basal ice samples are taken across discrete horizons. Instead, they propose the use of freezing envelopes that include all possible $\delta^{18}O$ and δD values of basal ice formed by freezing events in closed and open meltwater reservoirs. The height of such an envelope is defined by the range of variation in values of δR_I along the Global Meteoric Water Line, while the width of the envelope is dependent on the sensitivity of K included in individual basal ice samples (Hubbard and Sharp, 1995). A theoretical freezing envelope for a representative range of δR_I is presented in Figure 1C.

The co-isotopic freezing slopes and the envelope shown in Figures 1A, B, and C work best for basal ice formed from reservoirs that show either little or well-constrained variation in the value of K and in δR_I along the Global Meteoric Water Line. However, this approach may not be appropriate for basal ice formed in reservoirs characterized by large fluxes of both meltwater and rainfall runoff, like might occur in some low-altitude valley glaciers. In such open two-component reservoirs, F would be <<I and would result in a situation where O and I are essentially equal and where the value of K would be near zero. Therefore, the changing isotopic composition and volumetric flux of both meltwater and rainfall runoff would result in highly variable δR_I values. Thus, the relationship between basal ice

and δR_I could not be inferred from freezing slopes defined by progressive freezing of a subglacial reservoir with a relatively stable isotopic composition. Ice formed from freezing in a open two-component reservoir would simply reflect the $\delta^{18}O$ and δD value of O (since I and O are equal) plus the maximum fractionation coefficients (α and β) 2.91‰ and 21.2‰ for $\delta^{18}O$ and δD, respectively.

Values of $\delta^{18}O$ and δD for ice produced from a open two-component reservoir can be calculated from the equations:

$$\delta^{18}O = \delta_L^{18}O + 2.19‰ \quad (5)$$

and

$$\delta D = \delta_L D + 21.2‰ \quad (6)$$

where $\delta_L^{18}O$ and $\delta_L D$ are initial values of dR_I and 2.91‰ and 21.2‰ are the isotopic enrichment factors (a and b) of ice relative to δR_I (Lehmann and Siegenthaler, 1991). The equations are independent of freezing slopes and the value of K, but allow for maximum variation in the values of δR_I.

Athropogenic ^3H can also be incorporated into basal ice formed in a open reservoir that receives significant amounts of rainfall runoff. The amount incorporated, however, is a function of the concentration of ^3H within the reservoir and fractionation of $^3H/^1H$ during freezing. According to Strasser (1996), fractionation due to freezing results in only about 4% enrichment from the liquid to solid phase. Thus, ^3H activity of basal ice, produced from variable reservoir concentrations of ^3H, will generally reflect the ^3H activity of O (since I and O are essentially equal).

FIELD SITE

The Matanuska Glacier is located in south-central Alaska about 150 km northeast of Anchorage and flows northwest out of icefields in the Chugach Mountains. It is approximately 45 km in length and ranges in width from 2.2 km at its source in the icefields to 5 km at the terminus (Fig. 2A). Over its 45-km length, the elevation of the glacier surface descends from 3,500 to 500 m, and at the terminus rainfall during the ablation season can exceed 20 cm. The northern portion of the terminus is stagnant and covered with debris, while the western portion is characterized by active ice and a lack of debris cover (Fig. 2B). In general, the terminus has remained in its current position for the last 200 yrs (Williams and Ferrians, 1961).

Subglacial hydrologic activity is most evident at the western portion of the terminus where ice flows through a small basin or overdeepening (Arcone et al., 1995). Along the ice margin numerous discharge vents can be observed in the form of upwellings or fountains. Many of these emanate from directly under basal ice exposures that range in thickness from 1 m to more than 5 m. Collective discharge from the vents coalesces into a proglacial catchment that subsequently discharges into the headwaters of the Matanuska River (Fig. 2B).

METHODS

The discharge vents shown in Figure 2B represent the distal portion of the subglacial drainage system. During summer, flow from the vents is generally characterized by acute levels of suspended sediment, a characteristic of subglacially derived water (Lawson et al., 1996; Strasser, 1996). Samples of discharge were obtained from several of the largest vents throughout the early summer of 1995, either manually or by an ISCO^{Tm} water sampler equipped with a weighted sampling tube inserted into the throat of the vents. In most instances, sampling was done daily at times of high and low discharge determined from a gauging station located 300 m downstream from the vents. Immediately after collection, samples were sealed in 25 and 250 mls Nalgene^{Tm} bottles.

Approximately 10 mls of sample were allocated for $^{18}O/^{16}O$ and $^2H/^1H$ ratio measurements. Sample splits were sent to Coastal Labs, Inc., for analysis where the ratios of $^{18}O/^{16}O$ and $^2H/^1H$ were determined using the standard calculation method:

$$\delta(‰) = | R_{(sample)} - R_{(standard)} / R_{(standard)} | \times 1000 \quad (7)$$

where R is equal to the $^{18}O/^{16}O$ and D/H ratio for the experimental sample or the laboratory standard respectively (e.g., SMOW). Analytical accuracy was reported as 0.1‰ and 1.0‰ for $\delta^{18}O$ and δD, respectively.

Analysis for ^3H was conducted at the Michigan State University low level ^3H lab using a standard scintillation counting procedure (Kessler, 1988). For each sample, 250 ml was allocated for electrolytic enrichment and subsequent ^3H analysis. The activity of ^3H in a sample is expressed in ^3H units, where one TU is equivalent to 7.2 dpm/l or one ^3H atoms to every $^1H* 10^{18}$ atoms. Low level ^3H analysis has been shown to be accurate to ±1 TU if electrolytic enrichment is applied to samples (Ostlund and Werner, 1962).

RESULTS

The $\delta^{18}O$ and δD values measured in vent discharge are plotted in Figure 3A. Also plotted are (1) $\delta^{18}O$ and δD values for 25 samples from a 2-m-thick vertical section through basal ice (Strasser et al., 1996), (2) average $\delta^{18}O$ and δD values for englacial ice (Lawson, unpublished), and (3) a local meteoric water line derived for the Matanuska Glacier area (Lawson, unpublished).

The vent discharge $\delta^{18}O$ and δD values show a considerable amount of scatter about the local meteoric water line. Values of $\delta^{18}O$ vary from −21‰ to −24.6‰ and average −22.8‰. Values of δD vary from −172‰ to −189‰ and average −181.8‰. The data indicate that the isotopic composition of vent discharge is significantly more depleted with respect to ^{18}O and D than the average composition of basal ice and only slightly less depleted than the average composition of englacial ice.

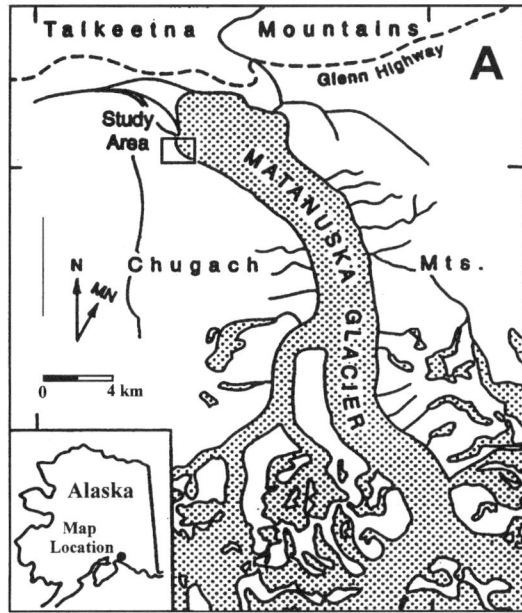
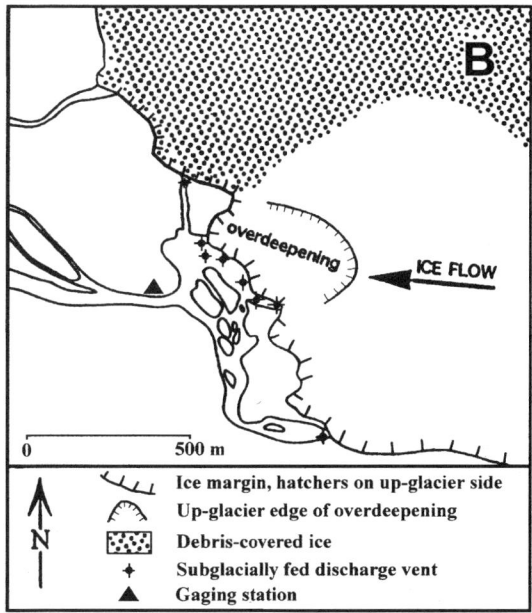

Figure 2. A, Map showing location of the Matanuska Glacier, Alaska; B, map showing portion of glacier terminus.

Activities of ^3H measured in vent discharge are presented in Figure 3B together with activities measured in englacial ice (Strasser et al., 1996) and in Anchorage precipitation during the summer of 1996 (U.S.G.S., unpublished). As in the case of vent discharge $\delta^{18}O$ and δD values, ^3H activities measured in vent discharge also show considerable variation. They range from 2.3 to 8.3 TU and average 4.6 TU. The activities appear to be 1 to 7 TU higher than activities measured in englacial ice. On the other hand, vent discharge activities appears about 2 to 10 TU less than activities measured in Anchorage precipitation. Also, the lowest observed activities are from vent discharge associated with protracted period of dry sunny days. The highest ^3H activities observed, however, are from vent discharge associated with major rainstorm events.

Results of the application of the open two-component reservoir model to the δO^{18} and δD values of vent discharge samples from 1995 are presented in Figure 4. The results show a definite but imprecise relationship between the calculated $\delta^{18}O$ and δD values for basal ice and those actually measured in basal ice. In addition, the calculated values show much more scattered about the local meteoric water line than the measured values. The average calculated $\delta^{18}O$ and δD value for basal ice, however, plots very close to the measured values that show the least depletion with respect to ^{18}O and D.

DISCUSSION

A couple of reasons might explain why the calculated $\delta^{18}O$ and δD values for basal ice in Figure 4 do not match the measured values as predicted by the open two-component reservoir model. They include (1) calculated values are reflective of discrete basal ice horizons whereas measured values represent composites of multiple freeze-on horizons within a single sample, and (2) calculated values do not reflect possible homogenization of multiple discrete basal ice horizons by subsequent regelation. The fact that the average of the calculated basal ice values plots very near to measured values that show the least depletion with respect to ^{18}O and D is not surprising, however, because the average effectively reflects what would be the homogenized value of multiple discrete freeze-on horizons. This is an important consideration since basal ice samples analyzed by Strasser et al. (1996) consisted of short horizontal cores measuring 2–3 cm in diameter and containing multiple layers of ice and debris. Melting of these samples prior to analysis clearly resulted in homogenization of the freeze-on horizons and generation of composite $\delta^{18}O$ and δD values.

Association of the average calculated basal ice value only with measured basal ice values that show the least depletion in ^{18}O and D is reasonable because rainfall at the Matanuska Glacier during the summer of 1995 was unusually high. This would have resulted in significant addition of isotopically "heavy" rainwater to the subglacial drainage system; thus producing vent discharge less depleted in ^{18}O and D than would occur during periods of low rainfall. The effect varying amounts of rainfall might potentially have on the isotopic composition of basal ice is also presented in Figure 5A. This figure compares $\delta^{18}O$ values measured in basal ice with summer precipitation deviations from normal recorded from 1964 to 1979 at Palmer, Alaska, which is located approximately 80 km west of the Matanuska Glacier. An accretion rate of 7 cm y^{-1} has been assumed for basal ice (Strasser, 1996) and a 1963 reference datum that has been assigned to one of the $\delta^{18}O$ values based on

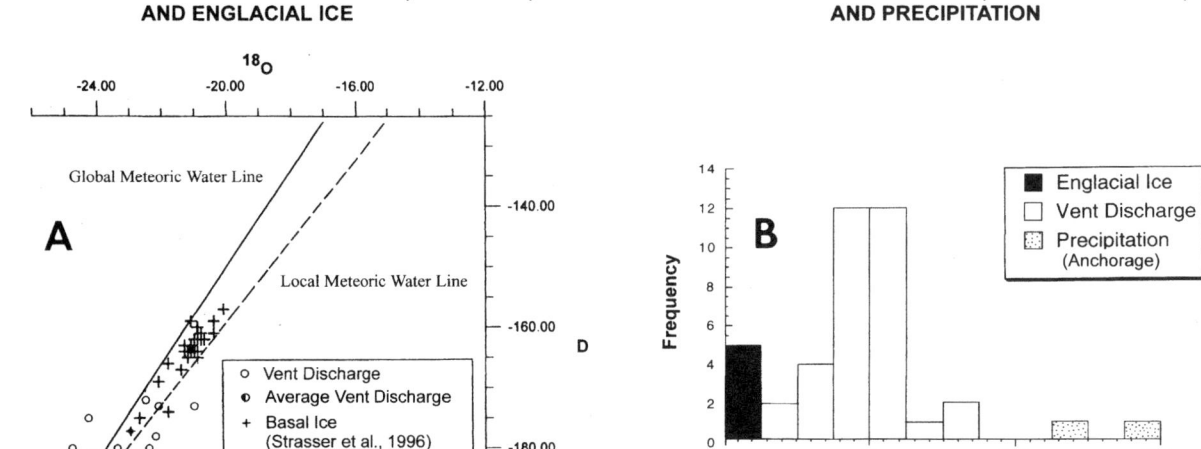

Figure 3. A, $\delta^{18}O$ and δD measured in vent discharge, basal ice, and englacial ice. Local meteoric water line after Lawson (unpublished); B, ^3H activity measured in englacial ice, vent discharge, and Anchorage summer precipitation.

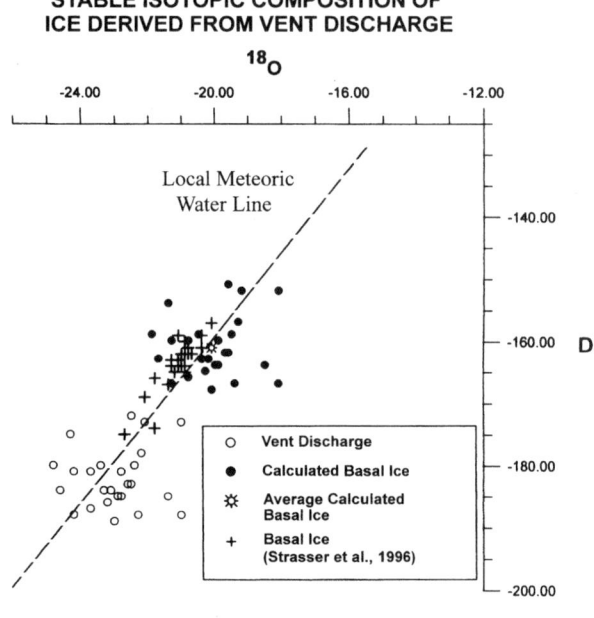

Figure 4. Calculated $\delta^{18}O$ and δD of ice derived from vent discharge.

the occurrence of a bomb tritium peak as seen in Figure 5B (Strasser, 1996; Lawson et al., in press). The pre-1963 $\delta^{18}O$ values show considerable depletion in ^{18}O and are probably the result of diffusion process(es) occurring along the contact between basal and englacial ice. The post-1963 values, however, are more representative of the basal ice and show variations that seem to generally coincide with precipitation deviations from normal. In effect, positive precipitation deviations appear to produce basal ice enriched in ^{18}O, whereas negative deviations appear to produce basal ice depleted in ^{18}O.

The ^3H activities measured in vent discharge (Fig. 3B) clearly indicate that a significant portion of the discharge is of recent meteoric origin, a conclusion also supported by the wide scatter in vent discharge $\delta^{18}O$ and δD values about the local meteoric water line (Fig. 3A). Since the Matanuska Glacier lies only 150 km from Anchorage and receives precipitation from the same relative source as Anchorage, it should be possible to approximate the percent of rainfall runoff in vent discharge by comparing the average summer ^3H activity measured in discharge (4.6 TU) against the average 1995 summer ^3H activity measured in Anchorage precipitation (13.2 TU). This approach assumes that the ^3H activity for vent discharge due to melting ice and snow alone is 2.3 TU (background) and that any increase above this value is due to rainfall. Such a calculation shows that approximately 21% of vent discharge can be attributed to rainfall.

Historical records of ^3H activity in Anchorage precipitation from 1956 to 1995 can also be used to estimate ^3H activity in the vent discharge of previous summers. Long-term summer month records (e.g., May–September) of ^3H activity measured in Anchorage precipitation and corrected to a reference date of December 31, 1995, are presented in Figure 6 (unpublished data from R. L. Snyder, U.S. Geological Survey). Assuming that precipitation represents 10% (dry year) to 21% (wet year) of vent discharge during most years, a range of possible historical ^3H

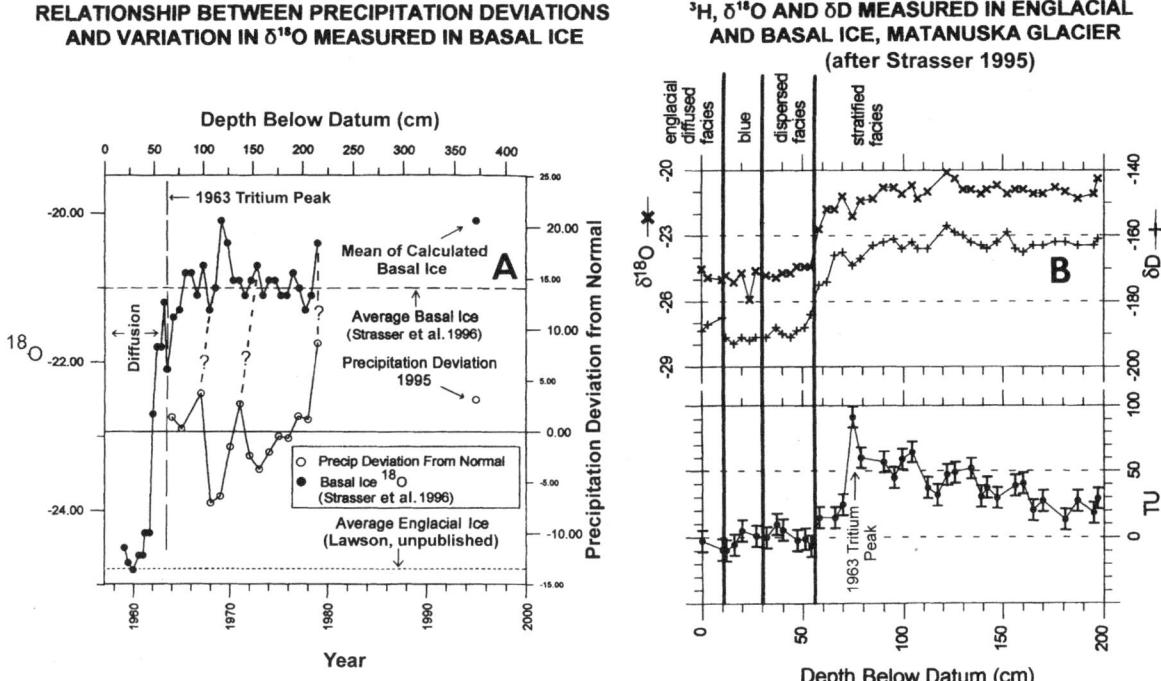

Figure 5. A, Relationship between variation in δ¹⁸O measured in basal ice and summer precipitation deviations from normal (inches) recorded at Palmer, Alaska, from 1964 to 1979 (NOAA Climatological Data-ALASKA); B, δ¹⁸O, δD and ³H measured in englacial and basal ice, Matanuska Glacier (modified from Strasser, 1996).

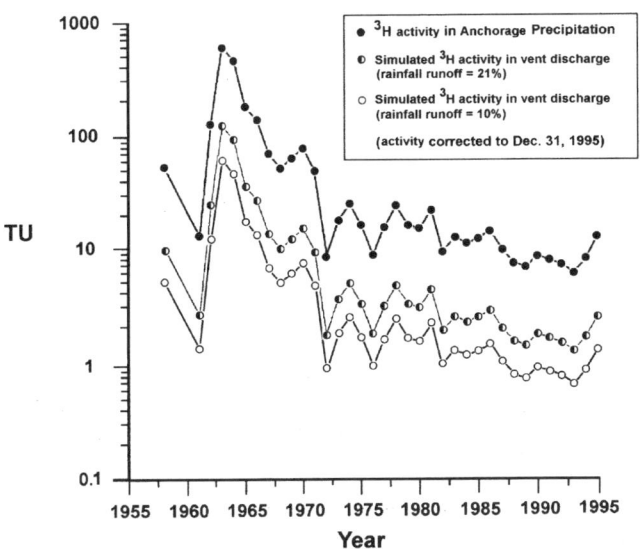

Figure 6. Simulated ³H activity in vent discharge versus ³H measured in summer precipitation for Anchorage, Alaska, from 1956 to 1995 (unpublished data from R. L. Snyder, U.S. Geological Survey). Note: activities corrected to December 31, 1995.

activity in discharge can be calculated. The calculated range shown in Figure 6 clearly corresponds to the ³H activity measured in basal ice by Strasser et al. (1996) and suggests that in the summer of 1963 rainfall runoff may have comprised about 17% of the subglacial discharge.

CONCLUSIONS

The objective of this study was to demonstrate that the ¹⁸O, D composition, and ³H activity of basal ice at the Matanuska Glacier reflects that of the subglacial vent discharge corrected for fractionation due to freezing. The results of this study show that (1) application of a open two-component reservoir model can explain the ¹⁸O, D composition, and ³H activity measured in basal ice, and supports the hypothesis that basal ice is being accreted to the glacier bed within overdeepenings through freeze-on; and (2) rainfall runoff appears to play an important role in the ¹⁸O, D composition, and ³H activity of basal ice at the Matanuska Glacier.

REFERENCES CITED

Arcone, S. A., Lawson, D. E., and Delaney, A. J., 1995, Short-pulse radar wavelet recovery and resolution of dielectric contrasts within englacial ice of the Matanuska Glacier, Alaska, U.S.A.: Journal of Glaciology, v. 41, p. 68–86.

Alley, R. B., Cuffey, K. M., Evenson, E. B., Strasser, J. C., Lawson, D. E., and Larson, G. J., 1996, How glaciers entrain and transport sediment at their beds: physical constraints: Quaternary Science Reviews, v. 16, p. 1017–1038.

Alley, R. B., Lawson, D. E., Evenson, E. B., Strasser, J. C., and Larson, G. J., 1998, Glaciohydraulic super cooling: A freeze-on mechanism to create stratified, debris-rich basal ice. II. Theory: Journal of Glaciology, v. 44, p. 562–568.

Boulton, G. S., 1972, The role of thermal regime in glacial sedimentation: Institute of British Geographers, Special Bulletin no. 4, p. 1–19.

Clarke, G. K. C., Collins, S. G., and Thompson, D. E., 1984, Flow, thermal structure, and subglacial conditions of a surge-type glacier: Canadian Journal of Earth Sciences, v. 21, p. 232–240.

Hubbard, B., and Sharp, M., 1995, Basal ice facies and their formation in the Western Alps: Arctic and Alpine Research, v. 27, p. 300–310.

Iverson, N. R.,1993, Regulation of ice through debris at glacier beds: implications for sediment transport: Geology, v. 21, p. 559–562.

Iverson, N. R., and Semmens, D. J.,1995. Intrusion of ice into porous media by regelation: a mechanism of sediment entrainment by glaciers: Journal of Geophysical. Research, v. 100, p. 10219–10230.

Jouzel, J., and Souchez, R. A., 1982, Melting-refreezing at the glacier sole and the isotopic composition of the ice: Journal of Glaciology, v. 28, p. 35–42.

Kamb, B., and LaChapelle, E. R., 1964, Direct observation of the mechanism of glacier sliding over bedrock: Journal Glaciology, v. 5, p.159–172.

Kessler, M. J., 1988, Effective use of low level liquid scintillation analysis: The Second International Seminar for Liquid Scintillation Analysis, Proceedings, June 8, 1988, Tokyo, Japan, p. 256–301.

Knight, P. G., 1994, Two-facies interpretation of the basal layer of the Greenland Ice Sheet contributes to a unified model of basal ice formation: Geology, v. 22, p. 971–974.

Lawson, D. E., Evenson, E. B., Strasser, J. C., Alley, R. B., and Larson, G. J., 1996, Subglacial supercooling, ice accretion, and sediment entrainment at the Matanuska Glacier, Alaska: Geological Society of America, Abstracts with Programs, v. 28, no 3. p. 75.

Lawson, D. E., Strasser, J. C., Evenson, E. B., Alley, R. B., Larson, G. J., and Arcone, S. A., 1998, Glaciohydraulic super cooling: A freeze-on mechanism t create stratified, debris-rich basal ice. I. Field evidence: Journal of Glaciology, v. 44, p. 547–561.

Lehmann, M., and Siegenthaler, U.,1991, Equilibrium oxygen and hydrogen isotope fractionation between ice and water: Journal of Glaciol.ogy, v. 37, p. 23–26.

Osterlund, H. G., and Werner, E., 1962, The electrolytic enrichment of tritium and deuterium for natural tritium measurements: Tritium in the physical and biological sciences: International Atomic Energy Association, v 1, p. 95–104.

Robin. G. deQ., 1976, Is the basal ice of a temperate glacier at the pressure melting point?: Journal of Glaciology, v. 16, p. 183–196.

Souchez, R. A., and Jouzel, J., 1984, On the isotopic composition in δD and $\delta^{18}O$ of water and ice during freezing: Journal Glaciology, v. 30, p. 369–372.

Strasser, J. C., 1996, Processes of subglacial ice growth and debris entrainment at the Matanuska Glacier, Alaska [Ph.D. thesis]: Bethlehem, Pennsylvania, Lehigh University, 137 p.

Strasser, J. C., Lawson, D. E., Larson, G. J., Evenson, E. B., and Alley, R. B., 1996, Preliminary results of tritium analysis in basal ice, Matanuska Glacier, Alaska, U.S.A.: evidence for subglacial ice accretion: Annals of Glaciology, v. 22, p. 126–132.

Weertman, J., 1961, Mechanism for the formation of inner moraines found near the edge of cold ice caps and sheets: Journal of Glaciology, v. 3, p. 965–978.

Weertman, J., 1964, Glacier sliding: Journal of Glaciology, v. 5, p. 287–303.

Williams, J. R., and Ferrians, O., 1961, Late Wisconsinan and recent history of the Matanuska Glacier, Alaska: Arctic, v. 14, p. 82–90.

MANUSCRIPT ACCEPTED BY THE SOCIETY OCTOBER 8, 1998

Microstructures of glacigenic sediment-flow deposits, Matanuska Glacier, Alaska

Matthew S. Lachniet* and Grahame J. Larson
Department of Geological Sciences, Michigan State University, East Lansing, Michigan 48824
Jeffrey C. Strasser
Department of Geology, Augustana College, Rock Island, Illinois 61201
Daniel E. Lawson
U.S. Army Cold Regions Research and Engineering Laboratory, Anchorage, Alaska 99505
Edward B. Evenson
Department of Earth and Environmental Sciences, Lehigh University, Bethlehem, Pennsylvania 18015
Richard B. Alley
Earth Systems Science Center and Department of Geoscience, The Pennsylvania State University, University Park, Pennsylvania 16802

ABSTRACT

Microstructures of glacigenic sediment gravity-flow deposits formed at the terminus of the Matanuska Glacier, Alaska, were analyzed to characterize flow type. These sediment flows have been classified into four types based primarily on water content and sedimentological characteristics (Lawson, 1979a, 1982). Thin sections of flow deposits show a variety of micro- and mesoscale characteristics that vary according to water content of the source flow. Wet-type flow deposits are characterized in thin section by a well-defined parallel and imbricated microclast fabric and thin laminations resulting from laminar to plastic flow regimes. Dry-type flow deposits are characterized in thin section by bi- or polymodal or random microclast fabrics, greater textural heterogeneity, and deformational microstructures associated with plastic to brittle flow regimes. Thin laminations and a "laminar flow fabric" in wet-type flow deposits may be characteristic of sediment gravity flow in a glacial environment. Characterization of these microstructures supports the contention that micromorphological analyses can be used to elucidate sediment flow genesis and the conditions of the flow just prior to deposition. Thus, micromorphology may also be useful for differentiating sediment-flow type in Pleistocene diamictons in other locations.

INTRODUCTION

Micromorphological analysis of glacial sediments has been used in previous studies to differentiate genetic types of till. In North Sea glacial deposits, for example, "flow tills" have been differentiated from a basal lodgement till on the basis of microstructures (van der Meer and Laban, 1990). Additionally, micromorphology has been used to characterize tectonic deformation of basal tills associated with a deforming bed (van der Meer, 1993), and to elucidate subglacial conditions and processes acting on tills (Menzies, 1990; Menzies and Maltman, 1992). The use of this technique to determine sedimentary genesis in Pleistocene and recent glacial deposits, however, has not been pursued extensively.

Here we use micromorphology to (1) analyze, inventory, and characterize microstructures found in contemporary subaerial glacigenic sediment-flow deposits; and (2) to differentiate the microstructures representative of deposits from dry-type sediment flows and wet-type sediment flows. Dry-type sediment-flow deposits, as used in this study, correspond approximately to Lawson type I and II flow deposits (low water content), and wet-

---

*Current address: Department of Earth Sciences, Syracuse University, Syracuse, New York 13244 type sediment-flow deposits correspond approximately to Lawson type III and IV flow deposits (high water content; Lawson, 1979a, 1982; see below for sediment flow type characteristics). Here we evaluate the use of micromorphological analysis to differentiate contemporary dry-type from wet-type sediment-flow deposits formed at the terminus of the Matanuska Glacier.

This study deals exclusively with the micromorphology of sediment-flow deposits; the study of the micromorphology of tills is beyond the scope of this study and has not been undertaken at the Matanuska Glacier. Future investigation on the micromorphology of known glacial sediments will allow the further distinction between sediment-flow deposits and true tills.

SITE DESCRIPTION

The Matanuska Glacier is located in south-central Alaska (Fig. 1), approximately 140 km north of Anchorage at 61°47'N, 147°45'W. The glacier flows northward approximately 40 km from ice fields in the Chugach Mountains and terminates at the east-west–trending Matanuska valley. The terminus of the glacier consists of a stagnant zone of debris-covered ice and an active clean-ice zone. Proglacial sedimentation occurs near the active ice zone in what is called the western terminus region (Lawson, 1979a).

Near the terminus the Matanuska Glacier flows out of an overdeepening and significant volumes of debris are incorporated into the ice mass as freeze-on occurs at the base of the glacier (Strasser et al. 1996; Lawson et al., 1998). Freeze-on produces debris-rich basal ice, with debris concentrations in individual strata and lenses as much as 74% by volume in the stratified basal ice facies (Lawson, 1979a). When basal ice ablates, water-saturated debris is released and it may undergo fluidization and/or liquefaction, and flow under the influence of gravity. The debris is generally resedimented several times after release from the ice; consequently, the majority of the diamictons deposited at the terminus are resedimented (Lawson, 1979a, 1982).

LITERATURE REVIEW

Glacigenic sediment flows have been investigated by several researchers, most notably by Hartshorn (1958), Boulton (1968, 1971), Marcussen (1973, 1975), Evenson et al. (1977), and at the Matanuska Glacier by Lawson (1979a, 1982). Many sediment-flow deposits were interpreted as "tills," and hence were given the name "flow tills" (e.g., Hartshorn, 1958). As the term "till" implies glacially derived sediment deposited in situ, the use of "sediment gravity flow" is a more accurate term for resedimented debris (Lawson, 1982). However, "flow till" will be presented in this summary when it was used by the original author cited, and is here used interchangeably with "sediment gravity flow" or "sediment flow" for short.

Hartshorn (1958) was the first to recognize the significance of flow tills produced by Pleistocene ice sheets. In the areas Hartshorn studied in southeastern Massachusetts, sediment-

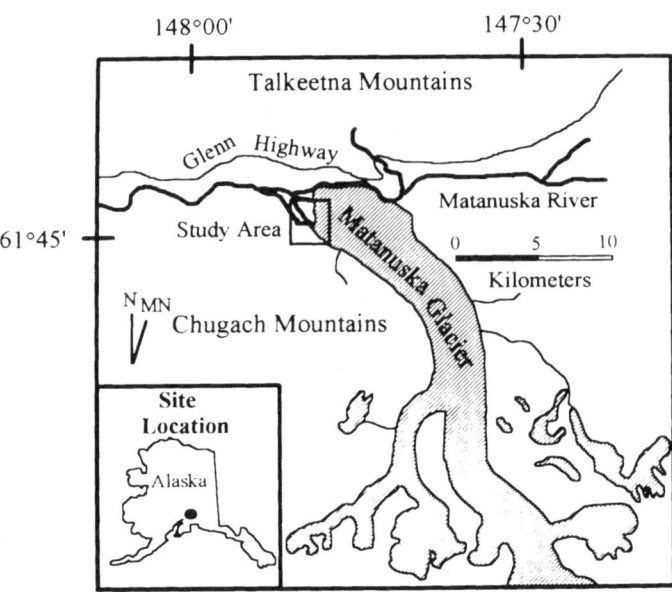

Figure 1. Location of Matanuska Glacier, Alaska (from Strasser et al., 1996).

flow deposits overlay outwash deposits, and were originally and incorrectly interpreted by other workers as basal till formed by lodgement during a readvance of the Laurentide Ice Sheet. These sediment-flow deposits appeared structureless at the macroscale. Boulton (1968) described flow tills deposited on proglacial outwash at the margins of Vestspitsbergen glaciers, in a manner that produced sequences similar to those described by Hartshorn (1958).

At the Vestspitsbergen glaciers, debris for sediment flow was supplied from englacial debris bands exposed at the termini. These debris bands originated from compressive flow and dipped steeply upglacier. They contain as much as 80% debris by volume and contain rounded and subrounded clasts, which led Boulton (1968) to interpret their source as the glacier bed. During ablation of englacial ice and debris bands, debris was released from the ice and formed saturated flows that were deposited on the ice, ice-cored moraine surfaces, or proglacial sediments.

Most significantly, Boulton (1968) proposed the idea that many tills previously interpreted to be subglacial may in fact be of proglacial sediment flow origin. Consequently, many multi-till sequences in Pleistocene glacial deposits, which were originally interpreted to represent several cycles of glacial advance and retreat, may in fact be sediment-flow deposits associated with a single advance and retreat cycle.

Lowe (1976, 1982) used the general term "sediment gravity flow" or "sediment flow" to describe sediment-water mixtures of various types, such as turbidity currents, fluidized or liquefied sediment flows, grain flows, and sediment flows. "Turbidity currents" are strictly subaquatic, and form as a result of turbulent fluid supporting suspended sediment. Liquefied sediment flows occur when the sediment settles downward and expels pore fluids from the matrix. Fluidized sediment flows, in which the upward force acting on sediment grains by escaping water is equal to the downward force of gravity, and the sediment is not displaced vertically, may not be common (Lowe, 1976). "Grain flows" occur when the sediment is entirely supported by intergranular contacts. A "debris flow" supports larger grains by the internal strength of the matrix material. The use of the term "debris" in this context does not refer to sediment transported in ice.

SEDIMENT-FLOW CLASSIFICATION AT THE MATANUSKA GLACIER

At the terminus of the Matanuska Glacier, the source of most sediment flows is the water and sediment released by the ablation of debris-rich basal ice. Sediment flowage is also initiated from the saturation of sediment on the active glacier ice or ice-cored moraine (Lawson, 1979a). Additionally, sediment flows can be initiated from the ablation of englacial debris bands on the glacier surface (similar to flows reported by Boulton, 1968).

Subaerial sediment gravity flows at the Matanuska Glacier have been classified into four types by Lawson (1979a, 1982), primarily as a function of water content (see below). Flow is initiated when the saturation of the sediment reduces the material's shear strength to failure. Flows are transported down slope, commonly in channels, and deposited when the slope approaches horizontal or the degree of saturation is sufficiently reduced. Lawson (1979a, 1982) has extensively studied and described characteristics of sediment gravity flows at the Matanuska Glacier, which are summarized in the following section and in Table 1. In these flows, grain size is generally larger than fine silt, shown in frequency curves for Lawson-type sediment flows in Figure 2.

Flow characteristics

Type I flows generally have a water content ranging from 8%–14% by weight, are nonchannelized, and are characterized by a cohesive plug flowing over a thin (a few centimeters) basal shear zone. Thicknesses range to as much as 2 m, sorting is poor, and clasts are supported in a fine-grained matrix. Surfaces of type I flows may have angles as large as 45° in the marginal and frontal slopes. This type is similar to a "debris flow" (Lowe, 1976, 1982).

Type II flows are characterized by a water content of 14%–19% by weight, and are generally channelized. As with type I, plug flow occurs over basal and lateral shear zones where laminar flow is dominant and which contain normally graded tractional gravels. Thicknesses are as much as ~1.5 m, sorting is poor, and texture is similar to type I flows. Surfaces of type II flows can hold angles similar to type I flows. Pore fluid expulsion channels (1–2 mm diameter) occur in clusters in the plug zone. If the water content is higher, flow can be more plastic than in type I flows. This type may be similar to a "grain flow" (Lowe, 1976, 1982), but could also have characteristics of a "debris flow" at lower water contents.

Type III flows have a water content by weight of 18%–25%, are channelized, and exhibit differential shear throughout, although thin discontinuous plugs may occasionally be present at the lower range of water content. These flows are thinner (0.5 m per lobe), and hold lower surface angles than type I and II flows. Grain size can decrease downflow, and particles may be imbricated up slope. As type III flows commonly occur in meltwater channels, small lenses of fluvial sediments may be intercalated. Clasts in low-viscosity sediment flows (such as wetter type II and type III) may sink and concentrate in distinct horizons (Boulton, 1971). Type III sediment flows most closely correspond to a "liquefied sediment flow" (Lowe, 1976, 1982).

A sediment flow with greater than 25% water by weight is considered a type IV sediment flow. This type of flow follows meltwater channels, appears laminar throughout, and may be fully liquefied. Individual flows are thinner than other flow types, and have near-horizontal slopes. A fine-grained flow body overlies similar fine-grained silt and sandy silt traction particles. After flow has ceased, loss of pore fluids and grain settling would have a tendency to destroy flow structures and fabric, but may allow for the grading of particles. If this type of flow becomes fluidized, it would be similar to a "fluidized sediment flow," although the possibility of true fluidized flows being found in nature has been questioned (Lowe, 1976).

Microstructures in sediment-flow deposits can be of at least four varieties: (1) structures formed during flow, (2) during deposition, (3) those structures inherited from the parent deposit, and (4) those formed from postdepositional changes. Distinguishing inherited structures from flow or depositional structures may be difficult in some dry-type flow deposits. For example, a type III flow may have been deposited on ice that subsequently melted, initiating a type I flow. If plastic deformation of the plug in the type I flow was minimal, the thin section would show type III structures. Lawson noted however, that the fabric of the source deposit is generally destroyed by mixing during mobilization (Lawson, 1982).

METHODS

Sampling protocol and thin section preparation

Sediment-flow deposits were sampled with metal Kubiena tins, rectangular boxes (750 mm × 500 mm × 400 mm) with two open ends covered by lids. A vertical face was cleared and an open end of a Kubiena tin was placed on the surface. Sediment surrounding the tin was carefully cut away with a knife, and the tin was slowly pushed over the remaining sediment block until it was filled tightly and completely by sediment. Tins were not forced into the sediment in order to avoid structural disturbance

TABLE 1. CHARACTERISTICS OF SEDIMENT FLOW DEPOSITS, MATANUSKA GLACIER, ALASKA. FROM LAWSON, 1979a

Sediment flow type	Bulk Texture type 1) Mean (f) 2) Std dev (f)	Internal Organization		
		General	Structure	Pebble fabric
I	Gravel-sand-silt, sandy silt 1) -1 to 2 2) 3 to 4.5	Clasts dispersed in fine-grained matrix	Massive	Absent to very weak; vertical clasts. S_1 @ 0.49-0.55
II	Gravel-sand-silt, sandy silt, silty sand 1) 2 to 3 2) 3 to 4	Plug zone; clasts dispersed in fine-grained matrix. Shear zone; gravel zone at base, upper part may show de-creased silt-clay and gravel content; overall, clasts in fine-grained matrix.	Massive, intraformational blocks. Massive; deposit may appear layered where shear and plug zones distinct in texture.	Absent to very weak; vertical clasts. Absent to weak; bimodal or multi-modal; vertical clasts. S_1 @ 0.50-0.65
III	Gravelly sand to sandy silt 1) -2.5 to 2.5 2) 3.5 to 2	Matrix to clast dominated; lack of fine-grained matrix possible; basal gravels.	Massive; intraformational blocks occasionally.	Moderate, multimodal to bimodal parallel and transverse to flow. S_1 @ 0.60-0.70
IV	Sand, silty sand, sandy silt 1) ³3.5 2) £2.5	Matrix except at base where granules possible.	Massive to graded (distribution, coarse-tail).	Absent
	Surface forms	Contacts and basal surface features	Pene-contemporaneous deformation	Geometry* and maximum observed dimensions (length X width, thickness, m)
I	Generally planar; also arcuate ridges, secondary rills and desiccation cracks.	Nonerosional, conformable contacts; contacts sharp; load structures.	Possible subflow and marginal deformation during and after deposition.	Lobe: 50X20, 2.5
II	Arcuate ridges; flow lineations, marginal folds, mud volcanoes, braided and distributary rills on surface	Nonerosional, conformable contacts; contacts indistinct to sharp; load structures.	Possible subflow and marginal deformation during and after deposition.	Lobe: 30X20, 1.5
III	Irregular to planar; singular rill development; mud volcanoes.	Nonerosional, conformable contacts; contacts indistinct to sharp	Generally absent; possible subflow deformation on liquefied sediments.	Thin lobe; 20X10, 0.5; fan wedge; 30X65, 3.5; rarely, sheet of coalesced deposits.
IV	Smooth, planar; mud volcanoes possible.	Contacts conformable; indistinct.	Absent.	Thin sheet; 20X30, 0.3; Fills surface lows of irregular size and shape.

* Length and width refer to dimensions parallel and transverse to direction of movement prior to deposition.

during sampling. The tin and sediment were then removed from the deposit, and excess sediment was trimmed away until covers could be put on the two open sides of the tin. Location, sample number, and orientation of the sample were marked on the tins and in a field notebook. Tins were numbered and placed in airtight plastic bags to prevent desiccation.

During sampling, an attempt was made to sample contacts between depositionally distinct units. Cobble-rich deposits could not be sampled using Kubiena tins, and therefore are underrepresented in this study. However, these deposits did not appear to differ from cobble-poor deposits in macro-texture, structure, or composition, and are therefore not considered to be genetically distinct from the types that were sampled.

Samples were impregnated under vacuum with a polyester resin using Cobalt Napthenate as an accelerator and Lupersol DDM-9 for a catalyst. The interested reader is referred to Bouma (1969) for further details of the impregnation process. Hardened sediment blocks were thin sectioned and photomicrographs prepared using a Petroscope with multiple magnifications and a 35-mm camera. Each sample block was thin sectioned into both a bottom (B) and a top (T) on a vertical face.

Sample locations

The majority of samples were taken from an unvegetated ice-cored moraine at the glacier margin (Fig. 1). The moraine

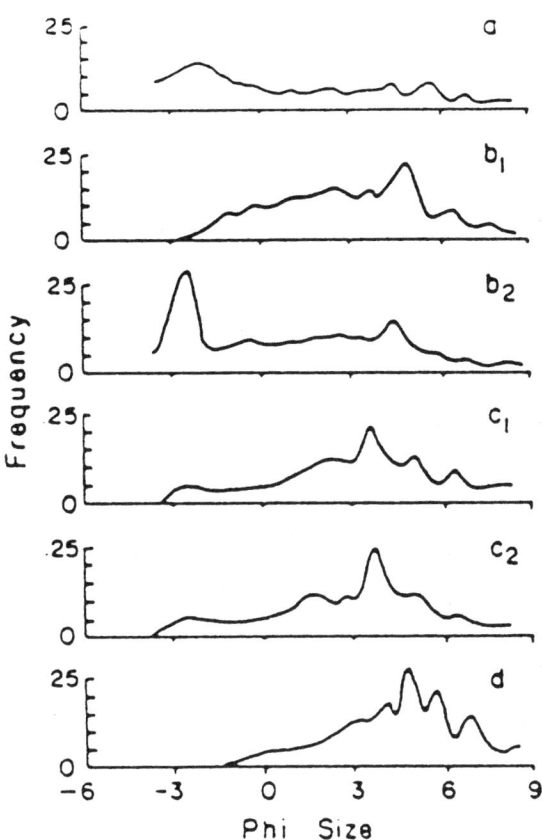

Figure 2. Typical grain size frequency curves for sediment gravity flows, in the body of type I (a), II (b_1), III (c_1), and IV (d) flows. b_2 and c_2 are samples from the basal zones of type II and III flows (after Lawson, 1982).

consists of several tall (5–10 m) eroding ridges, low areas, small stream channels, and seasonally or historically inundated lake areas. It is composed predominantly of older deposits in the ridges, deposited when the ice margin abutted the moraine, and of more recent resedimented flow deposits in low areas. Sediment-flow deposits in the ridges are highly variable in character and range from type I to type IV. Wet-type sediment-flow deposits are commonly found next to and intercalated with fluvial lenses (possibly cross-bedded) and meltwater silts produced by sheet flow. The meltwater silts are generally finely laminated (< 1–2 mm) and may contain injection structures. Where the silts were overridden by dry-type sediment flows (generally type I and II; Lawson, 1979a), they may be folded and faulted. Type III and type IV flows commonly occur in association with fluvial and meltwater deposits indicating they were formed under very wet conditions at or near ablating basal ice on the ice-cored moraine or near the glacier margin.

On the moraine, deposits are reworked frequently, often several times within a season (Lawson, 1979a, 1982). For this reason, sediment-flow deposits retaining a lobate surface shape on the moraine are probably recent in origin. These recently resedimented sediment-flow deposits observed during the early summer of 1996 are commonly flows of lower water contents (types I and II). In a general sense, the most common type of sediment flow now forming on the moraine represents a reworking of older morainal sediments. The morphology is typically lobate with compressive ridges at the toe. The upper surfaces of the sediment-flow deposits can have slopes as steep as 35° and contain vesicles, which may result from the expulsion of liquids or gases during flow and deposition. The deposits are supported by a silt matrix and contain assorted pebbles and cobbles that may or may not exhibit a clast macrofabric. The predominance of type I and II recent sediment flows may be attributed to the much drier conditions away from the ablating ice terminus.

Terminology

This study utilizes microstructural terminology and interpretations developed by van der Meer (1987, 1993, 1996), van der Meer and Laban (1990), Menzies (1990), Menzies and Maltman (1992), and the primary author to describe thin sections, and are summarized from these sources in this section. Microstructure types can be grouped into fabric, laminar, plastic, brittle, and miscellaneous structures. Some structures observed in this study are illustrated in Figure 3 (adapted from van der Meer, 1993, with observations from this study).

Microfabric terminology was originally developed by Brewer (1976) to describe plasma (clay-sized sediment) fabrics. In the Matanuska sediment-flow deposits, plasma-sized material is generally not present (see Fig. 2) and plasmic fabrics were not observed. In Brewers' soil terminology, "skeleton" grains are clasts generally larger than fine silt and are distinguished from the fine silt and clay plasma. Circular orientations of smaller grains and clasts parallel to the surface of a larger "skeleton" clast are called turbate or a "galaxy" structures (van der Meer, 1993). They are attributed to develop during rotational movement of the "skeleton" or "core" clast under plastic conditions. Bimodal clast fabrics in thin section consist of the apparent long axes of clasts dipping in two directions and may develop under plastic to semiplastic conditions. Parallel clast fabrics consist of most or all of the apparent long axes of clasts exhibiting a unidirectional aspect and are interpreted to form under laminar conditions. When formed during laminar flow, this fabric is here proposed to be called "laminar flow fabric," after Hampton (1975). Due to the two-dimensional nature of thin sections, what appears to be a parallel clast fabric may in reality consist of clast orientations both parallel and transverse to flow, and a bimodal clast fabric may actually be polymodal, which is consistent with the observations of clast fabrics in the Matanuska Glacier sediment-flow deposits (Lawson, 1979b).

Laminar microstructures are formed during flow under wet conditions and are expressed as thin (<3 mm) laminations within a deposit. Basal shear zones form under wet conditions as the body is separated from and flows over a traction gravel or underlying deposit (Lawson, 1979a, 1982). Silt wisps form as a silt

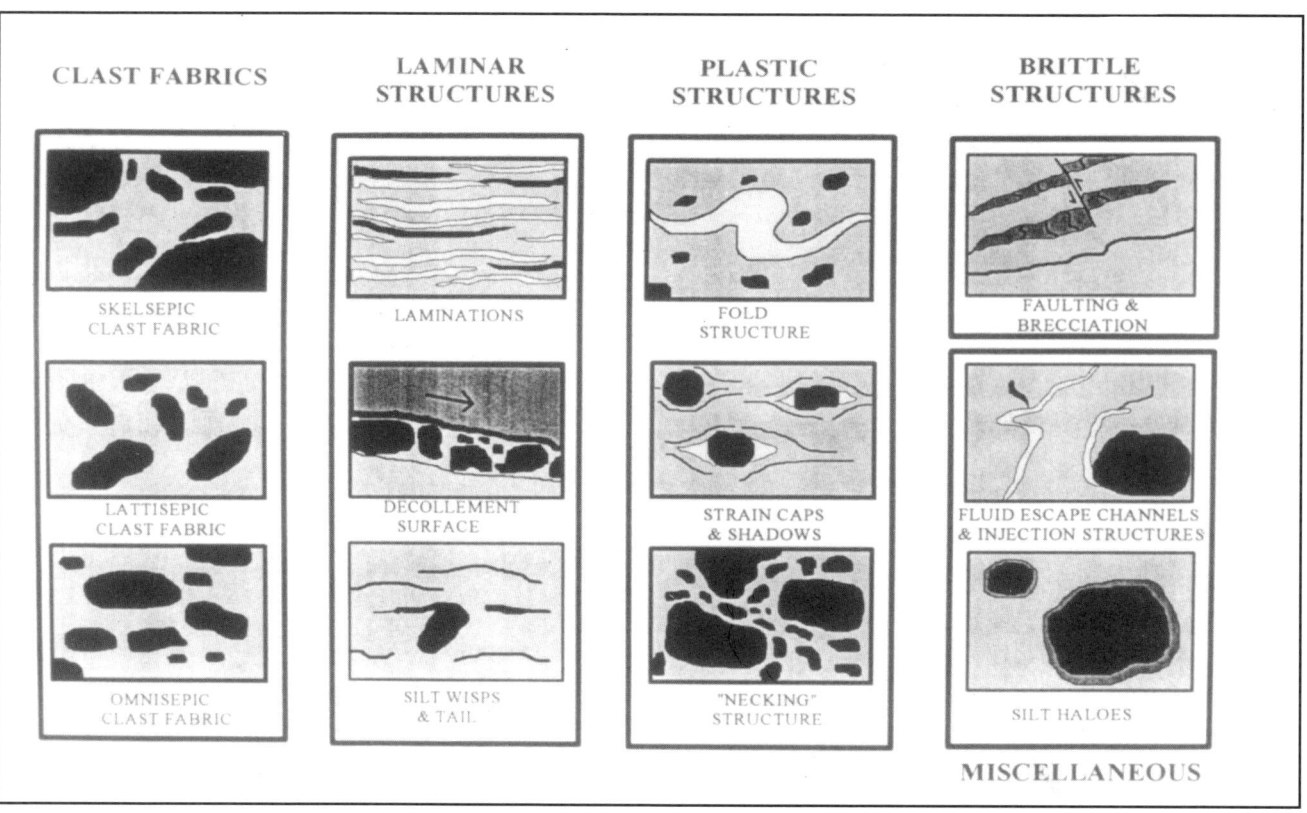

Figure 3. Sediment flow deposit microstructures. Adapted from van der Meer (1993).

clast or inclusion becomes elongated into a thin lineation during laminar or plastic flow. Silt tails are a variety of silt wisps that occur in association with a rotating silt-coated clast.

Plastic (or ductile) structures form in sediments with intermediate water contents. Examples are fold structures, turbate structures ("milky way structures"), strain caps and shadows, and necking structures. Fold structures are formed under compression of cohesive sediments. Strain caps and shadows are believed to be formed from rotation due to shearing (van der Meer, 1993). Necking structures are the orientation of smaller clasts parallel to the surface of a few close larger clasts, and are essentially a variation of a turbate structure. They may also be formed by fluid expulsion through channels between clasts (van der Meer, 1993).

Brittle structures form as a result of stresses on a dry or cohesionless sediment. Faulting, shearing, and brecciation are the most common types of brittle structures in the Matanuska sediment-flow deposits, and are most commonly postdepositional. These structures may also form in dry-type flows, but were not noted in this study.

Miscellaneous structures are represented by fluid-escape channels, which are common in sediment-flow deposits of the Matanuska Glacier. Fluid escape channels are generally vertical to subvertical, but can be horizontal if water movement is redirected by a relatively impermeable bed within a flow deposit. These channels may form as fines are removed by "washing," or as sand grains are deposited after injection into the channel. Silt-rich water can be injected into fractures, planes of weakness, or pores within the sediment, and results in a concentration of fines. Structures formed in this manner are here called "fluid injection structures." Many pebble- to cobble-sized clasts show haloes, or layers of silt around their edges. These may form as a result of rotation of the clast within the matrix (van der Meer, 1993), or as a coating of wet silt on the clast. Observations in this study indicate that higher water content flows generally have thinner or nonexistent haloes, as the cohesion of the sediment/water slurry decreases with increasing water content. The significance of fissility planes in thin section is worth consideration. In our samples, fissility planes were not observed in the field; when observed in thin section, they tended to mimic the angle of the deposit surface. These observations suggest that the fissility planes may be produced or enhanced during sample desiccation and processing, but may occur along preexisting planes of weakness.

RESULTS

Dry-type flow deposits

Dry-type sediment-flow deposits were identifiable in the field by several characteristics. Primarily, a high angle of surface slope and lobate shape are diagnostic, in addition to textural heterogeneity and a poorly defined macro-clast fabric. Most of the dry-type sediment-flow deposits sampled on the moraine were identified according to characteristics described by Lawson (1979a, 1982). The thin-section samples (14, 31, 45, 53, 70) discussed below were collected from five different dry-type sediment-flow deposits and show microstructures characteristic of this type of deposit.

Sample 14 was taken from a recent sediment-flow deposit originating about 1 m from older slumped deposits on the moraine. The surface of the deposit is lobate, dips about 10°, and holds a steeper angle at the nose. Surface relief is about 0.5 cm. The upper portion of the deposit contains vesicles, which probably originated from fluid expulsion during flow and deposition. The sample contains rounded pebbles and cobbles (<4 cm along c-axis) some of which were removed from the sides or back to permit sampling. The lobate morphology and sediment source are consistent with a type I sediment flow (Lawson, 1979a).

Thin sections made from this sample are oriented on a vertical plane parallel to flow and show two units. A silty lower unit which dips and is bounded by folded sand layers at the top and bottom. The upper folded sand layer (Fig. 4) shows a microclast fabric parallel to the trend of the layer. This layer is well sorted, and the lack of fines suggests a fluvial origin. The upper unit consists of silt and sand and shows fissility planes dipping parallel to the flow surface, which terminate at the contact with the lower unit (Fig. 5). Small phyllite clasts in the top unit are vertical to subvertical, and a small silt intraclast with vertical silt layers is present.

The lower unit was probably deformed and folded as the upper unit plastically overrode it. The fissility planes in the upper unit probably formed as the result of shearing during flow, and the presence of a silt intraclast is probably due to incomplete mixing of the source material, a characteristic of a low water content flow with a plug of limited deformation (Lawson, 1979a, 1982).

Sample 31 was also taken from a recent sediment-flow deposit originating near the moraine and shows a contact between a clast-rich sediment flow and underlying meltwater silts. The sediment-flow deposit is composed of rounded to subrounded pebbles and cobbles in a silty matrix. Sorting is poor and texture is heterogeneous; direction of movement is unknown. The sediment flow appears to have truncated the underlying faulted and deformed silts to form an erosional unconformity. Fine laminations at the base of the deposit were visible in the field. We interpret this deposit as a dry-type sediment flow, probably a type II.

Above the contact with the meltwater silts in 31B, there is a wavy layer of fine silt surrounding a few pebbles, interpreted to be the basal shear zone of the flow. Strain caps and shadows and necking structures surround the pebbles (Fig. 6). The body of the sediment-flow deposit contains many clasts with no consistent long-axis fabric. Fissility and a discontinuous silty layer show a deformational fold around a pebble clast in 31T.

The wavy morphology of the basal silt layers probably formed as the flow overrode the irregular surface of the meltwater silts. The fold in 31T probably formed during compressive flow as the body of the deposit encountered a "locking zone"—a zone of high friction and resistance, possibly caused by the slower moving sand/silt intraclast (present due to incomplete mixing of source sediments) in the right-center of the thin sections.

Sample 45 was taken on debris-covered ice from a sediment-flow deposit with some subhorizontal stratification, a sandy layer, and a high silt content. The sediment is underlain by basal ice. In the field, only one deposit was recognized, but two distinct units are observable in thin section. The lower unit is a homogeneous fine silty sand, which contains discontinuous and irregular silt layers and silt wisps. The main silt layer appears brecciated and is discontinuous (Fig. 7). Linear sandy layers and pores are visible throughout this facies. No microfabric was obvious in thin section. The upper surface of the lower unit is irregular and slightly folded (Fig. 8). Above this surface is a sand and gravel layer with intermixed silt and no apparent clast fabric that dips about 5°, overlain by a silty sediment with subhorizontal silt wisps and one obvious folded sand layer (Fig. 9).

There are several small clasts (<10 mm) shown in Figure 8 that do not appear to have a preferred orientation. The largest clast has a discontinuous halo of fine-grained sediment, while the other smaller clasts have weak or no haloes.

The upper unit in Fig. 8 is probably a dry-type sediment flow (I or II) with a tractional gravel at its base. The folded sand layer indicates compressive flow and a lack of a well-defined clast fabric may be an indication of the absence of laminar flow during transport.

The bottom unit shown in Figures 7 and 8 is probably a type IV flow deposit, which is characterized by silty sand and a lack of macroscale structure. The pores in this unit follow sandier layers that were most likely formed as fluid escape channels, one of which crosses the brecciated silt layer (Fig. 7). The discontinuous silt layer may be a relic flow structure that was not disaggregated

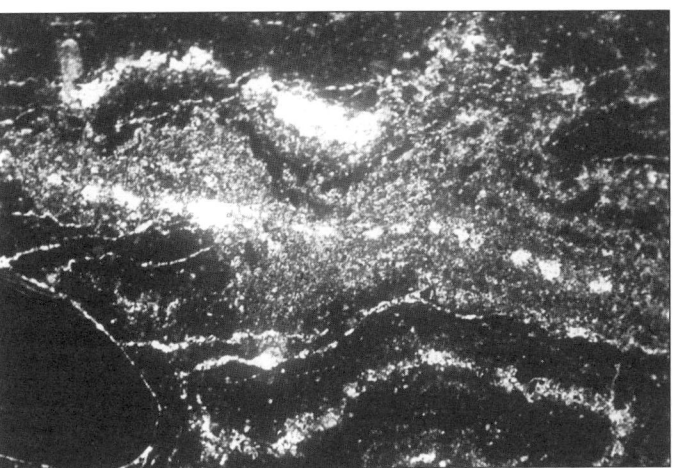

Figure 4. Folded sand layer in sample 14. Width of view is ~22 mm (Photo 3.20).

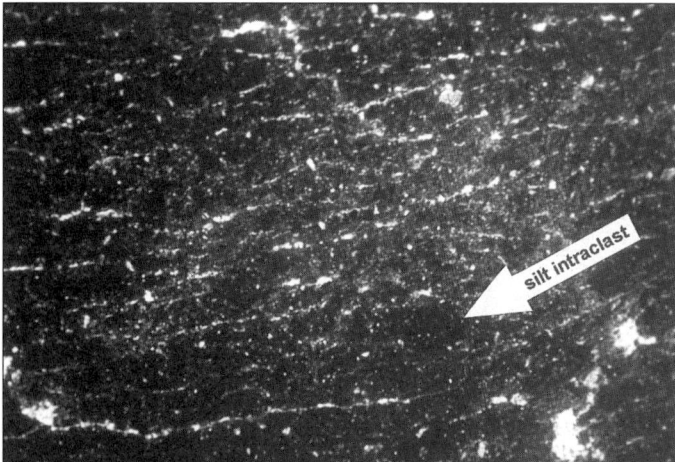

Figure 5. Sample 14, a dry-type deposit, showing fissility planes. There is a small silt intraclast near the center of the photo. Width of view is ~22 mm (Photo 3.24).

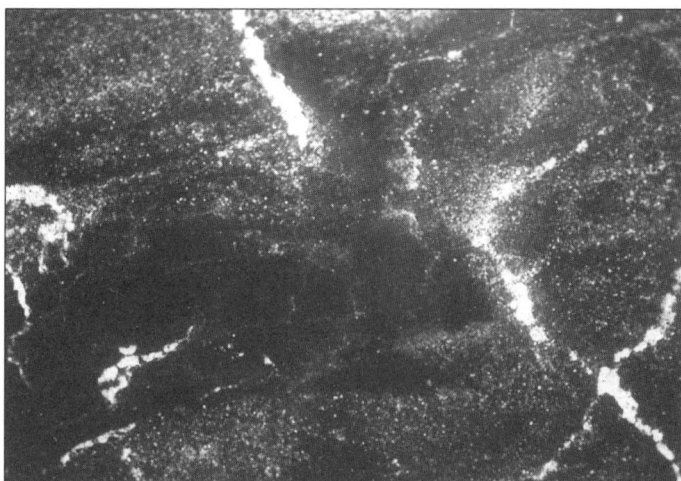

Figure 7. Discontinuous and brecciated silt layer in sample 45. Light areas are fluid escape channels. Width of view is ~22 mm (Photo 4.8).

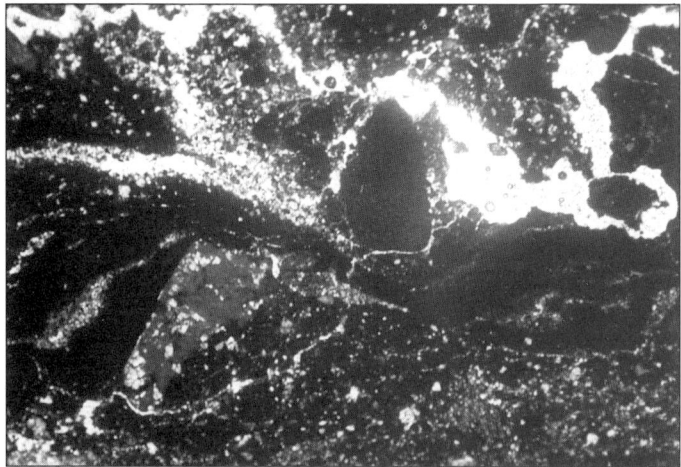

Figure 6. Contact between lower meltwater silts and upper dry-type sediment flow deposit in sample 31. The silt layers are a basal shear zone; strain caps and shadows are present around clasts. Width of view is ~22 mm (Photo 3.30).

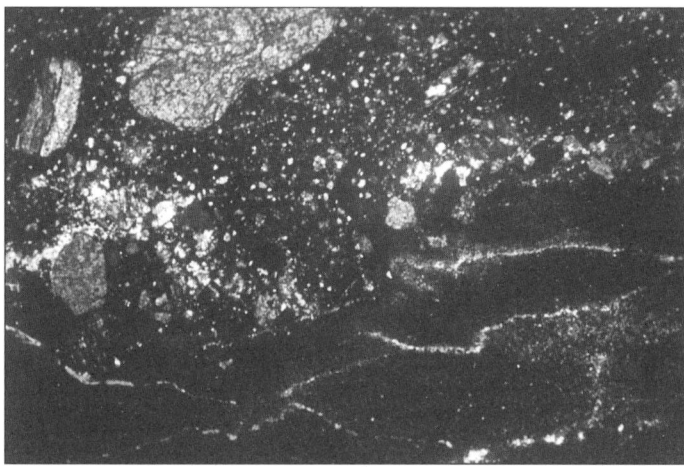

Figure 8. Contact between two deposits in sample 45. Note deformed silt below a traction gravel of the overlying sediment flow deposit. Width of view is ~22 mm (Photo 4.15).

during transportation or deposition. It should be noted, however, that the identification of this deposit as a possible type IV is based solely on micromorphology and not field identification.

Sample 53 was taken from a ridge of silty sediments on the moraine area near debris-covered ice. The unit was sampled normal to the strike of the ridge; flow direction could not be ascertained. The sediment appeared structureless in the field. The deposit is texturally heterogeneous and composed of silty sand with cobbles ranging as large as 20 cm in diameter. Only small pebbles were visible where the deposit was sampled.

Thin sections elucidate many structures not visible in outcrop. Angular metamorphic clasts smaller than 1 cm are interspersed throughout the sandy silt matrix along with prolate rounded phyllite clasts less than 0.5 cm long. A rounded phyllite clast (long axis is 4 cm) is present in 53B. The clasts show a

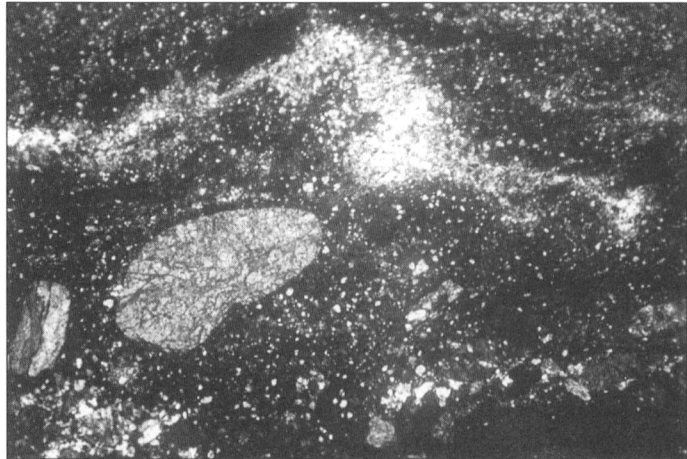

Figure 9. Folded sand lens in sample 45, above a locking zone. Note haloes around clasts. Width of view is ~22 mm (Photo 4.16).

poorly developed fabric in thin section (Fig. 10), with the two main axes dipping from ~45° to the left and ~40° to the right from the horizontal, and a turbate structure without a core stone. The larger clasts have weakly defined haloes, while around the largest phyllite clast there is a more strongly defined turbate structure (detail in Fig. 11).

The reorientation of clasts around the large phyllite clast and the turbate structures are evidence that internal deformation of the sediment was occurring. Some of the clasts have silt tails, which are formed from rotation of the clasts. Channels of sand are present around some of the clasts, and probably formed as fluid expulsion channels during deposition (Fig. 12).

Sample 70 was taken on the moraine area from a sediment-flow deposit that originated from slumping of morainal sediments. The appearance of the flow deposit is nonchannelized, and has a ropy surface formed from compressive deformation. There are several overlapping sublobes of the flow apparent on the surface, and their surfaces dip about 30°. The flow morphology and sediment source indicate this is a type I flow deposit (Table 1). The deposit was sampled parallel to flow direction from near the center of the flow body. No stratification was visible in outcrop, but a macrofabric was present in which the long axes of the clasts dipped downflow approximately parallel to the surface slope of the sediment flow. The deposit was matrix supported but contained many pebbles and cobbles. Some sandy areas in the matrix were also present.

Analysis of thin sections reveals a microclast fabric dipping about 30° to the left. The clast fabric is generally parallel (Fig. 13) near the base and grades to weakly bimodal higher in the flow. The transition from a parallel fabric near the base to a weakly bimodal fabric higher in the flow is an indication of differential shear stresses in the flow deposit. Laminations are not distinguishable. The sediment-flow deposit is heterogeneous and poorly sorted, and the pores are elongated along dip planes.

This sediment flow was probably initiated when morainal sediments were wetted enough to reduce cohesion and initiate flow. Considering the high angle of slope upon which this deposit flowed, internal shear stresses were probably great enough to overcome the plasticity of the plug and semilaminar flow occurred near the base, while semiplastic flow occurred higher in the body.

Wet-type flow deposits

Wet-type sediment-flow deposits were identified in the field from characteristics outlined by Lawson (1979a, 1982; Table 1). Primarily, a homogeneous, well-sorted texture, and a well-defined clast fabric were diagnostic, in addition to the presence in some samples of thin laminations. Most of the wet-type sediment-flow deposits sampled from the moraine area were deposited under wet conditions at or near the ablating ice terminus. The thin-section samples (13, 50, 28) discussed below were collected from two different wet-type sediment-flow deposits and show microstructures characteristic of this type of deposit.

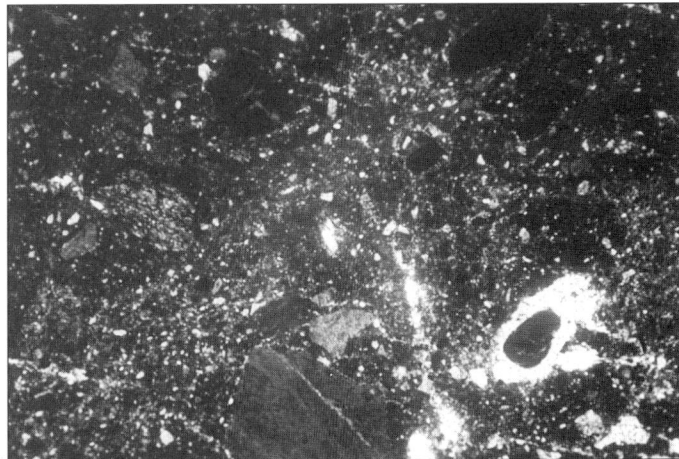

Figure 10. Poorly developed clast fabric in sample 53. Two directions of dip are ~45° to left and ~40° to right. Width of view is ~22 mm (Photo 4.31).

Figure 11. Reorientation of clasts around a phyllite "core" clast (turbate structure) in sample 53. Width of view is ~22 mm (Photo 4.29).

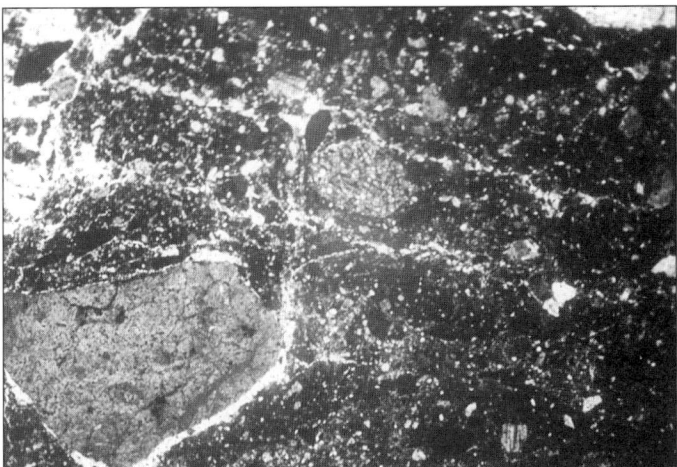

Figure 12. A fluid escape structure originating beneath a clast in sample 53. Width of view is ~22 mm (Photo 4.35).

Sample 13 was taken on the moraine from sediment-flow deposits in a morainal ridge. Flow direction was not apparent in the field. The deposit appears well sorted and generally massive with the exception of a few fine laminations (1–3 mm). It is composed of matrix-supported silt and contains rounded to subrounded clasts with a long axis length as much as 4 cm. Sand and gravel layers are present throughout the sediment block. No macro fabric was apparent in the field. The sorting and texture, along with the thin laminations indicate this is a wet-type sediment-flow deposit, probably a type III (Table 1).

Thin sections show one depositional facies, which is thinly laminated and contains silt, sand, and pebble sized clasts (<1 cm). Laminations dip to the right and the clasts exhibit a strong parallel fabric and are imbricated upslope (Fig. 14). Imbricated clasts were reported by Lawson (1979a) to be present in type III fan-type flows, and seems to be consistent with the observations of this deposit. Clast haloes are generally weakly defined, further indication of a high original water content of the flow.

The sediment in sample 13 was probably transported under laminar flow throughout, as shown by the high degree of sorting, and deposited under wet conditions, possibly in a fan-type setting. The topographic elevation in the morainal ridge and wet-type structures indicate the deposit was formed when the ice margin abutted the ridge. The thin sections most likely show characteristics of the body of the flow.

Sample 50 was taken from the same unit and stratigraphic position as sample 13. The unit appears massive, is composed of pebbles (<1 cm) in a matrix of silt and sand, with a clast fabric dipping slightly downwards and inwards from the outcrop face, which is oriented north–south. From this, movement was estimated to be approximately normal to the face.

In thin section, sample 50 (Fig. 15) shows a well-defined pebble fabric parallel to the trend of thin undulating laminations, and some of the clasts exhibit upslope imbrication. Other thin sections of this sample contained gravels overlain by silt wisps at the base of the slide. The strongly defined parallel clast fabric, upslope imbrication, and laminar flow fabric suggest the sediment-flow deposit formed as a wet-type flow, possibly in a fan setting. The microstructures of this sample share a resemblance to those of sample 13.

Sample 28 was taken 2 m from the active ice margin and was probably recently deposited. The sedimentary sequence consists of a 1-m-thick lower unit resting on basal ice, and a 10-cm-thick upper unit. The sample was taken at the contact between facies, and some cobbles were removed to facilitate sampling. Flow direction is not known.

The upper unit is a sandy silt, with thin undulating laminations, and small pebbles exhibiting a fabric parallel to bedding. The laminations consist of alternating sandy and silty layers, and are faulted in some locations. The pebbles are concentrated preferentially along laminations, which vary from horizontal to about 15°. Some large cobbles occur in this deposit (c-axis ~ 15 cm). A sand intraclast (1 cm diameter) is present near the bottom center of 28B (Fig. 16).

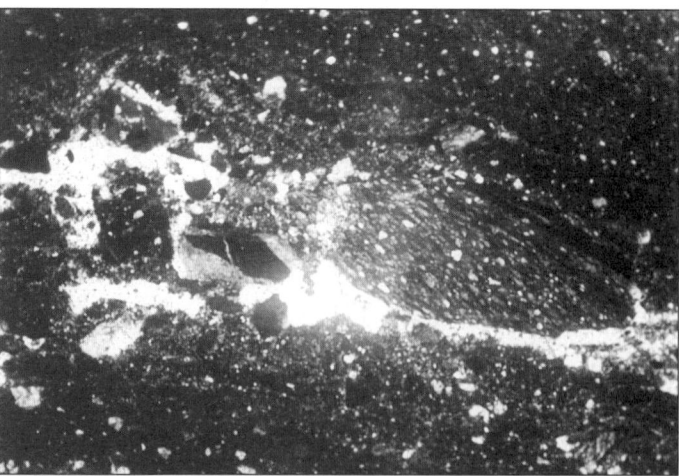
Figure 13. Parallel clast fabric in sample 70. Width of view is ~22 mm (Photo 5.27).

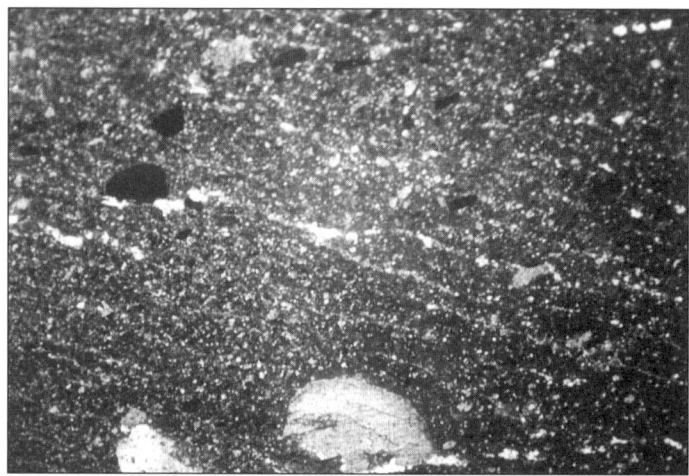
Figure 14. Wet-type flow deposit showing fissility, weak laminations, and upslope imbrication of small phyllite clasts, sample 13. Width of view is ~22 mm (Photo 5.15).

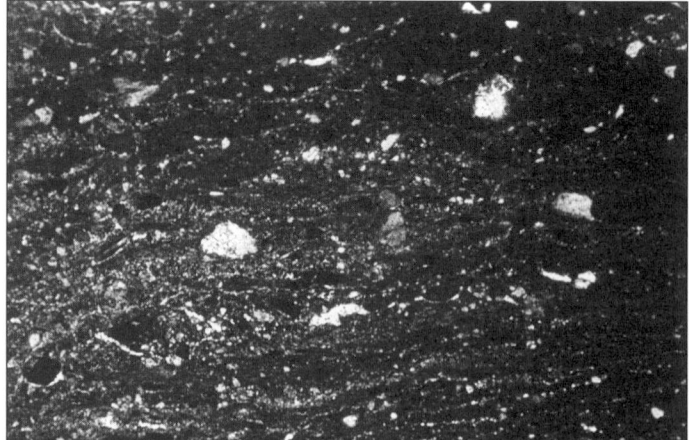
Figure 15. Laminar flow fabric in a wet-type sediment flow deposit, sample 50. Note the thin laminations, parallel clast fabric, and upslope imbrication of small phyllite clasts. Width of view is ~22 mm (Photo 7.21).

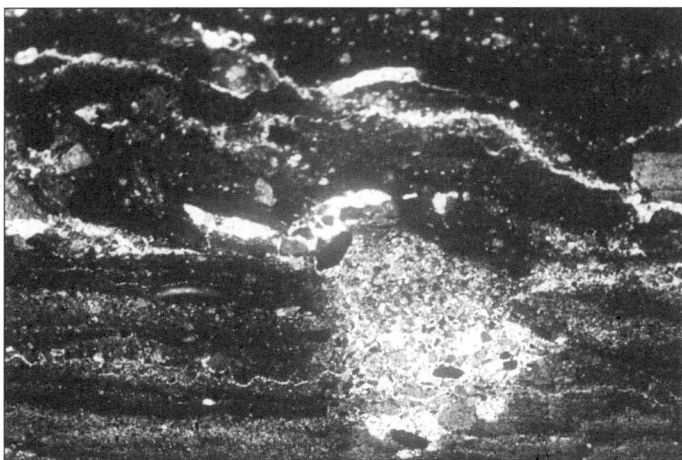

Figure 16. Thin laminations and sand intraclast in sample 28. Lower portion shows a parallel clast fabric. Width of view is ~22 mm (Photo 6.5).

The upper unit is probably a wet end-member sediment-flow deposit as shown by the laminations and parallel clast fabric. The internal organization of the unit probably reflects the laminar nature under which it flowed, possibly as a continuum of meltwater and sediment-flow deposition, an observation consistent with a type III flow (Table 1). Also, the sand intraclast may be a relic of the original sediment source, which was not disaggregated during flow. Melting of underlying basal ice probably caused slight normal faulting of the sample. The contact between the two facies is indistinct in thin section.

DISCUSSION

Sediment gravity flow deposits of lower water content (approximately Lawson types I and II) have characteristic microstructures that allow differentiation from higher water content flow deposits (approximately Lawson types III and IV; see Table 2). Of primary interest in distinguishing dry-type from wet-type sediment-flow deposits in a sediment sequence is the contact between the sediment-flow deposit and underlying sediments, in addition to the nature of the traction gravel-plug interface. In thin section, the underlying units may be faulted and deformed when overridden by dry-type flows, whereas they exhibit smooth contacts when overridden by wet-type flows. Above the traction gravel, plastic deformation structures are indicative of dry-type flow deposits, while an abundance of thin laminations and a parallel clast fabric indicates a wet-type flow deposit. Other characteristic microstructures of dry- and wet-type deposits are presented in the following sections.

Microstructures and characteristics of dry-type sediment-flow deposits

Type I and II flows demonstrate plastic deformation of the plug when a "locking zone" is encountered in the lag gravel or on surface irregularities of the underlying sediment. Evidence of this deformation is shown by the folding of layers and turbate structures, and the orientation of smaller clasts parallel to the faces of larger clasts, which other workers have attributed to clast rotation or translocation (van der Meer, 1993, 1996). Compressive flow structures are observable both in the field on the macroscale, and in thin section on the microscale.

Microdeformation of underlying sediment is apparent in thin section as faulting, folding, mixing, and fluid escape structures. During flow, sediment from other sources can be incorporated into the flow mass (Lawson, 1979a, 1982) and be completely or incompletely mixed. Micro- and mesoscale sediment clasts of this type are commonly present in thin section. A poorly defined (bimodal or random) clast fabric is observed in some samples and results from internal deformation of the plug during flow or from mixing during mobilization of the source material (Lawson, 1982). Plasticity of the sediment flow generally does not allow the formation of a parallel fabric (with the exception of flows on high-angle slopes, see sample 70). Semiplastic flow of parts of the plug and a weak clast fabric may develop in type I and II flows if the slope of the sediment surface is sufficiently steep. The presence of microstructures in these deposits, it should be noted, may be inherited from the sediment source and not produced as the result of flow or deposition. The higher cohesion of sediment allows the formation of thicker haloes around clasts.

Microstructures and characteristics of wet-type sediment-flow deposits

Wet-type sediment-flow deposits show characteristics that contrast with those of dry-type sediment-flow deposits. Under shear flow common to wet-type flows, microlaminations develop where silt layers alternate with sand and fine gravel. These laminations may form as a result of differential shear within the flow. Well-defined parallel clast fabrics commonly develop in wet-type deposits, and we suggest "laminar flow fabric" (after Hampton, 1975) as a descriptive term. This clast fabric is most obvious in the apparent C-axes of sand-sized phyllite clasts. In samples 13 and 50, imbrication of phyllite clasts was well developed, a characteristic of type III flows deposited as fans (Lawson, 1979a).

Consistent with laminar flow throughout the sediment, little internal plastic deformation of flow and underlying sediment is apparent in thin section. In contrast with the larger folds in the dry-type flow deposits, small-scale (<2 mm) laminar undulations are present in wet-type flow deposits where the flow overrode an irregular surface or encountered a friction zone of a lag gravel. Similar to dry-type flow deposits, pore fluid expulsion channels may be present.

In contrast with dry-type flow deposits, wet-type flow deposits typically develop smooth basal shear zones above a basal traction gravel. This surface is characterized by a planar zone of shear, often associated with an accumulation of silt in

TABLE 2. MICROMORPHOLOGICAL CHARACTERISTICS OF SUBAERIAL GLACIGENIC SEDIMENT FLOW DEPOSITS

	Dry-type	Wet-type
Viscosity regime and structures	brittle to plastic deformation; folding from compressive flow	plastic to laminar; thin flow laminations
Clast fabric in thin section	poorly developed, random	parallel; clasts may be imbricated upslope; "laminar flow fabric"
Flow/substrate interface	underlying sediment deformation	little underlying sediment deformation; smooth contacts
Fluid movement structures	fluid escape and injections structures	fluid escape and injections structures
Haloes	thick haloes	thin or nonexistent haloes

the lee of some gravels. Haloes are generally thin or nonexistent in wetter flows, which have less cohesion than dry-type flows.

Ice-marginal settings and sediment-flow deposits

Lawson (1979a, 1982) observed that sediment flows accounted for the majority of sediment deposition at the terminus of the Matanuska Glacier. Sediment flow formation at the Matanuska Glacier is partly dependent on melting of debris-rich basal ice. According to Strasser et al. (1996), the formation of debris-rich basal ice by freeze-on is hypothesized to occur in overdeepenings where supercooled water is expelled upward and downglacier, where it forms frazil ice as the pressure melting point is increased. The meshlike frazil ice incorporates sediment during nucleation and eventually becomes a dense mass, which is transported subglacially and ultimately exposed and ablated at the terminus

If this situation is analogous to depositional settings of Pleistocene ice sheets in some locations, the presence of large amounts of subaerial sediment-flow deposits in ice-marginal sediments may indicate that large amounts of debris-rich basal ice were present. The Great Lakes basins, the Finger Lakes area of western New York State, and other areas may have supported conditions of overdeepening. The relationship between sediment flows and debris-rich basal ice formation must be considered cautiously however, as substantial reworking of supraglacial material and ablation of debris bands may also permit sediment-flow formation (e.g., Boulton, 1968, 1971). It should be noted that the geologic situation at the Matanuska Glacier may be unique, and therefore may not serve as an analog for other glaciated locations.

CONCLUSIONS

Micromorphology can be used to distinguish wet-type from dry-type sediment-flow deposits occurring at the Matanuska Glacier, Alaska. Many characteristic structures in thin sections of sediment-flow deposits were often not visible in the field. Dry-type sediment-flow deposits are characterized by a lack of a well-defined clast fabric, generally exhibiting random or poorly defined clast fabrics and turbate structures, in addition to plastic and brittle deformation structures of the flow body and underlying sediments. Wet-type sediment-flow deposits are characterized by laminar flow structures, such as a well-defined parallel clast fabric, thin laminations, and a lack of underlying sediment deformation. These microstructures are consistent with the physics and rheology of flow occurring during transport and deposition. Laminar flow structures and a "laminar flow fabric" may be unique to wet-type sediment-flow deposits.

In a glacial environment, whereas plastic and brittle microstructures in sediments may be polygenetic and postdepositional (van der Meer, 1993; Menzies, 1990), flow structures are generally formed when sediment becomes saturated and flows under the influence of gravity. At a glacier's base, it is theoretically possible to develop structures having laminar characteristics, but most evidence of the subglacial conditions in a deforming bed (van der Meer, 1993; Menzies, 1990) indicate that brittle and plastic structures develop. Micromorphological analysis of unconsolidated sediments of other origins will improve our ability to interpret depositional sequences from recent and ancient glaciations.

ACKNOWLEDGMENTS

The authors would like to thank the useful criticisms of several reviewers, Nelson Ham, John Attig, and an anonymous reviewer; this paper benefited from their suggestions. Many thanks go also to the pioneers of glacial sediment micromorphology, J. J. M van der Meer, John Menzies, and Jim Rose, whose previous work served as an impetus for this study.

REFERENCES CITED

Bouma, A. H., 1969, Methods for the study of sedimentary structures: New York, John Wiley and Sons, 458 p.

Boulton, G. S., 1968, Flow tills and related deposits on some Vestspitsbergen glaciers: Journal of Glaciology, v. 7, no. 51, p. 391–412.

Boulton, G. S., 1971, Till genesis and fabric in Svalbard, Spitsbergen, in Goldthwait, R. P., ed., Till, a symposium: Columbus, Ohio State University Press, p. 41–72.

Brewer, R., 1976, Fabric and mineral analysis of soils: Huntingdon, New York, Krieger, 482 p.

Evenson, E. B., Dreimanis, A., and Newsome, J. W., 1977, Subaquatic flow tills: a new interpretation for the genesis of some laminated till deposits: Boreas, v. 6, p. 115–133.

Hampton, M. A., 1975, Competence of fine-grained debris flows: Journal of Sedimentary Petrology, v. 45, no. 4, p. 834–844.

Hartshorn, J. H., 1958, Flowtill in southeastern Massachusetts: Geological Society of America Bulletin, v. 69, p. 477–482.

Lawson, D. E., 1979a, Sedimentological analysis of the western terminus region of the Matanuska Glacier, Alaska: Hanover, New Hampshire, U.S. Army Cold Regions Research and Engineering Laboratory, Report 79-9, 112 p.

Lawson, D. E., 1979b, A comparison of the pebble orientations in ice and deposits of the Matanuska Glacier, Alaska: Journal of Geology, v. 87, p. 629–645.

Lawson, D. E., 1982, Mobilization, movement and deposition of active subaerial sediment flows, Matanuska Glacier, Alaska: Journal of Geology, v. 90, p. 279–300.

Lawson, D. E., Strasser, J. C., Evenson, E. B., Alley, R. B., Larson, G. J., and Arcone, S. A., 1998, Glaciohydraulic supercooling: a freeze-on mechanism to create stratified, debris-rich basal ice: I. Field evidence: Journal of Glaciology, v. 44, p. 547–562.

Lowe, D. R., 1976, Subaqueous liquefied and fluidized sediment flows and their deposits: Sedimentology, v. 23, p. 285–308.

Lowe, D. R., 1982, Sediment gravity flows: II. Depositional models with special reference to the deposits of high-density turbidity currents: Journal of Sedimentary Petrology, v. 52, no. 1, p. 0279–0297.

Marcussen, I., 1973, Studies on flow till in Denmark: Boreas, v. 2, p. 213–231.

Marcussen, I., 1975, Distinguishing between lodgement till and flow till in Weichselian deposits: Boreas, v. 4, p. 113–123.

Menzies, J., 1990, Brecciated diamictons from Mohawk Bay, S. Ontario, Canada: Sedimentology, v. 37, p. 481–493.

Menzies, J., and Maltman, A. J., 1992, Microstructures in diamictons–Evidence of subglacial bed conditions: Geomorphology, v. 6, p. 27–40.

Strasser, J. C., Lawson, D. A., Larson, G. J., Evenson, E. B., and Alley, R. B., 1996, Preliminary results of tritium analyses in basal ice, Matanuska Glacier, Alaska, U.S.A.: evidence for subglacial ice accretion: Annals of Glaciology, v. 22, p. 126–132.

van der Meer, J. J. M., 1987, Micromorphology of glacial sediments as a tool in distinguishing genetic varieties of till: Geological Survey of Finland, Special Paper 3, p. 77–89.

van der Meer, J. J. M., 1993, Microscopic evidence of subglacial deformation: Quaternary Science Reviews, v. 12, p. 553–587.

van der Meer, J. J. M., 1996, Micromorphology, in Menzies, J., ed., Glacial environments, Volume 2: Oxford, Butterworth-Heinemann, p. 335–355.

van der Meer, J. J. M., and Laban, C., 1990, Micromorphology of some North Sea till samples, a pilot study: Journal of Quaternary Science, v. 5, p. 95–101.

MANUSCRIPT ACCEPTED BY THE SOCIETY OCTOBER 8, 1998

Need for three-dimensional analysis of structural elements in glacial deposits for determination of direction of glacier movement

Aleksis Dreimanis
Department of Earth Sciences, University of Western Ontario, London, Ontario N6A 5B7, Canada

ABSTRACT

Direction of glacier movement is an important criterion in glacial stratigraphy and in indicator tracing. It is obtained locally from the orientation of erosional and depositional directional features and deformation structures that have been produced by moving glacier ice. However, in several subglacial and ice-marginal landforms these glaciokinetic features are formed by local secondary stresses that differ from the direction of glacier movement, yet they are still related to glacial flow. Such observations have been made particularly in drumlins, flutes, interlobate moraines, and end moraines formed by many small glacial tongues.

In order to determine the direction of glacier movement from secondary stress patterns, I suggest a three-dimensional kinetoarchitectural investigation, particularly when studying older glacial deposits where the upper parts of landforms have been eroded.

INTRODUCTION

Glacial stratigraphy should be based upon multiple criteria. Determination of the direction of glacier movement is particularly important in stratigraphic studies because a regionally consistent pattern of glacier movement characterizes each glacial advance, and successive glacial advances often come from different directions. Direction of glacier movement is also very important when doing indicator tracing in glaciated terrain.

A selective glacial stratigraphy, based only upon the directional pattern of glacier movement was proposed by Berthelsen (1973). Berthelsen (1973, 1978) concluded from his own and his co-workers' studies in Denmark, that the differences in the direction of successive glacial advances were very pronounced. As a result, he could distinguish four main Weichselian glacial advances in southeastern Denmark based upon the direction of glacier movement. He called this new type of stratigraphy kinetostratigraphy, ". . . where the main emphasis is placed on the study of the directional elements that reflect the movement patterns (kinetics) of former ice sheets" (Berthelsen, 1978, p. 25). The kinetostratigraphic drift units were distinguished by studies of orientation of the following glaciokinetic features: glacial striae on bedrock and boulder pavements, till fabric, sedimentary structures in stratified drift, and particularly by structural analysis of glacially induced deformation structures. For discussions of kinetostratigraphy see also Aber et al. (1989), Ehlers (1996), Rose and Menzies (1996), and Benn and Evans (1998). Berthelsen (1978, p. 25) referred to investigation by 12 authors who had tested the kinetostratigraphic approach in southeastern Denmark. Subsequently, Houmark-Nielsen and Berthelsen (1980) and Houmark-Nielsen (1987) successfully combined kinetostratigraphic and lithostratigraphic criteria in their studies of the Pleistocene stratigraphy in central Denmark. Houmark-Nielsen (1988) also emphasized the glaciotectonic unconformities as marker horizons on regional scale in stratigraphic studies of Denmark.

Kinetostratigraphy was applied successfully also outside Denmark. Thus Kluiving et al. (1991) distinguished four different Older Saalian glacial events by structured analysis of a continuous glacial cover in The Netherlands. In North America, Albino and Dreimanis (1988) applied kinetostratigraphic criteria in the north shore area of Lake Erie by determining the orientation of glaciotectonic deformation structures. They distinguished a sequence of three glacial advances from different directions. Hicock (1992) confirmed their conclusions. Hicock and Fuller (1995) investigated the interaction of the Cordilleran Ice Sheet and the glaciers of the Queen Charlotte Islands by applying mainly kinetostratigraphic criteria. Benn and Evans (1998, p. 557) in their discussion of kinetostratigraphy state that "kinetostratigraphy and related methods provide very powerful tools for unravelling the dynamic history of ice sheets."

The kinetostratigraphic approach, without using other criteria, was applied occasionally even before Berthelsen's (1973) proposal of the term "kinetostratigraphy." For example, Dreimanis (1935) related the orientation of glaciotectonic deformation structures to five till layers, and MacClintock (1958), MacClintock and Dreimanis (1964), and MacClintock and Stewart (1965) differentiated successive till layers by their fabric patterns. More commonly, glaciokinetic criteria have been combined with various lithologic, textural, and geomorphologic criteria in glacial stratigraphic studies since the second half of last century.

Several investigators of glacigenic deposits have also pointed out problems in the determination of direction of glacier movement. Thus Aber et al. (1989, p. 150–151) suggest that kinetostratigraphy's prerequisite—a consistent pattern of ice movement—may be the most difficult to deal with because the direction of ice movement may vary markedly across a region. Even in an idealized ice lobe the ice flow diverges along its flanks and also shifts the flow direction during advance and during retreat. However, the divergence and the pattern of changes are gradual and they may be deciphered by systematic regional investigation, unless there are erosional, nondepositional, or nondeformational gaps.

A more serious problem is the discovery, by several investigators of Pleistocene and modern glacial deposits, that the orientation of the supposed indicators of regional directions of glacier movement is governed by local stresses at the sole of glacier ice and that the local stresses may differ even from the direction of movement of the glacier ice over the site or the bedform investigated. When discussing these differences I will use the following terms:

1. "local stress direction": stress direction at the glacier sole at the site investigated;

2. "direction of glacier movement" without specification: direction of dominant ice movement at the site or the landform investigated; and

3. "direction of regional glacier movement": direction of ice movement along the axis of an ice stream, a lobe, or a valley glacier, or the dominant direction over an area of several hundreds of square kilometers covered by an ice sheet.

Before reviewing some published data on the differences among the local stress directions and the above two kinds of ice movement, I will briefly discuss the glaciokinetic criteria that may reflect either the direction of glacier movement or local stress direction.

Following the review of local subglacial stress directions that differ from the direction of glacier movement, I will propose a three-dimensional architectural investigation of the glaciokinetic elements in glacial landforms and particularly in the partly eroded buried remnants of older landforms. This proposal was discussed at two conferences (Dreimanis, 1997; Dreimanis and Zelčs, 1997a). Such a three-dimensional investigation should assist in overcoming the confusion that may be caused by the differences in regional ice flow direction, the directions of local subglacial stresses, and the direction of movement of overlying glacier ice.

GLACIOKINETIC CRITERIA AND THEIR ORIGIN

Glacier ice when moving is dynamic: it erodes or deforms underlying materials and may deposit primary till (Dreimanis, 1989, p. 41–42) by lodgement or subglacial drag (deformation till). Even the third variety of primary till, meltout till, though deposited from motionless basal ice, carries evidence of glacier movement prior to its melting out.

Erosional criteria

Parallel glacial striae and small grooves on bedrock or boulder pavements have been used for a couple of centuries as indicators of local direction of glacier movement. The sense of the ice movement may be determined by nailhead striae or forked striae, by stoss-lee features, or by transversely oriented crescentic marks, most of them dipping downglacier. For a brief discussion of the above glacial erosional features and further references see Ehlers (1996, p. 36–41), Menzies and Shilts (1996, p. 94–98), and Benn and Evans (1998, p. 178–192).

Linear subglacial erosional features also occur in soft sediment (Westgate, 1968; Ehlers and Stephan, 1979). They provide the same information as glacial striae. Occasionally, mylonitic fine striations occur in lodgement till (Berthelsen, 1978, p. 33).

The erosional criteria are considered to be most reliable for interpreting the local direction of glacier ice movement at its sole.

Glaciotectonic deformation structures

Though glaciotectonic structures have been recognized sporadically for several centuries (Aber et al., 1989, p. 1–6; Ehlers, 1996, p. 67–76), their detailed studies have been restricted to the last six decades (for references through 1993 see Aber, 1988, and 1993).

Glaciotectonic deformation occurs either in the subglacial shear zone or at the marginal compressive belt of a glacier (van der Wateren, 1995). The deforming stress is exerted by the weight and/or movement of the ice. If the cause of the deformation is merely the static weight of the ice, then the glaciotectonic structures do not reflect the glacier movement. Independent evidence is then is required for the direction of glacier movement. Houmark-Nielsen (1987) therefore compared the orientation of glaciotectonic structures with till fabric.

In most cases the strike of platy structural elements, for instance thrust and shear planes, but also axial planes of recumbent folds and till wedges, is transverse to the stress direction (Aber et al., 1989; van der Wateren, 1995). If the deformation features are formed in a compressive regime, they tend to rise in the direction of the stress vector, but those formed under extensional regime may dip in that direction (Banham, 1988; Hart and Boulton, 1991; Dreimanis, 1993; and references therein).

Some linear deformation features in subglacial till trend parallel to the direction of the local till-depositing glacier movement, for instance the so-called torpedo structures (Berthelsen, 1978, Fig. 9; similar raft structures are also in Figs. 90 and 91 in Lavrushin, 1976; and in Fig. 4 of Hart and Boulton, 1991). The orientation of upward-injected diapirs depends greatly upon local stresses rather than upon the glacier movement (Whittecar and Mickelson, 1979; Dreimanis, 1995; Dreimanis and Rappol, 1997).

Deformation structures also occur in flowtill and other mass movement deposits that are unrelated to glacial stresses. Dreimanis (1993) discussed how to distinguish glacial from nonglacial deformation structures, mainly by applying multiple criteria.

Till fabric

The orientation of stones or sand grains in till is usually called till fabric. Statistical till fabric measurements have become a standard method of reconstructing glacier ice movement (for a review see Benn, 1995; Benn and Evans, 1998; Ehlers, 1996, p. 41–45; and Hicock et al., 1996). Though in most subglacial tills the fabric maxima are parallel to the local direction of movement of ice and the clasts preferentially dip upglacier (Krüger, 1970), transverse maxima may dominate in till formed or deformed by compressive or squeeze flow (Boulton, 1971), for instance in deformation till. Also, till fabric maxima may be oblique to the direction of glacier movement. In fact, Richter (1932, 1936), who introduced the two-dimensional till fabric diagram, pointed out local variability of till fabric maxima in a regional study of the directions of glacier movement in northern Germany. Therefore, till fabric as an indicator of local ice movement should be used in conjunction with other glaciokinetic criteria as has been done, for instance, by Hicock and Dreimanis (1992) in the area of Toronto, Canada, and recommended by Hicock et al. (1996).

LOCAL GLACIAL STRESS DIRECTIONS DIFFERENT FROM THE DIRECTION OF GLACIER MOVEMENT

Though several authors have recorded evidence of local stress directions in glacial landforms that differ from the direction of glacier movement, Stephan (1971, 1985) paid particular attention to these differences because they create problems in reconstructing the ice flow direction in regional and/or stratigraphic studies. He pointed out that the orientation of glaciotectonic deformation structures that are commonly considered to strike transverse to the movement of the glacier, actually may have a different orientation caused by the load and very local stresses of the glacier itself. Other authors have also used till fabric as an indicator of either the direction of glacier movement or local stresses. Most of these studies have been done in the following landforms formed by ice sheets that by themselves are considered to be good indicators of the glacier movement:

1. subglacial streamlined landforms: drumlins and megaflutes; and
2. ice-marginal moraines: push and thrust moraines, lateral moraines, and interlobate moraines.

Subglacial streamlined landforms

Megaflutes. Gravenor and Meneley (1958) were among the first to propose the formation of megaflutes (large flutes) by lateral pressure from the erosional zone of high pressure in the grooves between fluting ridges to the depositional zone of low pressure in the flutes (Fig. 1). This proposal was based upon a few microfabric measurements in a flute at North Battleford, Saskatchewan. It was supported by the herringbone pattern of many more microfabric measurements in the megaflutes of the Athabasca area in Alberta, Canada, by Shaw and Freschauf (1973). Further confirmation was by Jones' (1982) study of micro- and macrofabric in flutes of east-central Alberta: a-axis maxima were oriented either oblique or perpendicular to the axes of flutes, and dipped upglacier (Fig. 2).

Drumlins. Drumlins and drumlinoids (very long narrow drumlins) may be of erosional, deformational, or depositional origin or their combination (Menzies, 1984; Hart, 1997). Boulton (1987, p.–53) emphasized deformation structures striking transverse to the direction of glacier movement, both in the core and its cover—the sheath. However, several other authors have recorded orientations of glaciotectonic deformation structures or till fabrics, particularly along the flanks of drumlins, that suggest local stress directions different from the direction of ice movement.

S. Jewtuchowicz was probably one of the first who reported stone alignment perpendicular to the direction of glacier movement in drumlins at Zbojno, Poland, in his talk at the 1957 INQUA Congress (Pippan, 1958, p. 147). About a decade later Andrews and King (1968) reported an orientation of the a and b axes of till fabric oblique to the trend of a drumlin in Yorkshire, United Kingdom, along its sides, suggesting that the drumlin grew by lateral accretion of till and some sandy layers. Lateral stresses from the interdrumlin areas towards the axes of drumlins have been also interpreted in several other studies of till fabric

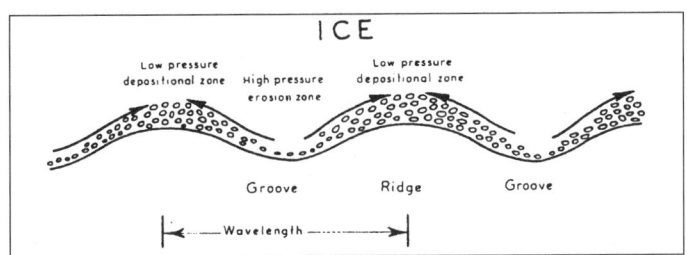

Figure 1. Hypothetical cross section through flutes at the base of a glacier. Arrows indicate the lateral component of glacier flow (Gravenor and Meneley, 1958, Fig. 6).

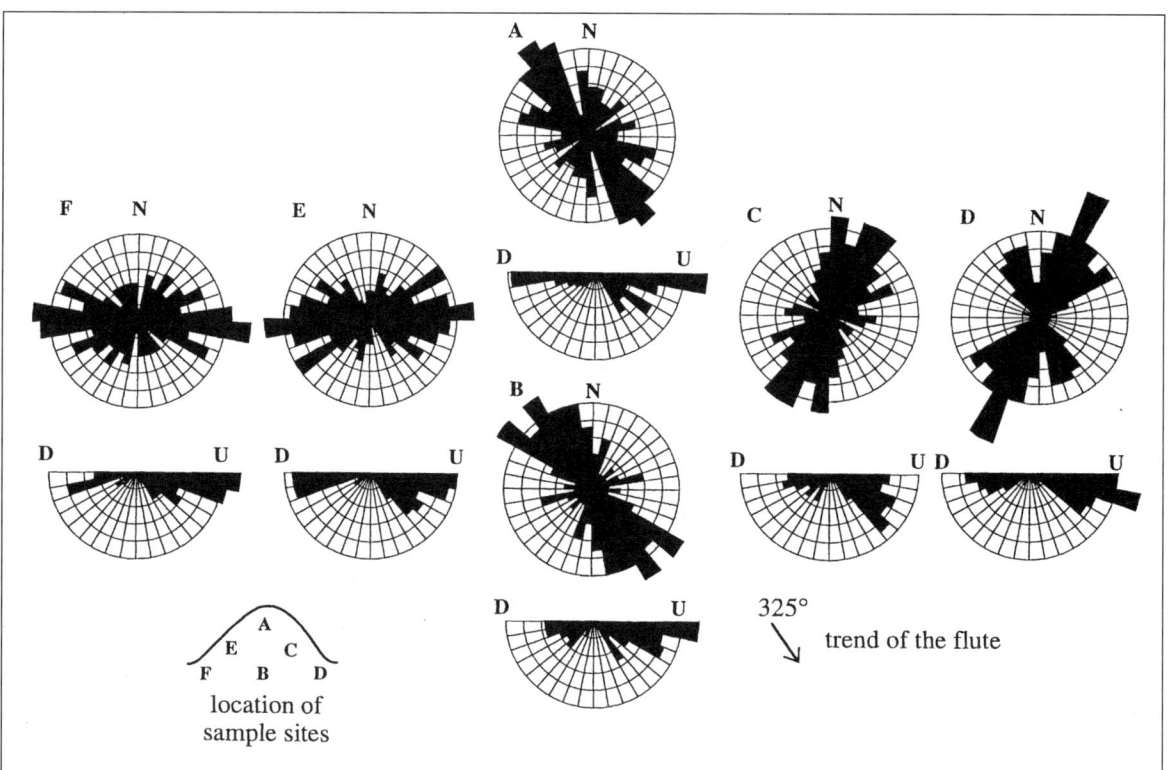

Figure 2. Pebble orientation and plunge in a fluting ridge, east-central Alberta: the two-dimensional rose diagrams show pebble orientation, and the half-a-circle diagrams show the plunge data along the pebble fabric maxima, with U at the upglacier side and D at the downglacier side. The long arrow is the direction of glacier movement, and the flute trends parallel to it. One hundred pebbles were measured at each site, and the rose diagram circles are at two pebble frequency. (After Jones, 1982, Fig. 6a.)

and/or glaciotectonic structures, for instance by Aario (1977) in Finland, Seret (1979) in northwest Ireland, Whittecar and Mickelson (1979), and Stanford and Mickelson (1985) in Wisconsin, United States; Zelčs (1987, 1993a, b) and Zelčs and Dreimanis (1997) in Latvia; Bluemle et al. (1993) in North Dakota, United States, and Wysota (1994) in Poland. If the Heiligenhafen moraine of northern Germany is a drumlinized palimpsest moraine, as proposed by Stephan (1971), this is one of the earliest detailed investigations of the orientation of glaciotectonic deformation structures in drumlins striking parallel and oblique to the direction of glacier movement. Gripp (1979) proposed that the Heiligenhafen moraine is an interlobate moraine with deformation structures pushed towards its axis, but Stephan (1985) disagreed. The lateral stresses towards the drumlin axes are most commonly explained by an interaction of the shear stress of the faster moving interdrumlin ice and its normal stress (Whittecar and Mickelson, 1979; Stephan, 1985; Zelčs, 1987, 1993a). Shaw and Freschauf (1973), Aario (1977), and Jones (1982), however, proposed hypothetical helicoidal ice flows in the interdrumlin areas or between the flute ridges.

A deflection of glacier flow around the drumlin core has been interpreted from till fabric studies in southwestern Saskatchewan, Canada (Kupsch, 1955), in northeast Ireland (Hill, 1971), in Bow Valley, Alberta, Canada (Walker, 1973), and in modern drumlins of Iceland (Krüger and Thomsen, 1984; Boulton, 1987) and Spitsbergen (Boulton, 1987).

Ice-marginal deformation structures related to ice streams, glacier lobes, tongues, or blocks

Glaciotectonic deformation structures in ice-marginal moraines, particularly in push or thrust moraines commonly strike parallel to the ice margin in outward convex belts (van der Wateren, 1995, Figs. 10.13 and 10.14), perpendicular to the direction of glacier movement at its terminus. However, in areas where the glacial terminus consisted of many small lobes or tongues, this relationship may be more complex as shown, for instance, in Schlesvig-Holstein, northwest Germany, by Stephan (1985, Fig. 1). The local glaciotectonic push direction, as concluded from the orientation of structures, is at right angles to the direction of ice flow at nine sites investigated. Stephan (1985) mentioned also several other studies in northern Germany, the earliest being Heerdt (1966), where similar orientation of glaciotectonic structures have been noted.

Stephan (1985, p. 48–51) proposed several interpretations for the deformation structures striking parallel to the regional direction of glacier movement. He explained most of them (six out of nine sites) by local "... lateral pushing at the flanks of ice tongues or lobes." Similar orientation of smaller structures may be produced subglacially in a shear zone between two ice streams. At two sites Stephan applied Pillewitzer's (1958) hypothesis of "block movement." According to this, the frontal position of a rapidly readvancing (surging?) glacial tongue that is constricted along its sides, may become separated in rigid blocks less than 100 m wide, and local transverse stresses produce deformation structures between these blocks.

Where ice streams in an ice sheet have been flowing faster along preexisting wide valleys, valley-side glaciotectonic structures striking parallel to the valley sides were formed by lateral stresses. Such lateral structures have been discussed by Ber (1987) from the Suwalki Lakeland in northeast Poland.

A NEED FOR THREE-DIMENSIONAL KNOWLEDGE

Most of the above discussed landforms are normally related to the direction of glacier movement: the streamlined subglacial ridges (flutes and drumlins) parallel to the ice flow, and the end moraines trending transverse to glacier movement at the glacier terminus. However, their internal structures and fabric may reflect stress directions that are different from the direction of glacier movement. These secondary stress directions may be related to (1) differences in the velocities of the movement of various parts of glacier ice, (2) differences in loading by adjacent parts of ice, or (3) a later glacial event with a different stress field.

In order to relate the internal structures to their external landform, and the local secondary stress directions to the direction of glacier movement, an architectural geometric reconstruction of a landform, including its interior, is required. It should include a three-dimensional construction of the lithostratigraphic units or lithofacies and the orientation of structures inside the landform, and also the orientation of various glaciokinetic features, such as fabric, striae, and stoss-and-lee features. The orientation measurements should be done in various parts of the landform, and the interpreted local stress directions should be plotted in relation to the orientation of the landform and the direction of glacier movement. If the orientation of the landform, for instance a flute, clearly indicates the ice-flow direction, then the features representing local secondary stress directions will assist in the deciphering of the process of formation of the landform.

Let us consider two case studies as examples. The first one, of megaflutes, will be based upon just one criterion, till fabric; the second one, of a drumlin, will be based upon several glaciokinetic criteria and a more complete kinetoarchitectural reconstruction.

Megaflutes

Jones (1982) investigated megaflutes in east-central Alberta, by measuring pebble fabric and microfabric at several levels in three flutes. The pebble fabric results (Fig. 2) were very consistent in all three flutes, but the microfabric orientations were more variable. Some of them showed transverse maxima and suggested that the till in flutes was probably deformation till. Jones (1982, p. 49) concluded the following origin of the flutes. Initial frozen-bed conditions and compressive flow caused glacial thrusting and plucking of blocks of basal debris near the margin of the ice. The blocks were lodged at the glacier bed and they resisted further movement. With continued glacier advance, thawed-bed conditions developed and deposition in a low-pressure zone was created in the lee of these obstacles. Lateral transport of debris in the lee of the blocks was accomplished by the presence of converging secondary flow cells in subglacial till, created by the basal pressure gradient. Till fabric analyses show a "herring-bone" fabric pattern (Fig. 2), supporting the existence of converging secondary flow.

Drumlin

A similar sequence of events is also suggested for the formation of drumlins in the Burtnieks drumlin field in Latvia by Zelčs (1993a) and Zelčs and Dreimanis (1997). It is based upon studies of the interior of several drumlins, considering the orientation of glaciotectonic deformation structures, pebble clast alignment in till, and deformed gravel, fracture patterns, striae on boulders, and boundinage lineation. The origin of the Sedaskalns drumlin (Fig. 3) will be discussed here briefly with some additional information to that in Zelčs and Dreimanis (1997), as an example of a kinetoarchitectural investigation, conditioned by the availability of sections, in this case in gravel pits along the northeast flank of the drumlin, that extend to its middle (Fig. 3A).

The formation of the Sedaskalns drumlin began by thrusting and downglacier translocation of megablocks of local, weakly lithified interbedded Devonian sandstone, shale, and weathered gravel, in the direction of glacier movement as evidenced by the orientation of glaciotectonic deformations in the bedrock slabs and lowermost gravel, as well as striae on bedrock slabs on the left side of the section (Fig. 4C). These translocated overthrust blocks became the core of the drumlin. As the velocity of glacier movement in the interdrumlin areas was higher than glacier movement over the more resistant core, lateral secondary pressure developed against the core, folding it into an anticlinal complex, as shown in the lower half of Figure 3B.

As evidenced by the orientation of glaciokinetic features (Fig. 4), the local stress along the northeast side of the drumlin gradually shifted from its initial direction parallel to glacier movement (from northwest to southeast) to local stresses from north-northwest, north, and even north-northeast. The latter one was maintained during the main drumlin-forming process, when the flanks of the drumlin core become covered by a sequence of imbricate overthrusts, first consisting of sand and gravel (Fig. 3B, layer d), later by interbedded till sheets a, b, and c and sand (Fig. 3B, right side, upper third, and 3C).

During the deposition of the uppermost till layer c, the stress direction finally returned parallel to the axis of the drumlin, coinciding with the direction of glacier movement (Fig. 4B). During

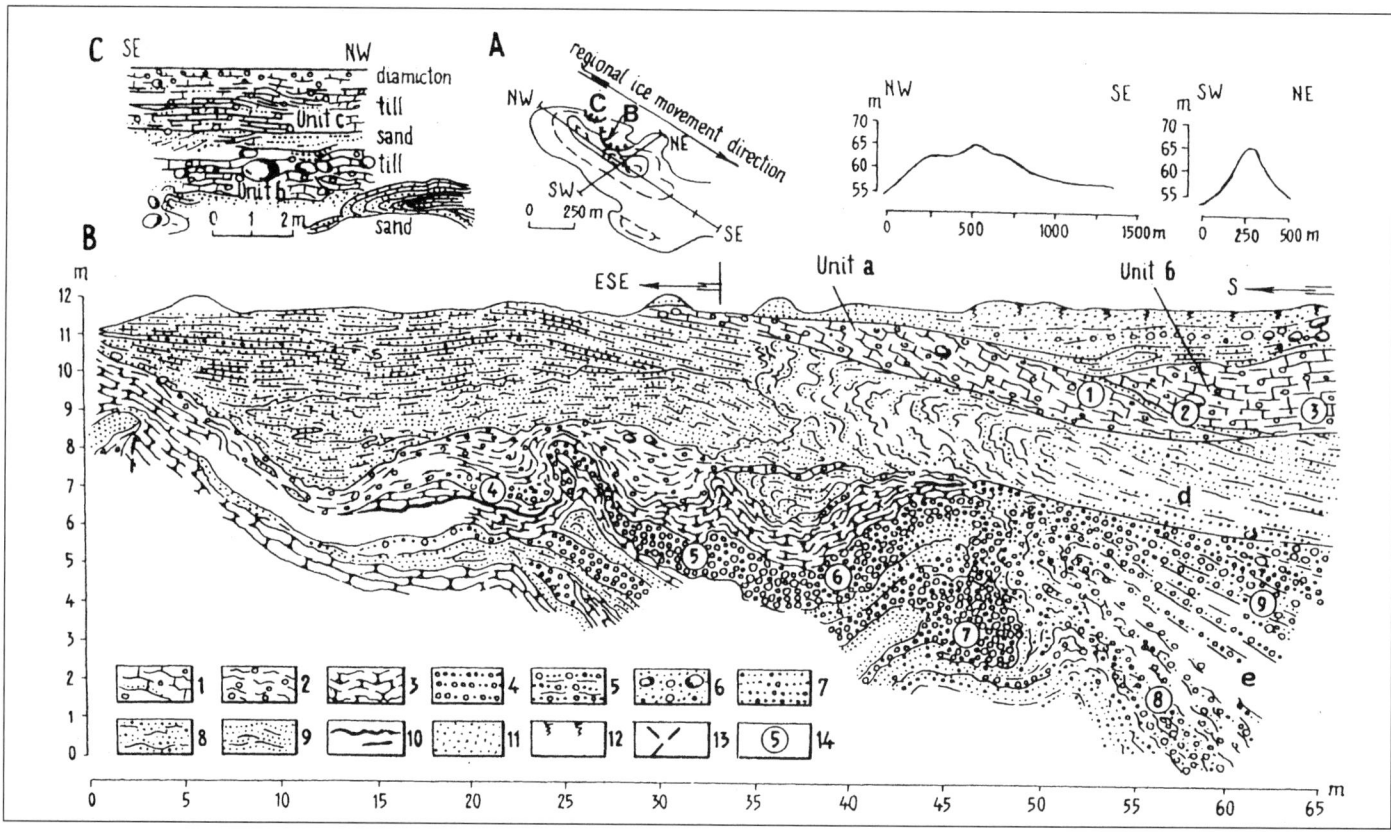

Figure 3. Sedaskalns drumlin. A, Outline of the drumlin, topographic profile sections, and location of the sections B and C along the walls of gravel pits. B, Glaciotectonic structures exposed along the southwest wall of main gravel pit; the right side of the section trends approximately transverse to the direction of glacier movement, but the left side approximately parallel to the glacier movement. C, A section parallel to the drumlin crest in a small gravel pit on the northeast flank of the drumlin. Legend: 1, reddish brown fissile basal silty sand till (units a and b); 2, local till; 3, deformed reddish sandstone with clayey silt interlayers; 4, pebble and gravel mixture; 5, gravelly pebble layers; 6, weakly graded sand with occasional pebbles and boulders; 7, sandy gravel; 8, gray coarse-grained sand; 9, reddish fine- and medium-grained sand; 10, clay and silt interlayers; 11, sand; 12, soil; 13, joints; 14, fabric measurement sites and numbers. Modified from Figures 7 and 8 of Zelčs and Dreimanis, (1997).

this final phase the crest of the drumlin (Fig. 3B, left side) became partly eroded, and sandy diamicton (meltout till) was deposited on the erosional surface (Fig. 3B and 3C).

Thus, in relation to the direction of glacier movement, three main directional phases can be distinguished in the Sedakalns drumlin, as represented by the orientation of glaciokinetic features:

1. formation of the core by stress parallel to the direction of glacier movement;
2. main building of the drumlin by secondary stresses from the interdrumlin areas towards the core in a herringbone pattern; and
3. return to stress direction parallel to the direction of glacier movement, involving partial erosion of the crest of the drumlin, but with some till deposition along its flanks.

Kinetoarchitectural reconstruction of landforms

Ideally, a complete kinetoarchitectural reconstruction of a landform would require several cross sections and at least one longitudinal section, and the investigation of several glaciokinetic criteria within them. The above two examples, the flutes and the drumlin, were given in order to show actual cases where the kinetoarchitectural reconstruction is partial, but still sufficient for drawing conclusions on the spatial arrangement of secondary stresses and their relation to the direction of glacier movement, an important criterion in glacial stratigraphy.

Thus, in the case of the Alberta flutes (Jones, 1982), the consistent sets of till fabric measurements in three cross sections of the flutes, with a herringbone pattern along their flanks pointing towards the axes of the flutes, suggested secondary local stress directions along this herringbone pattern. This minimal information was sufficient to reconstruct the kinetoarchitecture of the flutes investigated.

In the second case, the investigation of the Sedaskalns drumlin (Zelčs and Dreimanis, 1997) was conducted by studying several glaciokinetic criteria, and the lithologic composition of pebbles. The dominantly local pebble lithology suggested basal glacial transport for the tills and derivation of gravels from basally transported glacial drift, with incorporation of weathered gravels in the lowermost gravel layer.

Further examples of other case studies, with more or less

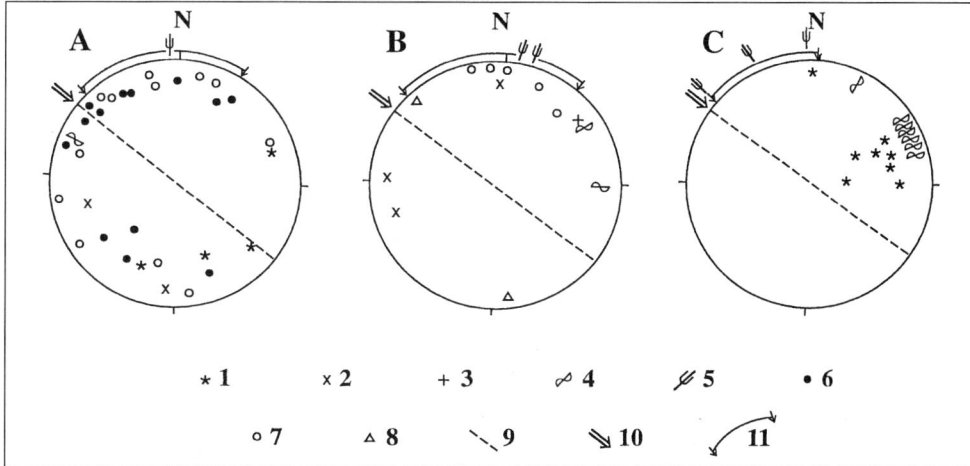

Figure 4. Sedaskalns drumlin, orientation of selected glaciokinetic features: A, in tills a and b of Figure 3B; B, in tills b and c of Figure 3C; C, in gravels and bedrock slabs underneath till a in Figure 3B. Legend: 1–3, mean vectors of clast fabric: 1, in till a and in gravel underneath till a; 2, in till b; 3, in till c. 4, hinges of drag folds in sand and gravel underneath till. 5, striae on boulders. 6–8, dip of shear planes: 6, in till a; 7, in till b; 8, in till c. 9, trend of the drumlin. 10, direction of glacier movement. 11, range of local deforming stress directions. (Selected features from Zelčs and Dreimanis, 1997, Fig. 8.)

complete kinetoarchitectural reconstructions in various landform areas, may be found in several references cited, particularly in Stephan (1971, 1985), Ber (1987), Aboltinsh (1989), Bluemle et al. (1993), Zelčs (1993a, b), Dreimanis and Zelčs (1997b), and Zelčs and Dreimanis (1997).

APPLICATION OF KINETOARCHITECTURE TO THE DETERMINATION OF THE DIRECTION OF GLACIER MOVEMENT

In many well-developed landform areas, for instance in drumlin fields, fluting fields, and in narrow push-or-thrust moraines, a detailed kinetoarchitectural investigation is not necessary when using these landforms as indicators of the direction of glacier movement. Their kinetoarchitecture may still be helpful in making conclusions about the origin of the landforms, particularly drumlins, as their internal structures and composition may reveal a considerable genetic variety (Menzies, 1987; Menzies and Shilts, 1996; Hart, 1997)

In areas of complex end moraines formed by many small ice tongues, as discussed by Stephan (1971, 1985), and in areas of partial drumlinization (Stephan, 1971), or where morainic ridges are oriented at right angles to each other (Zelčs, 1993a, b; Dreimanis and Zelčs, 1997b), kinetoarchitectural investigations will lead to the correct identification of the landforms and to the origin of structures within them. They will also determine the direction of glacier movement that existed during the formation of the landforms or glaciotectonic structures in question.

In cases where the upper parts of the landforms of an older drift layer have been eroded, the reconstruction of kinetoarchitecture in the remaining lower part of the landform may be particularly important for determining the direction of glacier movement during the deposition of the older drift unit. This would apply especially to partially eroded old drumlins, flutes, and lateral and interlobate moraines. Measurements of the orientation of glaciokinetic features in a single section or along one flank of the landform remnant, would lead to an erroneous conclusion about the direction of glacier movement during its formation. However, the kinetoarchitectural reconstruction of the landform remnant by measurements along several traverses may lead to correct identification of the landform and the direction of glacier movement. The identification by Stephan (1971, 1985) of the Heiligenhafen moraine in northwest Germany as a drumlinized older end moraine, using detailed regional investigation of various glaciokinetic features may serve as an example. The core of the moraine was formed by glacier advance from north-northeast, but the drumlinization was accomplished by the latest glacier advance from the east.

CONCLUSIONS

Determination of the direction of local or regional glacier movement may be erroneous if based upon single measurements of the orientation of glaciokinetic criteria (e.g., glaciotectonic deformation structures or till fabric) because some of the movement may be related to local secondary stresses. The local secondary stress directions have several definite patterns in most subglacially or ice-marginally developed landforms. These patterns may be recognized by detailed three-dimensional geometric kinetoarchitectural investigation of the landforms or their remnants in older glacial drift deposits. The known relationships of these patterns to the direction of glacier movement and other multiple criteria will assist in drawing correct interpretations of the direction of glacier movement at the site investigated.

ACKNOWLEDGMENTS

The technical expenses for this report have been covered by a grant from the Natural Sciences and Engineering Research Council of Canada. I am grateful to Lee Clayton, Stephen R. Hicock, Reet Karukäpp, and Vitālijs Zelčs for improving the manuscript by critical reading.

REFERENCES CITED

Aario, R., 1977, Flutings, drumlins and Rogen-landforms: Nordia, no. 2, p. 5–14.

Aber, J. S., 1988, Bibliography of glaciotectonic references, *in* Croot, D. G., ed., Glaciotectonic forms and processes: Rotterdam, A. A. Balkema, p. 195–210.

Aber, J. S., 1993, Expanded bibliography of glaciotectonic references, *in* Aber, J. S., ed., Glaciotectonics and mapping of glacial deposits: Regina, Canada, University of Regina, p. 99–137.

Aber, J. S., Croot, D. G., and Fenton, M. M., 1989, Glaciotectonic landforms and structures: Dordrecht, Netherlands, Kluwer Academic Publishers, Glaciology and Quaternary Geology Series, 200 p.

Aboltinsh, O. P., 1989, Glatsiostruktura i lednikovyi morfogenez (Glacial structure and glacial morphogenesis): Riga, Zinatne, 284 p. (in Russian).

Albino, K., and Dreimanis, A., 1988, A time-transgressive kinetostratigraphic sequence spanning 180° in a single section at Bradtville, Ontario, *in* Croot, D. G., ed., Glaciotectonics: Rotterdam, A. A. Balkema, p. 11–20.

Andrews, J. T. and King, C. A. M., 1968, Compressive till fabrics and till fabric variability in a till sheet and a drumlin: a small scale study: Proceedings of Yorkshire Geologic Society, v. 36, p. 435–461.

Banham, P. H., 1988, Thin-skinned glaciotectonic structures, *in* Croot, D. G., ed., Glaciotectonics, forms and processes: Rotterdam, A. A. Balkema, p. 21–25.

Benn, D. I., 1995, Fabric signature of subglacial till deformation, Breidamerkurjökull, Iceland: Sedimentology, v. 42, p. 735–747.

Benn, D. I., and Evans, D. J. A., 1998, Glaciers and glaciations: Arnold, London, 734 p.

Ber, A., 1987, Glaciotectonic deformation of glacial landforms and deposits in the Suwalki Lakeland (NE Poland), *in* van der Meer, J. J. M., ed., Tills and glaciotectonics: Rotterdam, A. A. Balkema, p. 135–143.

Berthelsen, A., 1973, Weichselian ice advances and drift successions in Denmark: Bulletin of the Geological Institute of University of Uppsala, v. 5, p. 21–29.

Berthelsen, A., 1978, The methodology of kineto-stratigraphy as applied to glacial geology: Bulletin of the Geological Society of Denmark, Special Issue, v. 27, p. 25–38.

Bluemle, J. P., Lord, M. L., and Hunke, N. T., 1993, Exceptionally long narrow drumlins formed in subglacial cavities, North Dakota: Boreas, v. 22, p. 15–24.

Boulton, G. S., 1971, Till genesis and fabric in Svalbard, Spitsbergen, *in* Goldthwait, R. P., ed., Till, a symposium: Ohio State University Press, p. 41–72.

Boulton, G. S., 1987, A theory of drumlin formation by subglacial sediment deformation, *in* Menzies, J., and Rose, J., eds., Drumlin symposium: Rotterdam, A. A. Balkema, p. 25–80.

Dreimanis, A., 1935, The rock deformations, caused by inland-ice, on the left bank of Daugava at Dole Island, near Riga in Latvia: Riga, A. Gulbis, 30 p. (in Latvian, with English summary).

Dreimanis, A., 1989, Tills: their genetic terminology and classification, *in* Goldthwait, R. P., and Matsch, C. L., eds., Genetic classification of glacigenic deposits: Rotterdam, A. A. Balkema, p. 17–83.

Dreimanis, A., 1993, Small to medium-sized glaciotectonic structures in till and its substratum and their comparison with mass movement structures: Quaternary International, v. 18, p. 69–79.

Dreimanis, A., 1995, Landforms and structures of the waterlain west end of St. Thomas moraine, S W Ontario, Canada: Geomorphology, v. 14, p. 185–196.

Dreimanis, A., 1997, Glaciodynamic criteria and kinetoarchitecture in glacial stratigraphy: CSPG-SEPM Joint Convention, June 1–6, 1997, Program with Abstracts, p. 83.

Dreimanis, A., and Rappol, M., 1997, Late Wisconsinan sub-glacial clastic intrusive sheets along Lake Erie bluffs, at Bradtville, Ontario, Canada: Sedimentary Geology, v. 111, p. 225–248.

Dreimanis, A., and Zelčs, V., 1997a, Supplementing glacial kineto-stratigraphy by kineto-architecture: The Peribaltic Group, INQUA Commission on Glaciation, Field Symposium on Glacial Geology at the Baltic Sea Coast in Northern Germany, University of Kiel, 7–12 September 1997, Abstracts of Papers and Posters, p. 6.

Dreimanis, A., and Zelčs, V., 1997b, Glaciotectonic deformations along the bluffs of the River Daugava, Daugmale Ribbed Moraine Area, Central Latvian Lowland: The Peribaltic Group, INQUA Commission on Glaciation, Field Symposium on Glacial Geology at the Baltic Sea Coast in Northern Germany, University of Kiel, 7–12 September 1997, Abstracts of Papers and Posters, p. 7–8.

Ehlers, J., 1996, Quaternary and glacial geology: Chichester, John Wiley & Sons, 578 p.

Ehlers, J., and Stephan, H. J., 1979, Forms at the base of till strata as indicators of glacier movement: Journal of Glaciology, v. 22, p. 345–355.

Gravenor, C. P., and Meneley, W. S., 1958, Glacial flutings in central and northern Alberta: American Journal of Science, v. 256, p. 715–728.

Gripp, K., 1979, Glazigene Press-Schuppen, frontal und lateral (Pressed scales in ice, in push moraines, in lateral moraines), *in* Schlüchter, C., ed., Moraines and varves: Rotterdam, A. A. Balkema, p. 157–166.

Hart, J. K., 1997, The relationship between drumlins and other forms of subglacial deformation: Quaternary Science Reviews, v. 16, p. 93–107.

Hart, J. K., and Boulton, G. S., 1991, The interrelation of glaciotectonic and glaciodepositional processes within the glacial environment: Quaternary Science Reviews, v. 10, p. 335–350.

Heerdt, S., 1966, Struktur und Endstehung der Stauchmoräne Kühlung (The structure and development of the Kühlung push-moraine): Geologie, v. 15, p. 1169–1213.

Hicock, S. R., 1992, Lobal interactions and rheologic superposition in subglacial till near Bradtville, Ontario, Canada: Boreas, v. 21, p. 73–88.

Hicock, S. R., and Fuller, E. A., 1995, Lobal interaction, rheological superposition, and implications for a Pleistocene ice stream on the continental shelf of British Columbia: Geomorphology, v. 14, p. 167–184.

Hicock, S. R., and Dreimanis, A., 1992, Sunnybrook drift in the Toronto area, Canada: re-investigation and reinterpretation, *in* Clark, P. U., and Lea, P. D., eds., The last interglacial-glacial transition in North America: Geological Society of America Special Paper 270, p. 109–118.

Hicock, S. R., Goff, J. R., Lian, O. B., and Little, E. C., 1996, On the interpretation of subglacial till fabric: Journal of Sedimentary Research, v. 66, p. 928–934.

Hill, A. R., 1971, The internal composition and structure of drumlins in North Down and South Antrim, Northern Ireland: Geografiska Annaler, v. 53A, p. 14–31.

Houmark-Nielsen, M., 1987, Pleistocene stratigraphy and glacial history of the central part of Denmark: Bulletin of the Geological Society of Denmark, v. 36, part 1–2, 189 p.

Houmark-Nielsen, M., 1988, Glaciotectonic unconformities in Pleistocene stratigraphy as evidence for the behaviour of former Scandinavian ice sheets, *in* Croot, D. G., ed., Glaciotectonics, forms and processes: Rotterdam, A. A. Balkema, p. 91–99.

Houmark-Nielsen, M., and Berthelsen, A., 1980, Kineto-stratigraphic evaluation and presentation of glacial-stratigraphic data, with examples from northern Samsø Denmark: Boreas, v. 10, p. 411–422.

Jones, N., 1982, The formation of glacial flutings in east central Alberta, *in* Davidson-Arnott, R., Nickling, W., and Fahey, B. D., eds., Research in Glacial, Glacio-fluvial and Glacio-lacustrine Systems, Proceedings of the 6th Guelph Symposium on Geomorphology, 1980: Norwich, Geo Books, p. 49–70.

Kluiving, S. J., Rappol, M., and van der Wateren, F. M., 1991, Till stratigraphy and ice movements in eastern Overijssel, The Netherlands: Boreas, v. 20, p. 193–205.

Krüger, J., 1970, Till fabric in relation to direction of ice movement. A study from the Fakse Banke, Denmark: Geografish Tidskrift, v. 69, p. 133–170.

Krüger, J., and Thomsen, H. H., 1984, Morphology, stratigraphy, and genesis of small drumlins in front of the glacier Myrdalsjökull, south Iceland: Journal of Glaciology, v. 30, p. 94–105.

Kupsch, W. O., 1955, Drumlins with jointed boulders near Dollar, Saskatchewan: Geological Society of America Bulletin, v. 66, p. 327–338.

Lavrushin, Y. A., 1976, Structure and development of ground moraines of continental glaciation: Moscow, Nauka, 237 p. (in Russian).

MacClintock, P., 1958, Glacial geology of the St. Lawrence Seaway and power projects: New York State Museum and Science Service Report of Investigation, v. 10, 20 p.

MacClintock, P., and Dreimanis, A., 1964, Reorientation of till fabric by overriding glacier in the St. Lawrence Valley: American Journal of Science, v. 262, p. 133–142.

MacClintock, P., and Steward, D. P., 1965, Pleistocene geology of the St. Lawrence Lowland: New York State Museum and Science Service Bulletin no. 394, 152 p., and 6 maps.

Menzies, J., 1984, Drumlins—a bibliography: Norwich, Geo Books, 117 p.

Menzies, J., 1987, Towards a general hypothesis on the formation of drumlins, in Menzies, J., and Rose, J., eds., Drumlin Symposium: Rotterdam, A. A. Balkema, p. 9–24.

Menzies, J., and Shilts, W. W., 1996, Subglacial environments, in Menzies, J., ed., Past Glacial Environments, Sediments, Forms and Techniques: Oxford, Butterworth/Heinemann, Glacial Environments, v. 2. p. 15–136.

Pillewitzer, W., 1958, Neue Erkenntnisse über die Blockbewegung der Gletscher (New knowledge about the block-movement of glaciers): Zeitschrift für Gletscherkunde und Glazialgeologie, v. 4, p. 23–33.

Pippan, T., 1958, Bericht über den V Internationalen Kongress der Inqua in Madrid-Barcelona vom 2. bis 16. September, 1957 (Report about the V International Congress of Inqua in Madrid-Barcelona, 2–16 September, 1957): Zeitschrift für Gletscherkunde und Glazialgeologie, v. 4, p. 145–150.

Richter, K., 1932, Bewegungrichtung des Inlandeises, rekonstruiert aus den Kritzen und Längstachsen der Geschiebe (Direction of movement of inland-ice, deciphered from striae and long axes of clasts): Zeitschrift für Geschiefeforschung, v. 8, p. 62–66.

Richter, K., 1936, Ergebnisse und Aussichten der Gefügeforschung im pommerschen Diluvium (Results and prospects of the clast-investigation in the diluvium of Pomerania): Geologisches Rundschau, v. 27, p. 196–206.

Rose, J., and Menzies, J., 1996, Glacial stratigraphy, in Menzies, J., ed., Past Glacial Environments: Glacial Environments, v. 2, p. 253–284.

Seret, G., 1979, La genése des drumlins (The origin of drumlins), in Schlüchter, Ch., ed., Moraines and varves, origin, genesis, classification: Rotterdam, A. A. Balkema, p. 189–196.

Shaw, J., and Freschauf, R. C., 1973. A kinematic discussion of the formation of glacial flutings: Canadian Geographer, v. 17, p. 19–35.

Stanford, S. D., and Mickelson, D. M., 1985, Till fabric and deformational structures in drumlins near Waukesha, Wisconsin, U.S.A.: Journal of Glaciology, v. 31, p. 220–228.

Stephan, H.-J., 1971, Glazalgeologische Untersuchungen im Raum Heiligenhafen (Ostholstein) (Glaciogeological investigations in the area of Heiligenhafen [East-Holstein]): Meyniana, v. 21, p. 67–86.

Stephan, H.-J., 1985, Deformations striking parallel to glacier movement as a problem in reconstructing its direction: Bulletin of the Geological Society of Denmark, v. 34, p. 47–54.

van der Wateren, F. M., 1995, Processes of glaciotectonism, in Menzies, J., ed., Modern Glacial Environments: Glacial Environments, v. 1, p. 309–335.

Walker, M. J. C., 1973, The nature and origin of a series of elongated redges in the Morley Flats area of the Bow Valley, Alberta: Canadian Journal of Earth Sciences, v. 10, p. 1340–1346.

Westgate, J. A., 1968, Linear sole markings in Pleistocene till: Geological Magazine, v. 105, p. 501–505.

Whittecar, G. R., and Mickelson, D. M., 1979, Composition, internal structures, and an hypothesis of formation for drumlins, Waukesha County, Wisconsin, U.S.A.: Journal of Glaciology, v. 22, p. 357–371.

Wysota, W., 1994, Morphology, internal composition and origin of drumlins in the southeastern part of the Chelmno-Dobrzyn Lakeland, North Poland: Sedimentary Geology, v. 91, p. 345–364.

Zelčs, V., 1987, Raznovidnosti gliatsiodislokatsiy i ikh reliefoobrazuyuschaya roly v prdelakh gliatsiodepressinnikh nizmennostey Latvii (Variety of glaciodislocations and their relief-formational role in the glaciodepressional lowlands of Latvia): Riga, Latviyskiy Gosudarstvennij Universitet, 35 p. (in Russian).

Zelčs, V., 1993a, Glaciotectonic landforms of divergent type glaciodepressional lowlands. Dissertation work synthesis: Riga, University of Latvia, 105 p. (in Latvian, English, and Russian).

Zelčs, V., 1993b, The Iecava drumlin field and the Daugmale ribbed moraine field, in IGCP 253, Termination of the Pleistocene, Peribaltic Group, Scientific Excursion in the Baltic States, 14–19 June, 1993: Tallinn, Institute of Geology, p. 17–21.

Zelčs, V., and Dreimanis, A., 1997, Morphology, internal structure and genesis of the Burtnieks drumlin field, northern Vidzeme, Latvia: Sedimentary Geology, v. 111, p. 73–90.

MANUSCRIPT ACCEPTED BY THE SOCIETY OCTOBER 8, 1998

Tunnel channels formed in Wisconsin during the last glaciation

Lee Clayton and John W. Attig
Wisconsin Geological and Natural History Survey, 3817 Mineral Point Road, Madison, Wisconsin 53705
David M. Mickelson
Department of Geology and Geophysics, University of Wisconsin, Madison, Wisconsin 53706

ABSTRACT

Tunnel channels are some of the more conspicuous landforms resulting from Pleistocene glaciation in Wisconsin. Many are several hundred meters wide, several meters to tens of meters deep, and several kilometers long. One group of about 80 tunnel channels occurs just behind the outermost moraine formed along the western edge of the Green Bay Lobe of the Laurentide Ice Sheet during the height of the Wisconsin Glaciation. They rise westward, in a downglacier direction, out of the Green Bay lowland, and end at a breach in the outermost moraine. Beyond each of these breaches is a fan of outwash.

Because they rise westward to end abruptly at the outermost moraine, we interpret at least these 80 to have been cut by subglacial rivers. Because they are shaped like river channels rather than valleys, we interpret them to be true channels, cut by rivers with water from bank to bank. Because they seem too big to have been cut by continually flowing rivers, requiring more water than should have been available, we tentatively interpret them to be the result of occasional floods of water that had accumulated in subglacial lakes, which perhaps were normally held in by a frozen-bed zone under the toe of the glacier.

INTRODUCTION

During the height of the Wisconsin Glaciation, more than 150 channel-like features were cut just behind the terminus of the Superior, Chippewa, Wisconsin Valley, Langlade, Green Bay, and Lake Michigan Lobes of the Laurentide Ice Sheet (Figs. 1, 2, and 3). They have been called tunnel channels or tunnel valleys, because, like eskers, they are thought to be the result of meltwater rivers flowing in subglacial tunnels, but the details of their origin remain obscure.

Tunnel channels were probably first recognized in the American Midwest (Fig. 2) by F. T. Thwaites. His field notes for July 10, 1927, state that a linear depression formed along the west side of the Green Bay Lobe "may have been the channel of a subglacial stream with reversed gradient" (Thwaites, 1927). Later, Thwaites (1943, p. 125; 1946, p. 48) suggested a similar interpretation for the whole series of "kettle chains" and moraine gaps in that area. However, he seems to have never elaborated on this idea.

The first detailed analysis of tunnel channels in the Upper Midwest was by Wright (1973), who described those formed under the western part of the Superior Lobe of the Laurentide Ice Sheet, in east-central Minnesota (Fig. 2). He thought they were cut by catastrophic discharge of subglacial meltwater breaking through the frozen-bed toe of the glacier. Mooers (1989) interpreted these trenches to have been cut by a noncatastrophic seasonal flow of subglacial meltwater. Patterson (1994, p. 16) agreed with Wright that they were cut by short-lived floods of subglacial meltwater.

Since they were recognized by Thwaites, the tunnel channels of Wisconsin have been only briefly mentioned in the geologic literature, and we have generally accepted Wright's interpretation. However, in this review, we will discuss problems with this and other theories that have been proposed.

Tunnel channels (or tunnel valleys) are a common but often unrecognized landform of the glacial landscape in many areas.

Clayton, L., Attig, J. W., and Mickelson, D. M., 1999, Tunnel channels formed in Wisconsin during the last glaciation, *in* Mickelson, D. M., and Attig, J. W., eds., Glacial Processes Past and Present: Boulder, Colorado, Geological Society of America Special Paper 337.

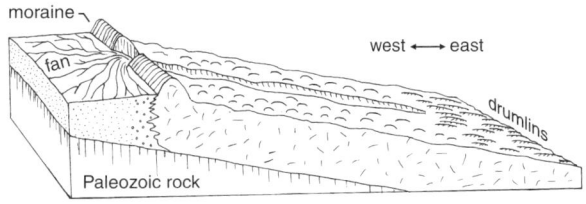

Figure 1. Stylized block diagram, showing a typical tunnel channel formed along the west side of the Green Bay Lobe of the Laurentide Ice Sheet. A tunnel channel begins on the downglacier edge of the drumlin zone, then rises westward, through a zone of hummocky till, to a breach in the outermost moraine, with an outwash fan beyond the breach. The dot pattern indicates outwash. The random-dash pattern indicates till and associated sediment. The scale is suggested by the size of the moraine, which is roughly 1 km wide and 5–20 m high.

An understanding of these features should contribute to a better understanding of conditions at the bed of the ice sheet. In Wisconsin they are an important source of construction aggregate and groundwater. In recent years, the Pleistocene geology of much of the tunnel-channel area in Wisconsin has been mapped at a scale of 1:24,000 and published at 1:100,000 (Mickelson, D. M., and Dolliver, P. N., 1979, unpublished report; Mickelson et al., 1983, p. 26–27; Mickelson, 1983, p. 17, 1986, p. 22–23; Johnson, 1986, p. 27–28; Clayton, 1986, p. 10; Attig, 1989, p. 10–11, 1993, p. 16–17; Clayton and Attig, 1989, p. 37, 1997, p. 29–31; Attig et al., 1988, 1989, p. 401–403, 1990, p. 10–11; and others in progress). As a result, we feel that a review of these features is timely. Our purpose here is to provide the first detailed discussion of the subglacial channels of Wisconsin.

Terminology

Following earlier usage, Wright (1973) used the term tunnel valley, which is a translation of the Danish term tunneldal, for landforms in Minnesota. However, he interpreted them to be the channels, not valleys, of subglacial rivers. Mooers (1989) used the term tunnel valley for those in Minnesota because he thought they were formed by the vertical and lateral migration of tunnel rivers that were much smaller than the valleys.

In this paper we will use the term channel in the hydrologic sense to mean a trench occupied bank to bank by a river. We use the term tunnel channel to mean a channel cut by a river flowing in a subglacial tunnel, where the size of the river matches the size of the channel (Fig. 4A). We will use the term tunnel valley in Mooers' rather than Wright's sense to mean a valley that was formed by lateral planation by a subglacial river that was considerably narrower than the valley. That is, a wide tunnel valley was formed when a narrow subglacial river moved laterally, much as a broad nonglacial valley bottom is formed by a river that moves laterally back and forth on its flood plain (Fig. 4B). Alternatively, a tunnel valley might also have formed when fluidized subglacial sediment flowed or squeezed into an already existing narrow river channel (Fig. 4C). We will use the term collapsed pipes for depressions resulting from the collapse of pipes that were well below the bed of the glacier and were initiated by groundwater piping, or sapping (Fig. 4D).

DESCRIPTION

The different groups of channels discussed here have somewhat different characteristics, and they could in fact be of different origin. For that reason, there is the potential for confusion resulting from the mixing of observations and interpretations of unrelated features. Therefore, where appropriate, we separately discuss specific groups of channels, but we concentrate on the most conspicuous and uniform set, formed along the west side of the Green Bay Lobe (Fig. 3). We also concentrate on that area, because the glacier, as well as the tunnel-channel rivers, there flowed uphill out of the Green Bay lowland—convincing evidence that these trenches were in fact eroded by subglacial rivers.

Tunnel channels formed along the west side of the Green Bay Lobe

About three-quarters of the 80 subglacial channels formed along the west side of the Green Bay Lobe are 0.15–0.45 km wide. A few are narrower, and several are as wide as 0.7 km or even wider. They maintain a fairly uniform width throughout their length; they do not become systematically wider in either the downstream or upstream direction. The length of the channels is difficult to determine because most of their upstream ends are obscure, at least some because they were buried by younger deposits. Few of the identified channels are shorter than 1 km. Most are between 2 and 7 km long, and the longest is about 15 km.

The depth of the channels at the time they formed is unknown because the bottoms of all of those studied in Wisconsin are apparently hidden by later sediment. Test drilling has been inconclusive because we have generally been unable to differentiate fluvial sediment deposited before, during, or after the time the channels were cut. Their present depth, not including the later sediment in their bottoms, is typically from 5 to 30 m, with some as deep as 50 m.

Most of the tunnel channels that formed beneath the Green Bay Lobe now slope upglacier (eastward) at 1–20 m/km (Fig. 5), some of them ascending more than 50 m from head to mouth (westward). Postglacial crustal rebound was only about 0.5 m/km in this area (Clayton and Attig, 1989) and therefore had little effect on the slope of the tunnel channels.

The channels have a radial pattern, with their mouths at the margin of the lobe when it was near its maximum extent during the last part of the Wisconsin Glaciation (Fig. 2). Those formed under the west side of the Green Bay Lobe are associated with the Hancock moraine and with the younger Almond moraine and the probably equivalent Johnstown moraine (Fig. 3). These moraines are about 0.5 km wide and 5–20 m high (Figs. 6, 7, and 8), and they formed during and just after the climax of the Wisconsin Glaciation. No tunnel channels are associated with the smaller moraines that formed after the Almond moraine.

Some channel mouths appear to be fairly uniformly spaced at

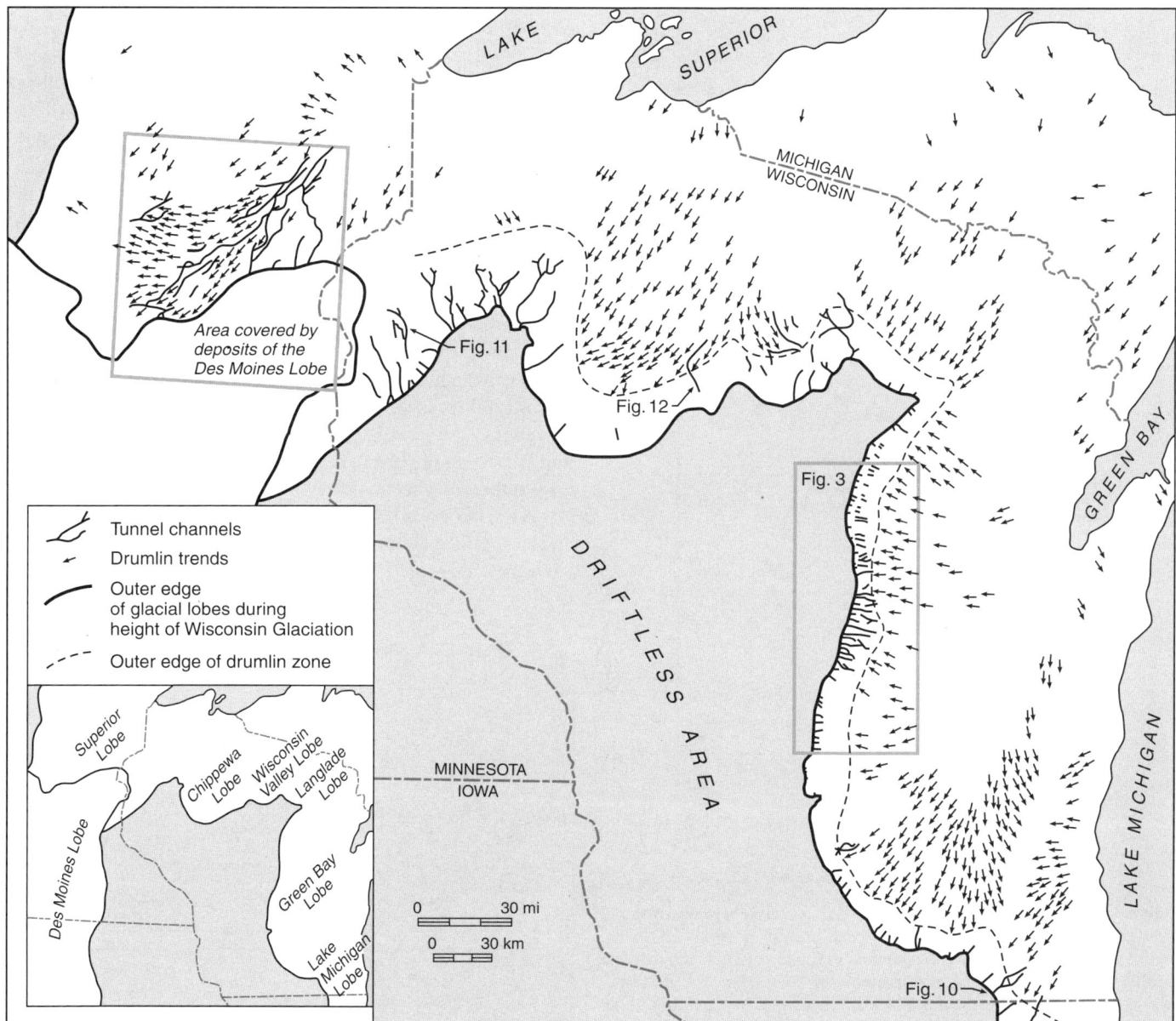

Figure 2. Tunnel channels in the Upper Midwest. Drumlin trends and lobe margins are from Farrand et al. (1984), Goebel et al. (1983), and Lineback et al. (1983). The tunnel channels in Wisconsin are from Johnson (1986, 1999), Mickelson (1983, 1986), Attig (1989), Clayton (1986, 1987), and Clayton and Attig (1990, 1997) or have been interpreted from 7.5-minute topographic quadrangles. Those within the rectangle in east-central Minnesota are from Figure 1 of Wright (1973). The drumlin arrows indicate the direction of ice movement. The Superior Lobe extended farther southwest; that area was later covered by deposits of the Des Moines Lobe.

about 3-km intervals, especially those formed along the middle of the west side of the Green Bay Lobe (Fig. 3). Elsewhere, channels of the Green Bay Lobe are from 0.5 to more than 20 km apart.

The channels that formed along the west side of the Green Bay Lobe generally occur in areas where the Pleistocene sediment is 15–100 m thick (Trotta and Cotter, 1973). Much of this material is sand deposited by glacial meltwater, and in most places less than half the total thickness is till, which here consists largely of sand. However, the material under the glacier when the tunnel channels formed was not necessarily highly permeable, because much of this sand may have been frozen. In the northern third of this area the sub-Pleistocene material is Precambrian igneous rock, in the middle third it is poorly cemented Cambrian sandstone, and in the southern third it is Cambrian and Ordovician sandstone and dolomite. The thickness and character of the geologic materials in the region do not seem to correlate with the spacing, size, or other characteristics of the channels.

Many of the supposed tunnel valleys or tunnel channels of northern Europe are completely buried by later deposits and have been identified by subsurface methods (e.g., Ehlers and a tunnel channel that was covered by outwash as shown in Figure 9. Well-construction reports in Wisconsin Geological and Natural History Survey files show several tens of meters of sand and gravel in the adjacent area, as well as in the channel, suggesting that the base of the blanket of sand and gravel extends below the bottom of the deepest pits.

Similarly, the Tamarack tunnel channel and the Patterson tunnel channel shown in Figure 7 are indicated by the gaps through the moraines, which are connected to lines of collapse pits in the outwash between the moraines. Well-construction reports show tens of meters of outwash on either side of the channels here.

The tunnel channels in Wisconsin are about the same size and shape as the spillways of proglacial lakes in the region (the largest of which are described by Kehew and Lord, 1987). Many of these spillways are bottomed with a lag of large boulders, and the subglacial channels might be expected to have a similar lag. None has been observed, perhaps because of the scarcity of outcrops rather than because the lag does not exist.

A fan of outwash sediment occurs at the mouth of most tunnel channels formed along the edge of the Green Bay and Lake Michigan Lobes (Figs. 5, 6, 7, 8, and 10). In most places the form of the

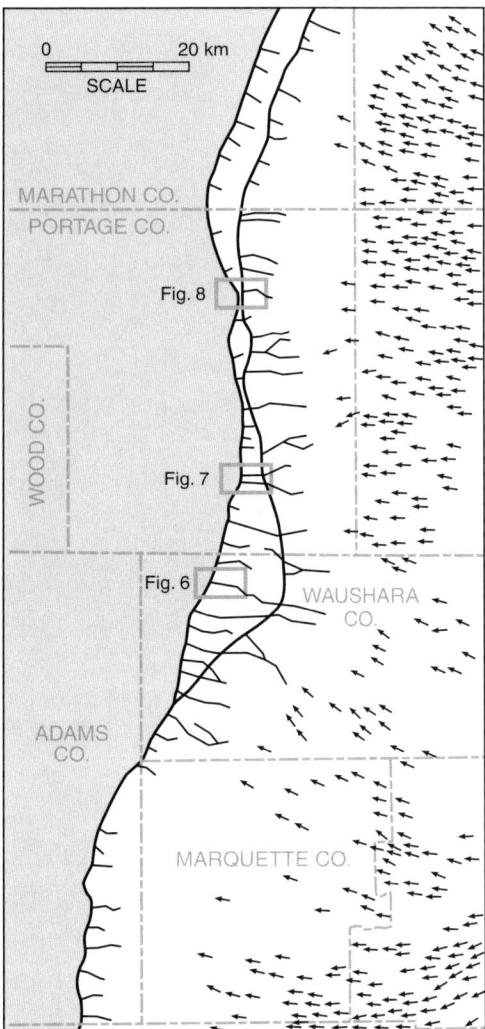

Figure 3. Tunnel channels (thin lines) formed in central Wisconsin along the west side of the Green Bay Lobe; location of map area is shown in Figure 2. The heavy lines are the Almond moraine (east), which merges southward with the Johnstown moraine, and the Hancock moraine (west). The arrows are drumlins, indicating ice-flow direction. The shaded area is the Driftless Area. Interpreted from 7.5-minute topographic maps and from geologic maps by Hadley and Pelham (1976), Attig (1989), and Clayton (1986, 1987).

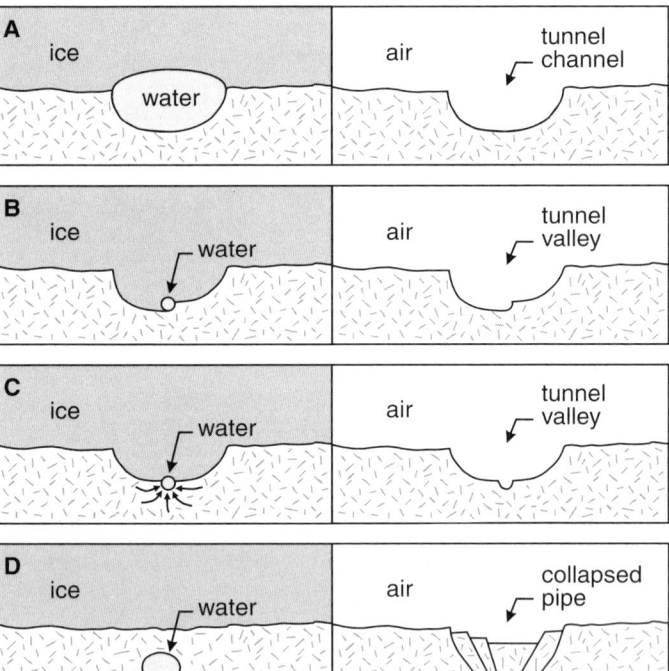

Figure 4. Cross sections showing the formation of different kinds of trenches formed by subglacial rivers; subglacial material is shown with dashes. A, *Tunnel channel:* a channel is eroded by a subglacial river that is as large as the channel. B, *Tunnel valley:* the channel is eroded by a subglacial river that is much smaller than the valley. C, *Tunnel valley:* a valley is formed by the flowage (indicated by arrows) of subglacial material into a subglacial river that is much smaller than the valley. D, *Subglacial pipe:* a tunnel is sapped into the subglacial material by groundwater and then enlarged by meltwater; the overlying sediment collapsed into the pipe after flow ceased.

Linke, 1989). Buried tunnel channels lacking any surface expression also exist in Wisconsin (Stewart, 1976, p. 34–83), but in most places we lack adequate subsurface information to identify them. However, some of those shown in Figures 2 and 3, including Thwaites' (1946, p. 48) "kettle chains," were completely buried but can still be recognized by the presence of collapse depressions.

The Plainfield-Huron tunnel channel, shown in Figure 6, for example, is covered by outwash to an unknown depth. All of the area east of the moraine in Figure 6 is an outwash plain, with only scattered collapse depressions in most places. A continuous string of collapse pits leads up to a gap in the moraine, suggesting this is fan is obvious, with its apex at the mouth of channel; however, where the channels are close together, the fans tend to merge together into a continuous outwash plain. Many of the fans are 2–4 km wide and have a longitudinal slope of 2–15 m/km near the apex, becoming less steep downstream. The fans formed along the middle and southern part of west side of the Green Bay Lobe are composed of sand, becoming gravelly within only a few hundred meters of the apex. The glaciofluvial sediment in this area is sandy because much of it was derived from till with 70%–90% sand, and the tunnel-channel sediment may be especially sandy because some of it was eroded directly from Cambrian formations consisting of sand or easily eroded sandstone. Farther north, where the glacier passed over Precambrian granite and other igneous and metamorphic rock, boulders larger than 1 m are abundant in at least one of the fans.

Many of the fans have small channels cut into their apexes (F in Figs. 6 and 7). These channels are generally no more than about 7 m deep and are typically 100–200 m wide, less than half as wide as the associated tunnel channel. Most are no more than 1 km long. These small channels were cut after the bulk of the fans were deposited, perhaps during the waning stage of a tunnel-channel flood. In addition, the general surfaces of the fans are covered with a pattern of braided channels that are no more than a few meters deep and a few tens of meters wide, apparently associated with the final stages of meltwater flow across the fans. Except for these fan-apex channels and braided channels, no channel extends beyond the outermost moraine formed along the west side of the Green Bay Lobe. All meltwater flow ceased on the fans when the ice had wasted eastward no more than a few hundred meters from the outermost moraine. As soon as a continuous trough opened between the moraine and the ice, meltwater flow shifted from the fans southward through the trough, although this may have been delayed if the way was blocked by buried stagnant ice. There is no evidence that any of this southward flow came from the tunnel channels.

The mouths of most tunnel channels consist of distinct gaps through the associated end moraine (Figs. 1, 6, 7, and 8). Most moraine gaps are the same width as the rest of the tunnel channel, but some are slightly narrower, perhaps because the end-moraine material was more prone to postglacial mass movement than the material behind the moraines. The tunnel channels that formed along the western side of the Green Bay Lobe occur in a zone of hummocky glacial sediment 10–20 km wide that lacks drumlins (Fig. 2 and 3; Attig et al., 1989). Immediately upglacier from the tunnel channels is a broad zone of drumlins. Eskers occur in this drumlin zone but are absent from most of the tunnel-channel zone.

Tunnel channels elsewhere in Wisconsin

The features interpreted to be tunnel channels in Dane County, which formed at the south end of the Green Bay Lobe (Mickelson, 1983, p. 17; Clayton and Attig, 1997, p. 29–31), differ from those formed along the west side. Most are short, narrow gorges cut through dolomite and sandstone of preglacial drainage divides; their mouths do not end at gaps in the outermost moraine; and none have prominent outwash fans at their mouths. The steepest measured upglacier slope is an 80-m rise within a distance of 0.5 km near the mouth of the tunnel channel west of Springfield Corners in northwestern Dane County (Mickelson, 1983, p. 17).

The tunnel channels formed under other lobes in Wisconsin tend to be somewhat wider than those of the Green Bay Lobe—commonly 1 km or wider. Those formed under the western part of the Superior Lobe in Minnesota range in width from 0.2 to 2 km (Wright, 1973, p. 253; Patterson, 1994, p. 72).

The tunnel channels of the Lake Michigan Lobe (lower-right part of Fig. 2) are about the same length as those of the Green Bay Lobe. Those of the Wisconsin Valley Lobe, the Chippewa Lobe, and the eastern part of the Superior Lobe are longer but tend to be ill defined and discontinuous; some continuous channels are tens of kilometers long. According to Wright (1973, p. 253), those formed under the Minnesota part of the Superior Lobe are as long as 150 km, but Mooers (1989, p. 24) considered them to consist of separate segments 10–20 km long, which is similar to the length of those in Wisconsin, and Patterson (1994, p. 72) said they are 5–45 km long.

The other tunnel channels of Wisconsin are similar in depth to those of the Green Bay Lobe. Wright (1973, p. 254) and Patterson (1994, p. 72) reported that those in Minnesota range in depth from a few meters to about 30 m deep and average about 10 m.

The lateral spacing of most channels in central and northwestern Wisconsin is wider and less regular than the spacing of those of the Green Bay Lobe. Spacings as wide as 10 km appear to be typical but some are spaced as much as a few tens of kilometers apart.

According to Wright (1973, p. 252, 266, Fig. 2), Mooers (1989, p. 26–27), and Patterson (1994, p. 81), eskers occur in many of the tunnel channels of Minnesota. Only a few occur in the Wisconsin channels, however, and none have been identified in channels formed along the west side of the Green Bay Lobe. The few eskers we have observed in tunnel channels are about a tenth as wide as the channels they occupy. Two of the most conspicuous are shown in Figures 11 and 12. Drillhole information indicates that the granite floor of the Mondeaux trench, shown in Figure 12, slopes downward to the north, opposite the direction of meltwater flow (Attig, 1993, p. 16–17).

Like the channels formed under the west side of the Green Bay Lobe, the other subglacial channels in Wisconsin tend to occur downglacier from the drumlin zone. In Minnesota, however, the situation is strikingly different. According to Wright (1973, Fig. 1 and 2), tunnel channels occur in the drumlin and esker zone, and according to Mooers (1989, p. 25), they are absent in the outer hummocky-till zone along the west side of the Superior Lobe (Fig. 2), although Patterson (1994, p. 81, fig. 2) observed that they are commonly present in this outer zone.

INTERPRETATION

Tunnel channels have been described in other parts of the world (e.g., Milthers, 1948; Woodland, 1970; Ehlers and Linke, 1989; Brennand and Shaw, 1994; Piotrowski, 1994, 1997;

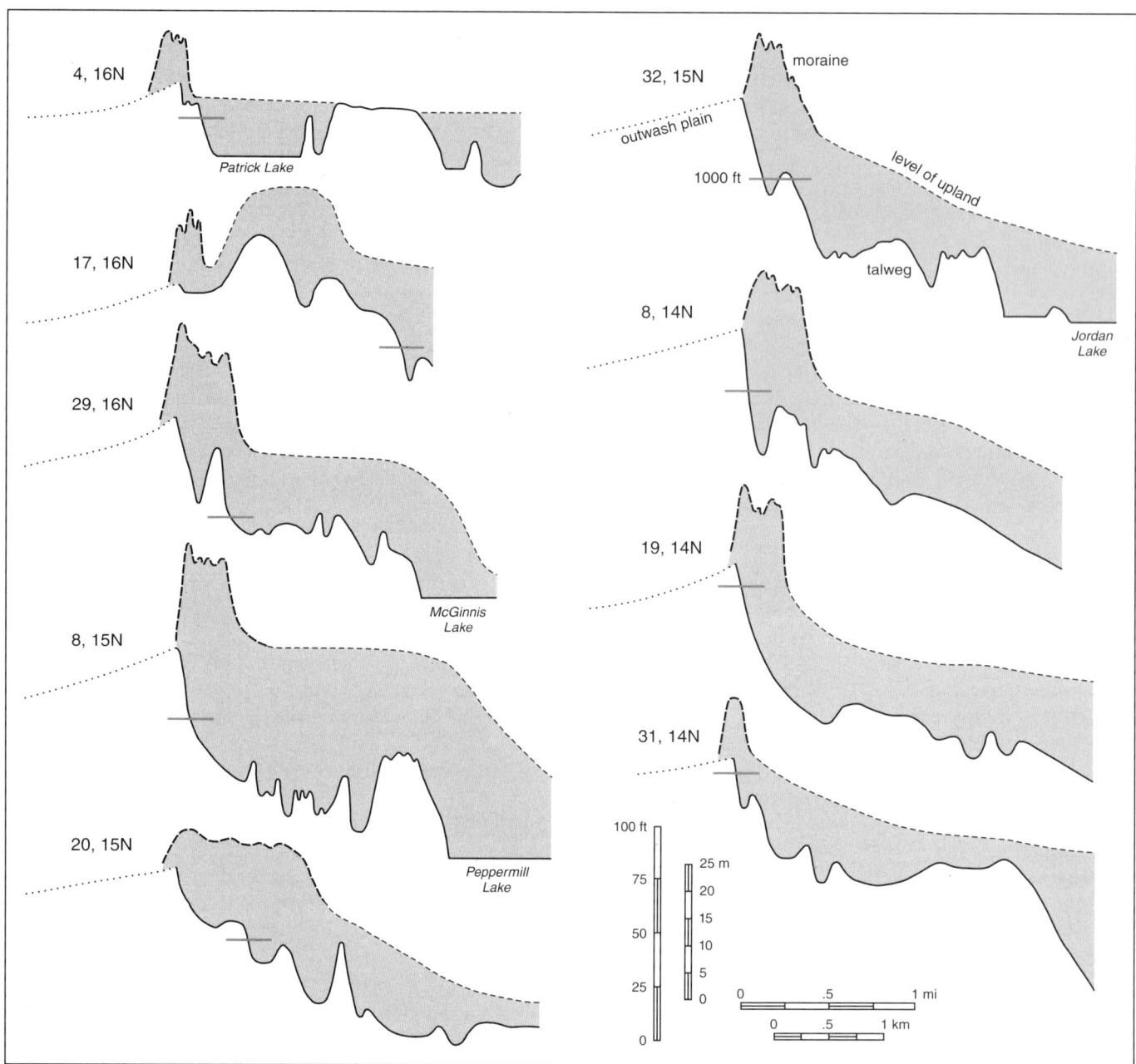

Figure 5. Longitudinal profiles of the lower ends of tunnel channels in central (top left) to southern (bottom right) Adams County, central Wisconsin (Fig. 3), showing their upglacier slope. The Green Bay Lobe here flowed westward, from right to left. The dotted line is the surface of the outwash plain west of the mouth of the tunnel channel. The solid line is the talweg of the present channel (because of post-channel fluvial deposition, the original talweg was lower). The horizontal cross line shows the level of the 1,000-ft contour. The dashed line shows the profile of the upland on either side of the tunnel channels; the heavy dashed line is the Johnstown moraine. The numbers give the location of the mouth of each tunnel channel (section and range north, Public Land Survey).

O'Cofaigh, 1996). However, some have different characteristics, and they may have had an origin different from the channels discussed here. For this reason they will not be discussed; instead we will focus on the tunnel channels of Wisconsin, which are more familiar to us.

A nonglacial origin for the channels shown in Figures 2 and 3 can be ruled out because they slope uphill and they are clearly associated with the outer margin of the Laurentide Ice Sheet during the height of the Wisconsin Glaciation. Nor were they cut by the glacier itself because they are the shape of river channels, not glacial valleys.

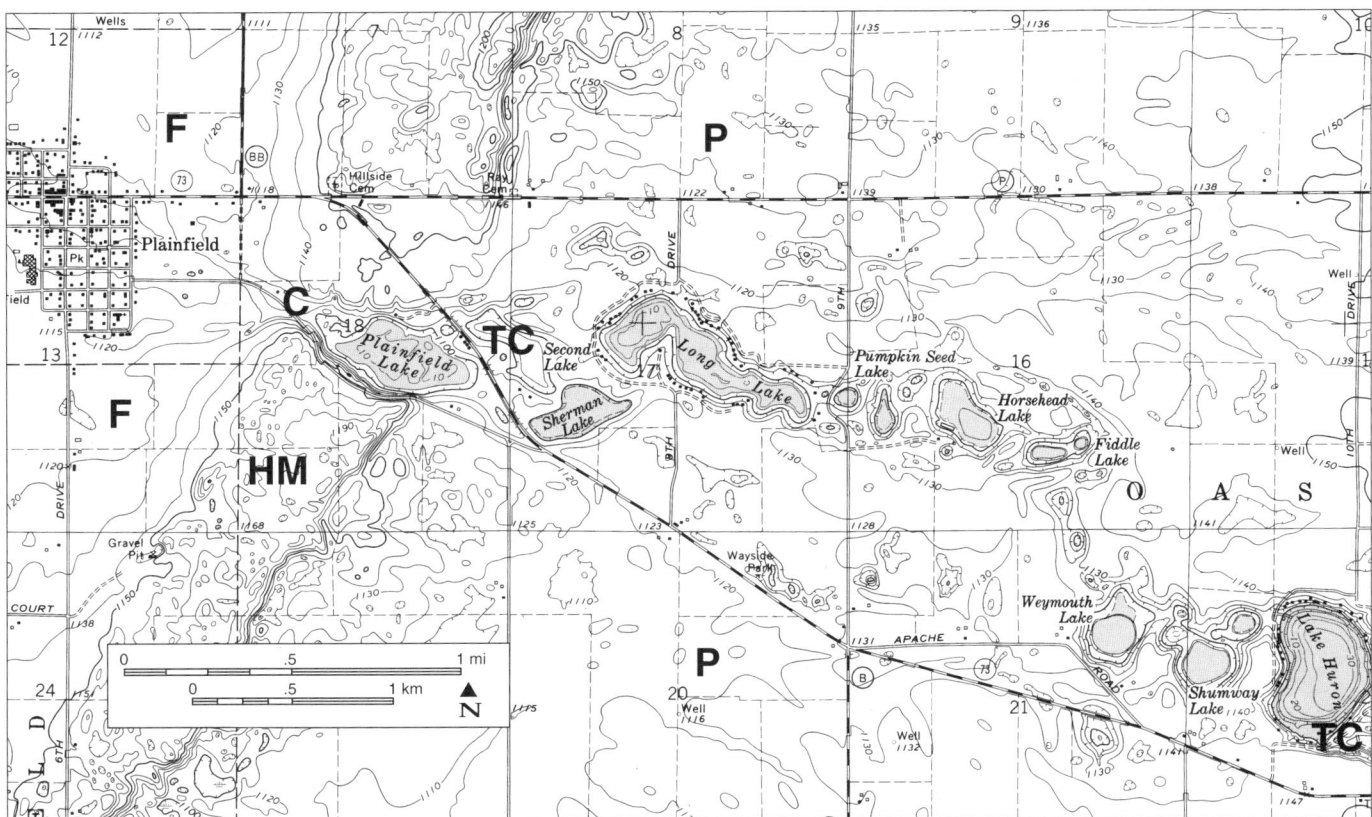

Figure 6. Tunnel channel breaching the Hancock moraine. HM, thick sandy till within the 1,150-ft contour; TC, the Plainfield-Huron chain of lakes in a series of collapse pits marking the location of the tunnel; F, a fan of outwash sand beyond the mouth of the channel, west of the moraine; P, a pitted outwash plain to the east of the moraine; C, a river channel cut into the apex of the channel-mouth fan. The Green Bay Lobe here flowed west-northwest. Figure is a part of U.S. Geological Survey Plainfield Quadrangle (7.5-minute series; 10-ft contour interval). The location of the map area is shown in Figure 3.

Palimpsest spillways

Some insight into the origin of the tunnel channels can be gained by comparing them with proglacial meltwater channels. Most proglacial meltwater channels in this region can be placed in one of two categories. Meltwater channels on proglacial outwash plains typically are braided, are no more than a few meters deep, are no more than a few tens of meters wide, and have a small depth/width ratio; they are generally too small to be obvious on the 1:24,000 topographic maps. These characteristics are the result of the abundant coarse bed load supplied directly from the glacier.

In contrast, the channels of spillways flowing out of proglacial lakes tend to meander (or at least are not highly braided), are several meters or more deep, are at least several tens of meters wide, and have a larger depth/width ratio; they are conspicuous landforms, easily identified on topographic maps (Fig. 13). These characteristics are the result a lack of bed load in water flowing out of lakes. In the Midwest, the largest spillways, which may be as wide as a few kilometers, are interpreted to have been the result of catastrophic floods from rapidly draining glacial lakes (Kehew and Lord, 1987).

The tunnel channels of the Upper Midwest fit into the second category, but with a somewhat smaller depth/width ratio, probably due to the post-tunnel sediment in their bottoms. They are not typical proglacial-lake spillways, however, for several reasons. Evidence of proglacial lakes is generally lacking around their upstream ends. Most lack the clear, fresh, unmodified cutbanks of known spillways; instead the banks are irregular and obscure in places. Similarly, the channels lack the flat bottoms and bars of typical spillways; the hummocky sand and gravel and occasional eskers in the bottom of the channels indicate that they contained ice after they formed.

In these respects, the tunnel channels resemble the outlets of proglacial lakes that were later overridden by the glacier—that is, palimpsest spillways. The thicker the blanket of till draped over the spillway, the more the form of the spillway is obliterated. An example of a spillway that has been modified but not totally obliterated when the ice advanced over it is shown in Figure 14.

However, the tunnel channels of the Green Bay Lobe could not be palimpsest spillways for several reasons. (1) The channels are radially oriented with respect to the margins of the glacier lobe rather than parallel to the slope of the land, as they should be if they were originally spillways of proglacial lakes. (2) Many slope upglacier (Fig. 5). (3) Most seem to end exactly at the outermost

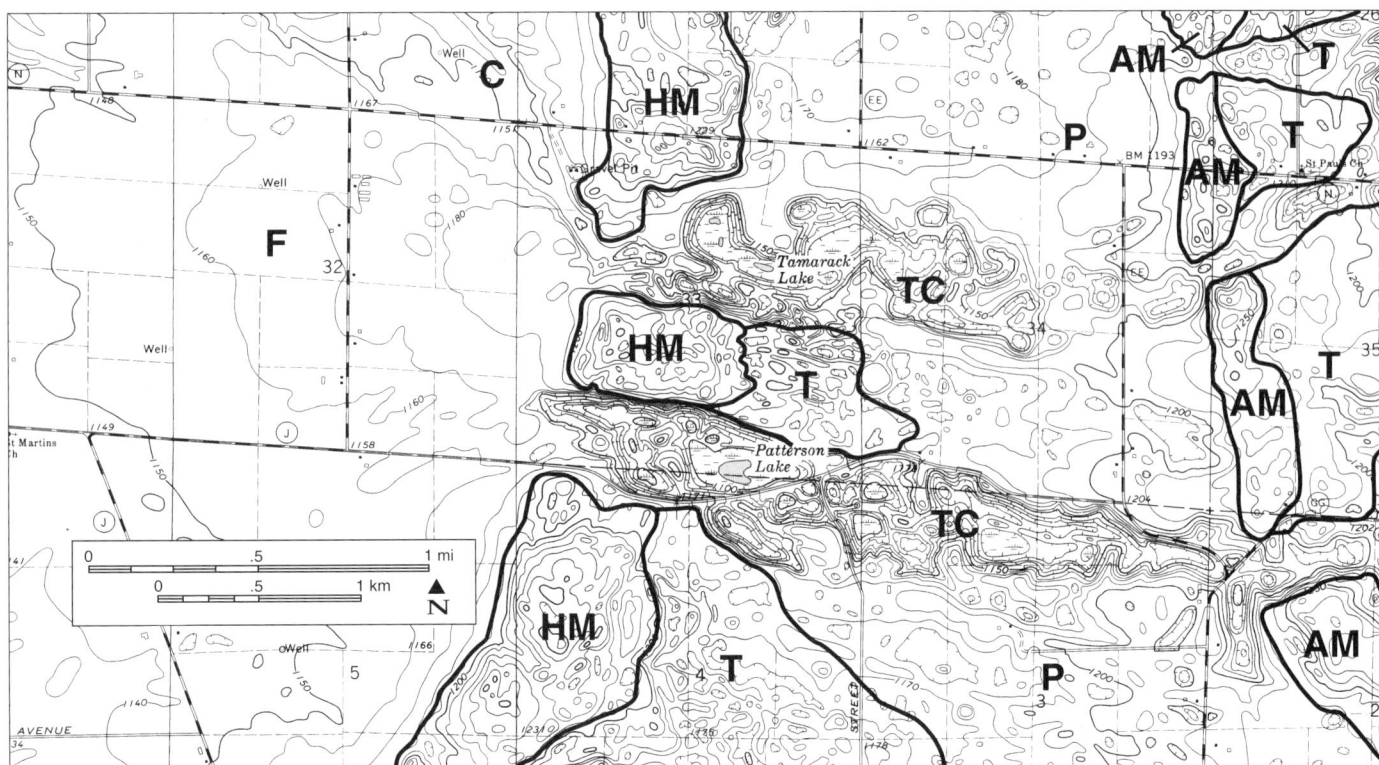

Figure 7. A pair of tunnel channels breaching a pair of moraines. HM, Hancock moraine; AM, Almond moraine; TC, the tunnels' positions marked by the two chains of collapse depressions; the northern one is occupied by Tamarack Lake and the other by Patterson Lake; T, thin till occurs behind the moraines; C, a river channel cut into apex of the channel-mouth fan; F, an unpitted outwash fan occurs west of the Hancock moraine; P, a pitted outwash plain east of the Hancock moraine. The Green Bay Lobe here flowed west. Figure is a part of U.S. Geological Survey Almond Quadrangle (7.5-minute series; 10-ft contour interval). The location of the map area is shown in Figure 3. Geology from Clayton (1986).

moraine; it is unlikely that the glacier would have advanced everywhere precisely to the mouths of a set of preexisting spillways.

For these reasons, we are convinced that the features shown on Figure 2, especially those formed along the west side of the Green Bay Lobe, were formed by rivers running under the glacier. There remains the question of the position and size of these rivers. Are they collapsed pipes, tunnel valleys, or tunnel channels?

Collapsed subglacial pipes

Could these trenches have resulted from an outburst of subglacial water through sediment beneath the bed of the glacier? Boulton and Hindmarsh (1987) suggested that groundwater movement at the glacier bed might have caused piping (hydraulic sapping). This process might have initiated a tunnel that was then enlarged by a catastrophic outbreak of meltwater stored in a thawed-bed zone upglacier from the terminal frozen-bed zone. If this occurred well below the glacier bed (Fig. 4D), a trench would have formed after the outburst, when the roof of the tunnel (pipe) collapsed.

Pipes should have begun to form in front of the glacier at spots where the permafrost was thin or lacking, such as under proglacial lakes. However, there is no obvious correlation between the position of tunnel channels and proglacial lake basins.

Pipes would be expected where the hydraulic potential gradient was steep over a subglacial aquifer that extended from a thawed-bed zone, then downglacier beneath a frozen-bed zone, to the edge of the glacier. However, we know of no correlation between tunnel channels and preexisting aquifers. With the right configuration of subglacial aquifers, some of the pipes might have discharged beyond the glacier margin, so that the collapse trenches extended out into the proglacial area, but we know of none that extended beyond the ice margin. In addition, piles of debris (sand volcanoes) might be expected around the pipe mouths, but we know of none near the trench mouths. We have seen outwash-filled pipes cut through soft early Paleozoic sandstone in Wisconsin, but they are no more than a few meters across, show no evidence of collapse, and have no surface expression.

Tunnel valleys

Mooers (1989) suggested that the Minnesota features are not tunnel channels but are tunnel valleys, cut by subglacial rivers that were much smaller than the valleys (Fig. 4B). Those of Wisconsin, however, probably are not tunnel valleys for the following reasons.

One objection to the tunnel-valley interpretation is that some look like large channels rather than valleys. The outlet channels

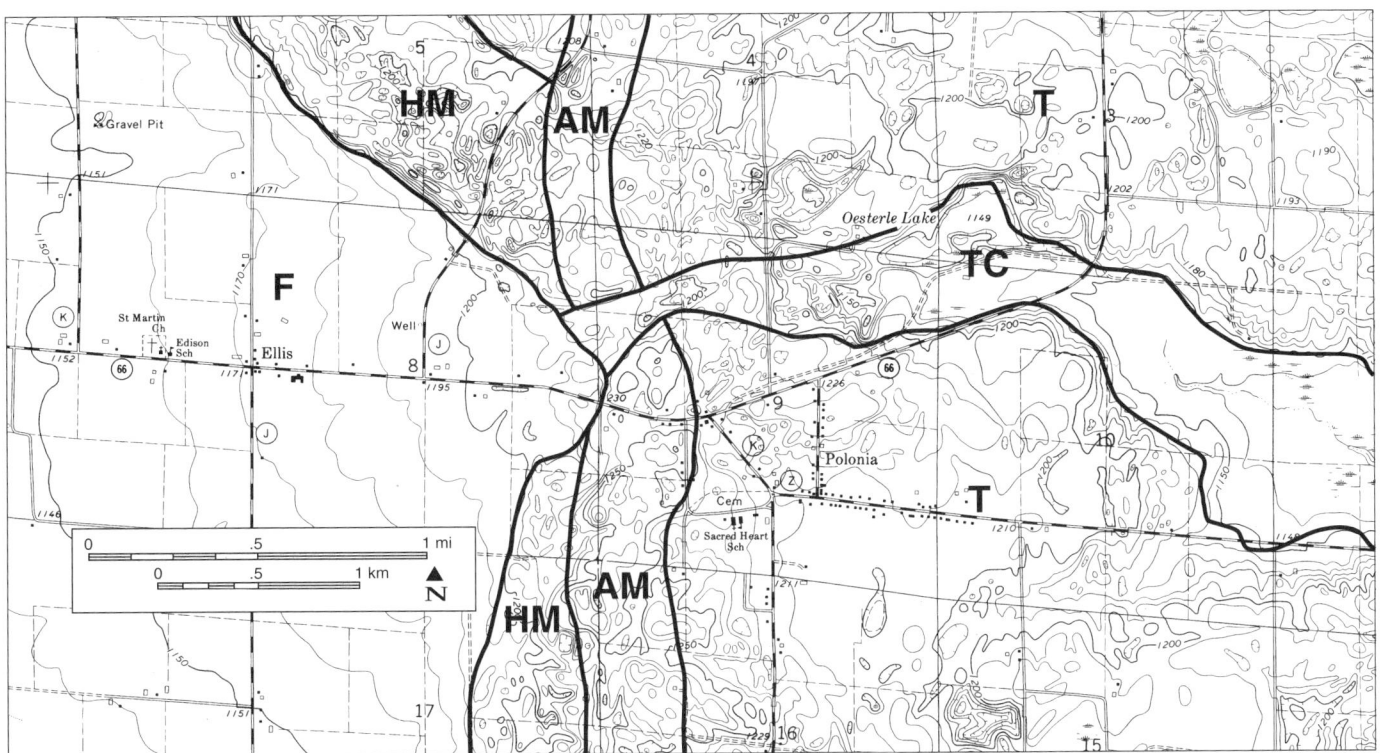

Figure 8. Tunnel channel occupied by Oesterle Lake. HM, thick till of the Hancock moraine; AM, thick till of the Almond moraine; T, thin till behind the moraines (the flat areas are offshore sediment of ice-walled-lake plains); F, an unpitted outwash fan west of the Hancock moraine; TC, collapsed supraglacial fluvial sand (overlain by flat peat to the east) in the tunnel channel. The Green Bay Lobe here flowed west. Figure is a part of U.S. Geological Survey Polonia Quadrangle (7.5-minute series; 10-ft contour interval). The location of the map area is shown in Figure 3. Geology from Clayton (1986).

of large proglacial lakes in the region (Kehew and Lord, 1987) have steep sides and a uniform size and cross-sectional shape, and their form tends to be slightly but smoothly meandering, as illustrated in Figure 13. In addition, they have few tributaries except for postglacial gullies. In contrast, valley bottoms are much more irregular, with variable widths and cross-sectional shapes where joined by tributary valleys. The Straight River tunnel channel is a conspicuous example of a trench that is shaped like a channel rather than a valley (Fig. 11).

Another objection to the tunnel-valley interpretation comes from a comparison with eskers. When the whole population of tunnel channels is compared with the whole population of eskers in Wisconsin, several factors indicate that the tunnel channels and eskers represent two significantly different populations of subglacial rivers. (1) The eskers represent subglacial rivers that were aggrading their beds to form ridges, and the tunnel channels represent subglacial rivers that were degrading their beds. (2) The tunnel channels are everywhere older than the nearby eskers. As a group the tunnel channels formed during the height of the Wisconsin Glaciation, whereas the eskers formed during the waning phases of the glaciation. (3) Although some eskers formed in the bottom of earlier-formed tunnel channels, there is no direct genetic relation between them. Nowhere have we observed tunnel channels grading into eskers, for example. (4) Few of the eskers are wider than 100 m and few of the tunnel channels are narrower than 200 m; typical tunnel channels are nearly an order of magnitude wider than the typical eskers. The esker rivers and the tunnel-channel rivers of Wisconsin therefore seem to have been the result of significantly different meltwater regimes.

The striking difference in size between the tunnel channels and the eskers indicates to us a striking difference in discharge through the tunnel rivers that formed them. The difference is doubly striking when meltwater production is considered. The tunnel channels formed when the glacier terminus was stable and when the climate was cold enough to form deep permafrost in front of the glacier (Attig and Clayton, 1986); little glacial meltwater would be expected. In contrast, the eskers formed when the glacier terminus was wasting back and the climate was warmer; more meltwater would be expected.

The size of the collapse depressions like those shown in Figures 6 and 7 provides further evidence that these are tunnel channels rather than tunnel valleys. Large tunnels should have resulted in large collapse pits, and small ones should have resulted in small pits. These pits are comparable in width to typical tunnel channels of the region; they are much wider than the eskers of the region.

These reasons make it seem likely that the Wisconsin features are not tunnel valleys, formed by rivers that were much smaller than the trench. Based on the difference in channel size, the discharge of

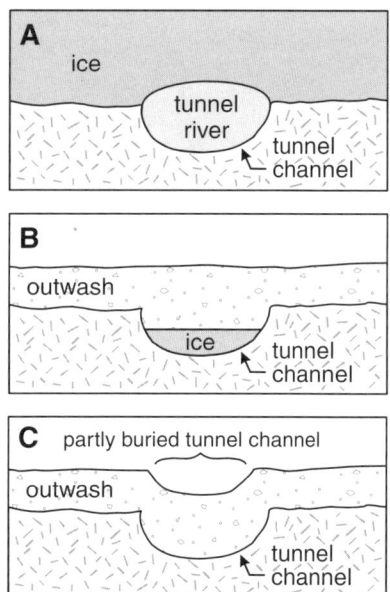

Figure 9. Cross sections showing the sequence of events responsible for a partly buried tunnel channel, such as the tunnel channels occupied by the Plainfield-Huron chain of lakes (Fig. 6) and Tamarack and Patterson Lakes (Fig. 7). A, A subglacial channel is cut. B, As the subglacial river stops flowing, the roof collapses, and outwash is deposited across the area before all the ice in the tunnel channel melts. C, The last ice melts, causing the overlying outwash to collapse, forming a partly buried tunnel channel.

tunnel-channel rivers might have been roughly two orders of magnitude larger than that of the esker rivers if the ratios of width to depth in each were comparable (Leopold et al., 1964, Fig. 7-21).

Wright (1973, p. 261) also thought the tunnel channels of Minnesota were cut by large rivers. He argued that the velocity normally present in small, esker-sized rivers would be too small to cause them to degrade their beds. In contrast, large rivers in tunnels would have high enough velocities to cause their channels to degrade, especially if they were supplied by water that had lost its bed load in subglacial lakes.

Boulton and Hindmarsh (1987) suggested that some tunnel valleys formed when mobile subglacial sediment was squeezed into a tunnel river as a result of elevated pore pressure under the glacier (Fig. 4C). We see little to indicate that such squeezing occurred in Wisconsin. The cross-sectional size and shape of a tunnel valley formed by squeezing should have varied, depending on the nature of the subglacial material. Instead, the tunnel channels of Wisconsin are fairly uniform in width, and the variations can not be correlated with the nature of the bank material—with the exception of a slight narrowing where some pass through the outer moraine. If the subglacial sediment were frozen, as we think, squeezing was unlikely.

Tunnel channels

Wright (1973) concluded that the Minnesota channels were cut during a cold climate, when the Superior Lobe stood at its outermost moraine. His paleoecological reconstructions for that time indicated a climate that was too cold to produce significant amounts of surface meltwater and a glacier that was too cold for the surface meltwater to penetrate to its base. Instead he thought that the meltwater originated at the base of the glacier well behind its toe, where it was stored until large rivers periodically and catastrophically burst through the glacier's frozen-bed toe.

Mooers (1989) disagreed, arguing that the Minnesota features are not tunnel channels, because the calculated supply of meltwater was inadequate to cut channels that big. He argued that they formed later than suggested by Wright, when the climate was warmer and most of the meltwater came from the surface of the glacier; he argued that the mouths of the valleys are associated with moraines formed at a time when the climate was milder, far behind the outermost moraine. With such a climate, the toe of the glacier would be thawing, so subglacial water would flow continually rather than be stored and periodically discharged.

In contrast, the tunnel channels along the west side of the Green Bay Lobe formed when the glacier stood at its outermost moraine, which formed while the area still had a cold tundra climate. Ice-wedge polygons and ice-wedge casts occur in many places in Wisconsin, including on outwash plains associated with the outer moraine of the Green Bay Lobe. Many also occur on the plain of proglacial Lake Wisconsin, which did not drain until the glacier had wasted back 20 km from the outer moraine (Clayton, 1987, p. 8–9, 1989, p. 13; Clayton and Attig, 1989, p. 44–45). Similarly, there is evidence for continuous permafrost after the other lobes of northern Wisconsin wasted a considerable distance back from their outermost positions (Johnson, 1986, p. 20; Attig and Clayton, 1986; Clayton and Attig, 1990, p. 56). It therefore is likely that the tunnel channels of Wisconsin formed during a relatively cold period, when the glacial toe probably had a frozen bed.

Mooers (1989) also argued against a frozen-bed glacial toe when the Minnesota tunnel valleys formed because they occur in the zone of drumlins and eskers, which require a glacier sliding over a thawed bed. In contrast, the tunnel channels formed along the western side of the Green Bay Lobe (Fig. 3), and generally elsewhere in Wisconsin (Fig. 2), occur between the outer moraine and the drumlin and esker zone; drumlins and eskers are lacking in the tunnel-channel zone, presumably because the glacier was frozen to its bed there (Attig, et al., 1989).

In summary, Mooers' paleoclimatologic objections to a tunnel-channel interpretation of the Minnesota features do not apply to those in Wisconsin.

Shaw and others (Brennand and Shaw, 1994; and the references listed therein) have suggested that many presumed drumlins were really formed by catastrophic discharge of subglacial water. The association of Midwest drumlins and tunnel channels might be taken as supporting evidence for the catastrophic origin of the channels. However, we believe that the drumlins in this area were molded by direct glacial action, not by subglacial water (Whittecar and Mickelson, 1979; Colgan and Mickelson, 1997).

Nevertheless, we agree with Wright (1973) and Mooers (1989) that tunnel channels could not have flowed continuously, because not enough meltwater would have been available to

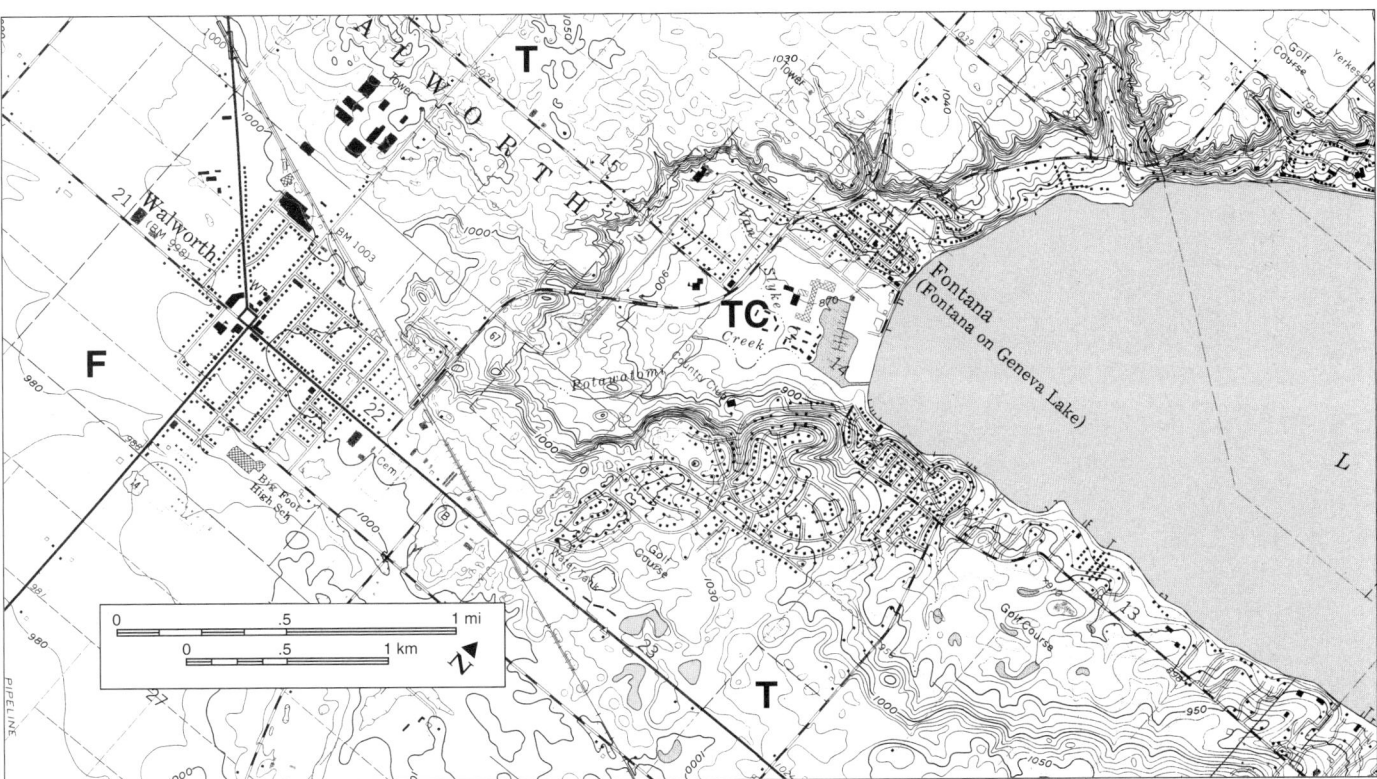

Figure 10. Tunnel channel occupied by Lake Geneva. F, area to the southwest of the 1,000-ft contour occupied by an outwash fan; T, till in the area east of the fan, according to the soil maps of the area (Haszel, 1971); TC, collapsed outwash sand in the tunnel channel. Lake Geneva is at an elevation of 864 ft, 130 ft (40 m) below the apex of the fan. The Lake Michigan Lobe here flowed southwest. Figure is a part of U.S. Geological Survey Walworth Quadrangle (7.5-minute series; 10-ft contour interval). The location is shown in Figure 2.

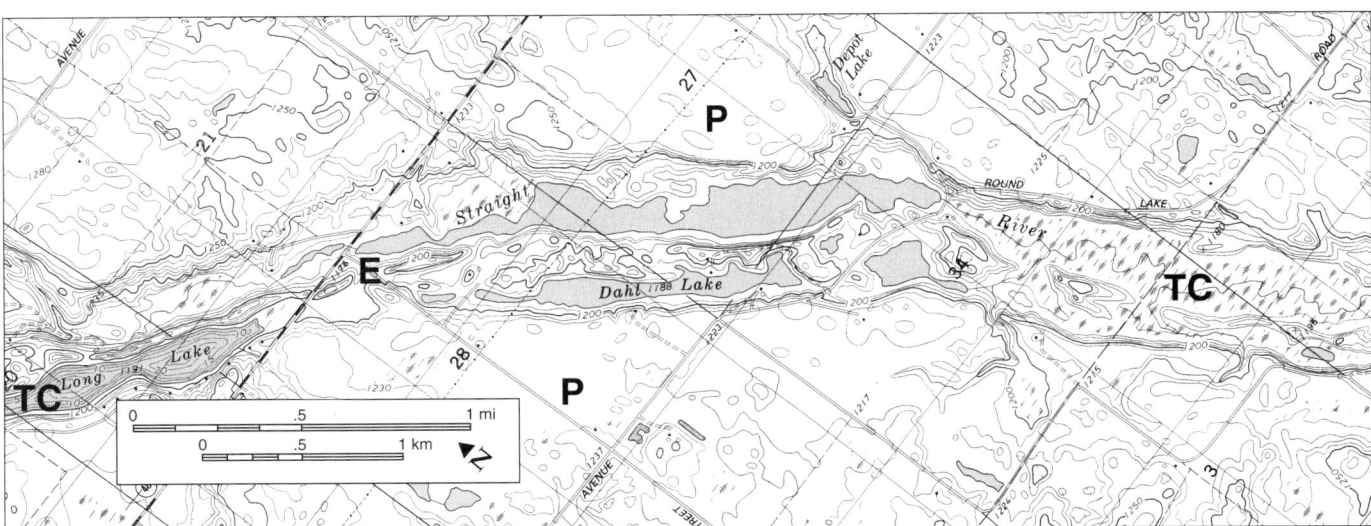

Figure 11. Tunnel channel (TC) occupied by Straight River. E, esker in bottom of tunnel channel; P, outwash plain on either side of the tunnel channel. During the final phases in the formation of the channel, buried ice melted, causing the outwash to collapse into the channel. The Superior Lobe here flowed southeast. Figure is a part of U.S. Geological Survey Luck and Big Round Lake Quadrangles (7.5-minute series; 10-ft contour interval). The location is shown in Figure 2. Geology from M. D. Johnson (1999).

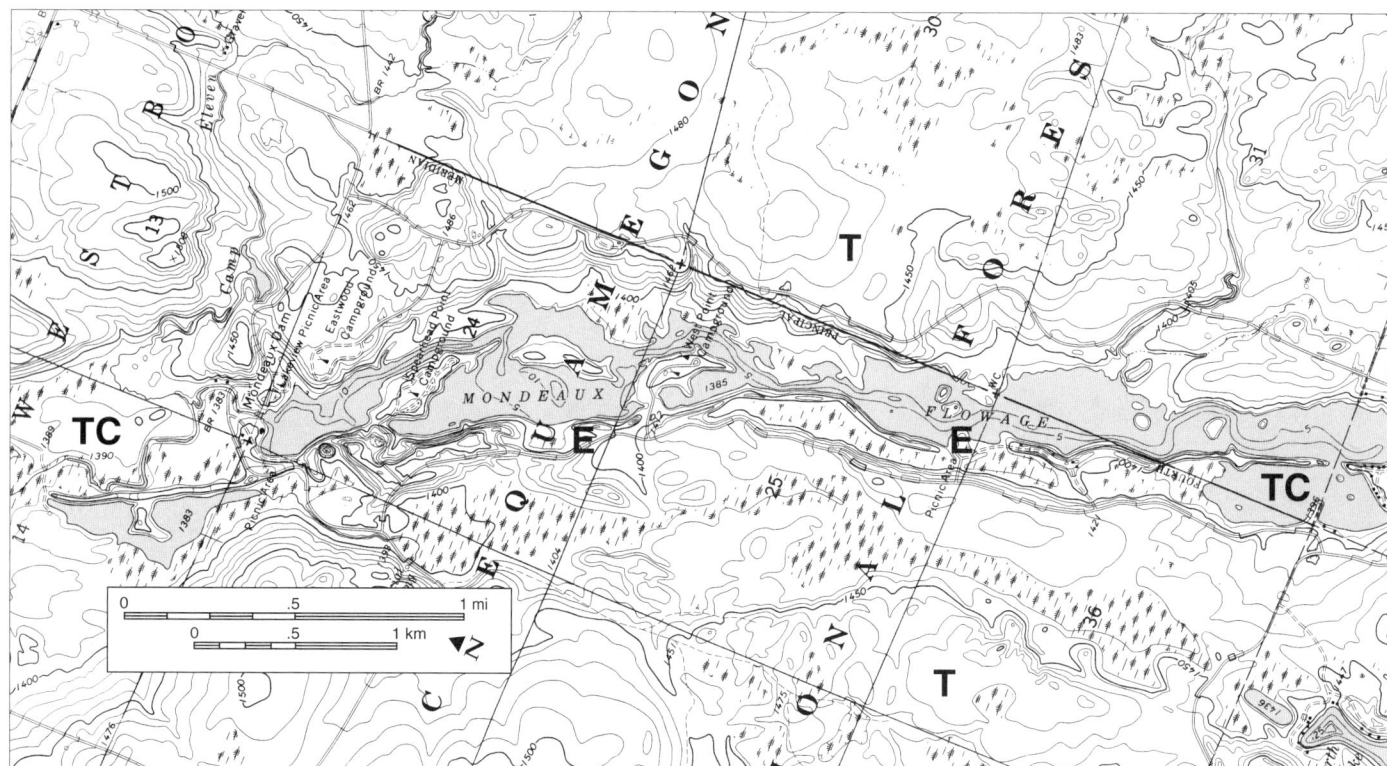

Figure 12. Tunnel channel occupied by Mondeaux River. T, the upland (roughly above the 1,450-ft contour) underlain by till; TC, the channel bottomed with fluvial sand and gravel; E, esker in the bottom of the tunnel channel. The Chippewa Lobe here flowed south-southeast, but just to the north of here it flowed to the southwest. Figure is a part of U.S. Geological Survey Mondeaux Dam Quadrangle (7.5-minute series; 10-ft contour interval). The location is shown in Figure 2. Geology from Attig (1993).

supply such large rivers. If they are tunnel channels, they must have flowed discontinuously, discharging large amounts of stored water during short, widely spaced intervals.

This conclusion is in harmony with the conclusion that most of the large glacier-lake spillways in the Midwest (as in Fig. 13) were cut during floods caused by the failure of a glacial dam or by the overtopping of a drainage divide, rather than by normal continuous flow of meltwater (Kehew and Lord, 1987). The tunnel channels are similar in size to these spillways, suggesting that they too were cut by floods.

A variety of evidence indicates that the glacier was frozen to its bed in an outer zone about as wide as the tunnel-channel zone and that it was thawing in a broad zone behind that (Mickelson, 1983, 1986; Attig et al., 1988, 1989). A catastrophic outbreak of a subglacial lake from the thawed-bed zone through the frozen-bed zone is just what might be expected to produce tunnel channels.

Some channels may have been used more than once, but after the last flood ceased in each, most were partly buried by younger sediment. When flow ceased, the roof of the tunnel collapsed, forming a depression on the glacier surface, which directed the flow of supraglacial meltwater. The supraglacial streams covered the fallen ice with outwash, which collapsed when the ice melted, producing the hummocky topography characteristic of the bottoms of tunnel channels in most parts of Wisconsin.

CONCLUSION

The subglacial trenches shown in Figure 3 are a coherent group of landforms with uniform morphology, which indicates that they are of similar origin. They are remarkably different from the eskers of the region, which indicates that they formed under different conditions from the eskers.

We are convinced that they were in fact formed by rivers flowing in subglacial tunnels. Their orientation with respect to the outer margin of the ice sheet and the uphill gradient of many, including those along the west side of the Green Bay Lobe, allows no other interpretation. The channel-like shape of the trenches indicates that they are tunnel channels (not tunnel valleys), probably cut by catastrophic floods of subglacial meltwater, perhaps when it broke through the frozen-bed toe of the glacier, when the glacier had reached its maximum extent during the coldest part of the Wisconsin Glaciation.

ACKNOWLEDGMENTS

We thank Kent Syverson, Jürgen Ehlers, and especially Carrie Patterson and Mark Johnson, who provided information on tunnel channels in their field areas and reviewed the manuscript at least once.

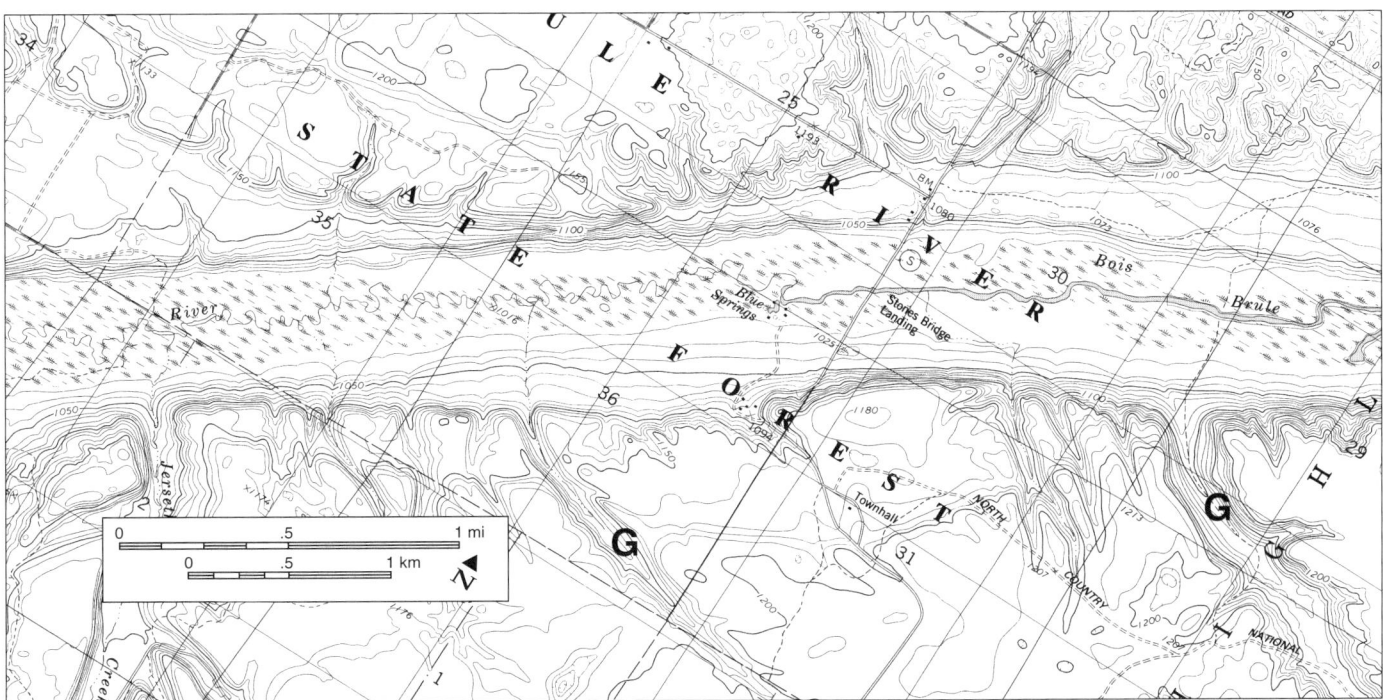

Figure 13. Typical large proglacial meltwater channel, the outlet of glacial Lake Superior, in northwestern Wisconsin. G, gullies eroded through the cutbanks after flow had ceased in the channel. It never was covered by the glacier after it formed; compare its smooth bottom with the irregular bottoms of the channel shown in Figure 11, which formed under a glacier, and the channel shown in Figure 14, which was a proglacial channel subsequently overrun during a glacial readvance. Figure is a part of U.S. Geological Survey Lake Minnesuing Quadrangle (7.5-minute series; 10-ft contour interval).

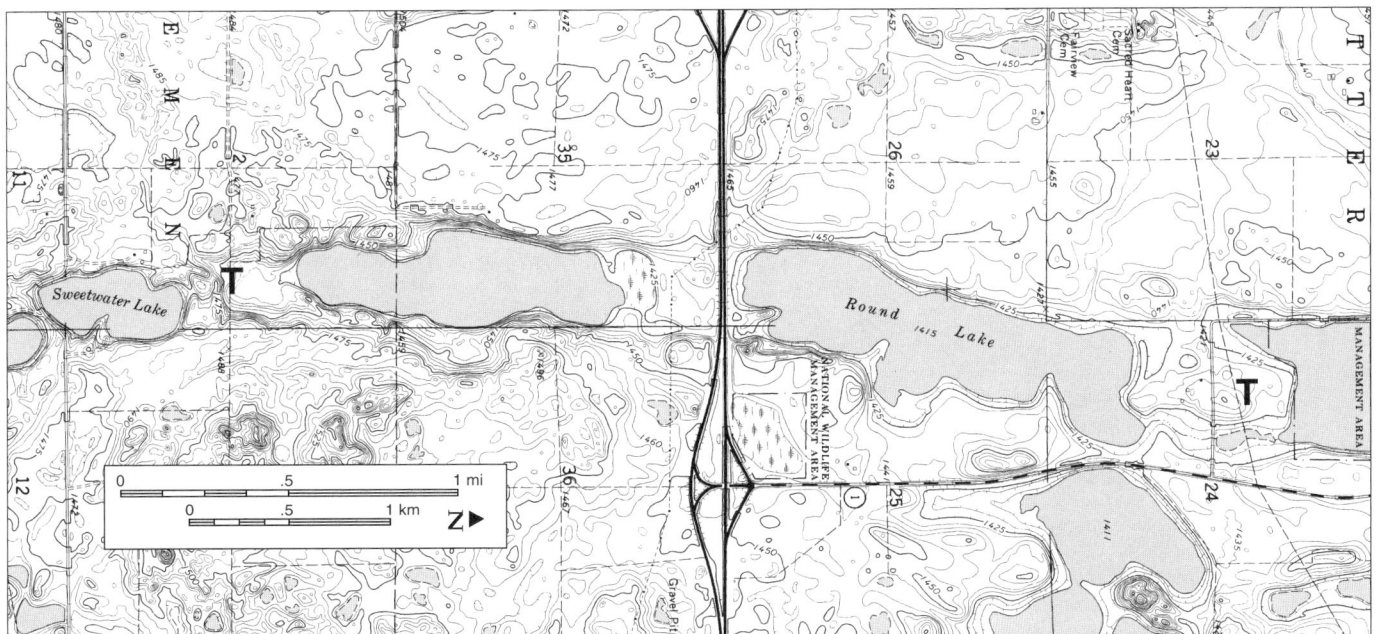

Figure 14. A palimpsest spillway, occupied by Round and Sweetwater Lakes, behind the maximum position of the late Wisconsin Kensal glacial readvance, in western Barnes County, North Dakota. T, mounds of till between the lakes indicating this is either a tunnel channel or a palimpsest spillway. It is probably not a tunnel channel because, like others in the region, it extends south well beyond the maximum Kensal position, where it is a proglacial spillway with no evidence it was ever under the ice (Kelly and Block, 1967; Clayton, 1980). Figure is a part of U.S. Geological Survey Sanborn Quadrangle (7.5-minute series; 5-ft contour interval).

REFERENCES CITED

Attig, J. W., 1989, Pleistocene geology of Marathon County, Wisconsin: Wisconsin Geological and Natural History Survey Information Circular 65, 27 p.

Attig, J. W., 1993, Pleistocene geology of Taylor County, Wisconsin: Wisconsin Geological and Natural History Survey Bulletin 90, 25 p.

Attig, J. W., and Clayton, L., 1986, Late Wisconsin continuous permafrost in northern Wisconsin ended about 13,000 years ago: Geological Society of America Abstracts with Programs, v. 18, p. 278.

Attig, J. W., Clayton, L., and Mickelson, D. M., 1988, Late Wisconsin permafrost history and tunnel channel formation in Wisconsin: American Quaternary Association Program and Abstracts, p. 104.

Attig, J. W., Mickelson, D. M., and Clayton, L., 1989, Late Wisconsin landform distribution and glacier-bed conditions in Wisconsin: Sedimentary Geology, v. 62, p. 399–405.

Attig, J. W., Clayton, L., Lange, K. I., and Maher, L. J., 1990, The Ice Age geology of Devils Lake State Park: Wisconsin Geological and Natural History Survey Educational Series 35, 28 p.

Boulton, G. S., and Hindmarsh, R. C. A., 1987, Sediment deformation beneath glaciers: rheology and geological consequences: Journal of Geophysical Research, v. 92, p. 9059–9082.

Brennand, T. A., and Shaw, J., 1994, Tunnel channels and associated landforms, south-central Ontario: their implications for ice-sheet hydrology: Canadian Journal of Earth Science, v. 31, p. 505–522.

Clayton, L., 1980, Geologic map of North Dakota: U.S. Geological Survey, scale 1:500,000, 1 sheet.

Clayton, L., 1986, Pleistocene geology of Portage County, Wisconsin: Wisconsin Geological and Natural History Survey Information Circular 56, 19 p.

Clayton, L., 1987, Pleistocene geology of Adams County, Wisconsin: Wisconsin Geological and Natural History Survey Information Circular 59, 14 p.

Clayton, L., 1989, Pleistocene geology of Juneau County, Wisconsin: Wisconsin Geological and Natural History Survey Information Circular 66, 16 p.

Clayton, L., and Attig, J. W., 1989, Glacial Lake Wisconsin: Geological Society of America Memoir 173, 80 p.

Clayton, L., and Attig, J. W., 1990, Geology of Sauk County, Wisconsin: Wisconsin Geological and Natural History Survey Information Circular 67, 68 p.

Clayton, L., and Attig, J. W., 1997, Pleistocene geology of Dane County, Wisconsin: Wisconsin Geological and Natural History Survey Bulletin 95, 64 p.

Colgan, P. M., and Mickelson, D. M., 1997, Genesis of streamlined landforms and flow history of the Green Bay Lobe, Wisconsin, U.S.A.: Sedimentary Geology, v. 111, p. 7–25.

Ehlers, J., and Linke, G., 1989, The origin of deep buried channels of Elsterian age in northwest Germany: Journal of Quaternary Science, v. 4, p. 255–265.

Farrand, W. R., Mickelson, D. M., Cowan, W. R., and Goebel, J. E., 1984, Quaternary geologic map of the Lake Superior 4° × 6° Quadrangle, United States and Canada: U.S. Geological Survey Miscellaneous Investigations Series I-1420(NL-16), scale 1:1,000,000.

Goebel, J. E., Mickelson, D. M., Farrand, W. R., Clayton, L., Knox, J. C., Cahow, A., Hobbs, H. C., and Walton, M. S., Jr., 1983, Quaternary geologic map of the Minneapolis 4° × 6° Quadrangle, United States: U.S. Geological Survey Miscellaneous Investigations Series I-1420(NL-15), scale 1:1,000,000, 1 sheet.

Hadley, D. W., and Pelham, J. H., 1976, Glacial deposits of Wisconsin: Wisconsin Geological and Natural History Survey Map 10, scale 1:500 000, 1 sheet.

Haszel, O. L., 1971, Soil survey of Walworth County, Wisconsin: U.S. Department of Agriculture Soil Conservation Service, 107 p.

Johnson, M. D., 1986, Pleistocene geology of Barron County, Wisconsin: Wisconsin Geological and Natural History Survey Information Circular 55, 42 p.

Johnson, M. D., 1999, Pleistocene geology of Polk County, Wisconsin: Wisconsin Geological and Natural History Survey Bulletin 92 (in press).

Kelly, T. E., and Block, D. A., 1967, Geology and groundwater resources, Barnes County, North Dakota: North Dakota Geological Survey Bulletin 43, Part 1, 51 p.

Kehew, A. E., and Lord, M. L., 1987, Glacial-lake outbursts along the mid-continent margins of the Laurentide ice-sheet, *in* Mayer, L., and Nash, D., eds., Catastrophic flooding: Boston, Allen & Unwin, p. 95–120.

Leopold, L. B., Wolman, M. G., and Miller, J. P., 1964, Fluvial processes in geomorphology: San Francisco, Freeman, 522 p.

Lineback, J. A., Bleuer, N. K., Mickelson, D. M., Farrand, W. R., and Goldthwait, R. P., 1983, Quaternary geologic map of the Chicago 4° × 6° Quadrangle, United States: U.S. Geological Survey Miscellaneous Investigations Series I-1420(NK-16), scale 1:1,000,000, 1 sheet.

Mickelson, D. M., 1983, A guide to the glacial landscapes of Dane County, Wisconsin: Wisconsin Geological and Natural History Survey Field Trip Guide Book 6, 52 p.

Mickelson, D. M., 1986, Glacial and related deposits of Langlade County, Wisconsin: Wisconsin Geological and Natural History Survey Information Circular 52, 30 p.

Mickelson, D. M., Clayton, L., Fullerton, D. S., and Borns, H. W., Jr., 1983, The late Wisconsin glacial record of the Laurentide Ice Sheet in the United States, *in* Wright, H. E., Jr., Late-Quaternary environments of the United States: Minneapolis, University of Minnesota Press, v. 1, p. 3–37.

Milthers, V., 1948, Det danske Istidslandskabs Terrænformer og deres Opstaaen (The morphology and genesis of the glacial landscape in Denmark): Danmarks Geologiske Undersøgelse, series 3, no. 28, 233 p.

Mooers, H. D., 1989, On the formation of tunnel valleys of the Superior Lobe, central Minnesota: Quaternary Research, v. 32, p. 24–35.

O'Cofaigh, C., 1996, Tunnel valley genesis: Progress in Physical Geography, v. 20, p. 1–19.

Patterson, C. J., 1994, Tunnel-valley fans of the St. Croix moraine, east-central Minnesota, U.S.A., *in* Warren, W. P., and Croot, D. G., eds., Formation and deformation of glacial deposits: Rotterdam, A. A. Balkema, p. 59–87.

Piotrowski, J. A., 1994, Tunnel-valley formation in northwest Germany—geology of mechanisms of formation and subglacial bed conditions for the Bornhöved tunnel valley: Sedimentary Geology, v. 89, p. 107–141.

Piotrowski, J. A., 1997, Subglacial hydrology in north-western Germany during the last glaciation: groundwater flow, tunnel valleys and hydrological cycles: Quaternary Science Reviews, v. 16, p. 169–185.

Stewart, M. T., 1976, An integrated geologic, hydrologic, and geophysical investigation of drift aquifers, western Outagamie County, Wisconsin [Ph.D. thesis]: Madison, University of Wisconsin, 65 p.

Thwaites, F. T., 1927, Field Notes, July 10, 1927: unpublished field notes, Madison, Wisconsin Geological and Natural History Survey, small-format loose-leaf file, section 14, T. 22 N., R. 9 E.

Thwaites, F. T., 1943, Pleistocene of part of northeastern Wisconsin: Geological Society of America Bulletin, v. 54, p. 87–144.

Thwaites, F. T., 1946 [1963], Outline of glacial geology: Ann Arbor, Michigan, Edwards Brothers, 142 p.

Trotta, L. C., and Cotter, R. D., 1973, Depth to bedrock in Wisconsin: Wisconsin Geological and Natural History Survey map, scale 1:1,000,000.

Whittecar, G. R., and Mickelson, D. M., 1979, Composition, internal structures and an hypothesis of formation for drumlins, Waukesha County, Wisconsin, U.S.A.: Journal of Glaciology, v. 22, p. 357–372.

Woodland, A. W., 1970, The buried tunnel-valleys of East Anglia: Yorkshire Geological Society Proceedings, v. 37, p. 521–578.

Wright, H. E., Jr., 1973, Tunnel valleys, glacial surges, and subglacial hydrology of the Superior Lobe, Minnesota: Geological Society of America Memoir 136, p. 251–276.

MANUSCRIPT ACCEPTED BY THE SOCIETY OCTOBER 8, 1998

Spooner Hills, northwest Wisconsin: High-relief hills carved by subglacial meltwater of the Superior Lobe

Mark D. Johnson
Department of Geology, Gustavus Adolphus College, St. Peter, Minnesota 56082

ABSTRACT

A 15-km-wide band of high-relief hills (area 1–15 km^2; relief as much as 60 m) occurs concentric to and 15–30 km behind the terminal ice-margin position of the Superior Lobe in northwestern Wisconsin. I refer to these hills as the Spooner Hills (named after nearby Spooner, Wisconsin). Drill holes, exposures, and well logs show the Spooner Hills to be composed of several layers of till and sorted sediment that are in turn covered with the till of the most recent ice advance. The hills are roughly equidimensional, but a number of them are elongate parallel to the direction of regional ice flow. Hill-top elevations increase toward the former ice-margin position.

The hills contain a variety of sediment types representing a variety of depositional environments. Because much of this sediment predates the most recent glaciation, the Spooner Hills are considered erosional features. The interhill valleys form a branching, anastomosed network that, in some places, connects up with prominent tunnel channels. I suggest that the hills are erosional remnants formed when these valleys were excavated by subglacial meltwater. The tunnel channels represent the outlets for meltwater and sediment. The water and sediment released at the margin was deposited proglacially in large outwash fans and plains, while some of the sediment may have been frozen to the bed of the glacier.

This band of hills is distinct from any other landscape region in northwestern Wisconsin and represents a landform type not commonly described in glaciated regions (Clayton, 1986; Booth and Hallet, 1993). Features similar to these occur associated with the Green Bay Lobe of eastern Wisconsin, the Lake Michigan Lobe of northern Michigan, and the Itasca Moraine of the Wadena Lobe, Minnesota.

INTRODUCTION

Subglacial meltwater has long been understood to be an important factor in glacial geomorphology and sedimentology. Several common landforms found in glaciated landscapes are clearly understood to be formed by the depositional or erosional action of subglacial meltwater, including eskers, tunnel channels/valleys, and some end moraines (e.g., Price, 1973; Wright, 1973; Attig et al., 1989; Mooers, 1989; Fyfe, 1990; Booth, 1994; Cofaigh, 1996; Pair, 1997). In recent years, subglacial meltwater erosion has been called upon to explain other glacial landforms, namely drumlins, hummocks, and Rogen moraine whose origin has been traditionally explained by other mechanisms, such as glacier erosion, glaciotectonism, and ice stagnation (Shaw and Kvill, 1984; Shaw and Gilbert, 1990; Fisher and Shaw, 1992; Rains et al., 1993; Shaw et al., 1996; Munro and Shaw, 1997). These latter studies have stimulated interest in reevaluating the role of subglacial meltwater.

In this paper I describe the geomorphology, stratigraphy, and origin of a band of hills that occur in northwestern Wisconsin (Fig. 1). This band of hills is distinct from any other landforms in northwestern Wisconsin and represents a landform type not com-

Johnson, M. D., 1999, Spooner Hills, northwest Wisconsin: High-relief hills carved by subglacial meltwater of the Superior Lobe, *in* Mickelson, D. M., and Attig, J. W., eds., Glacial Processes Past and Present: Boulder, Colorado, Geological Society of America Special Paper 337.

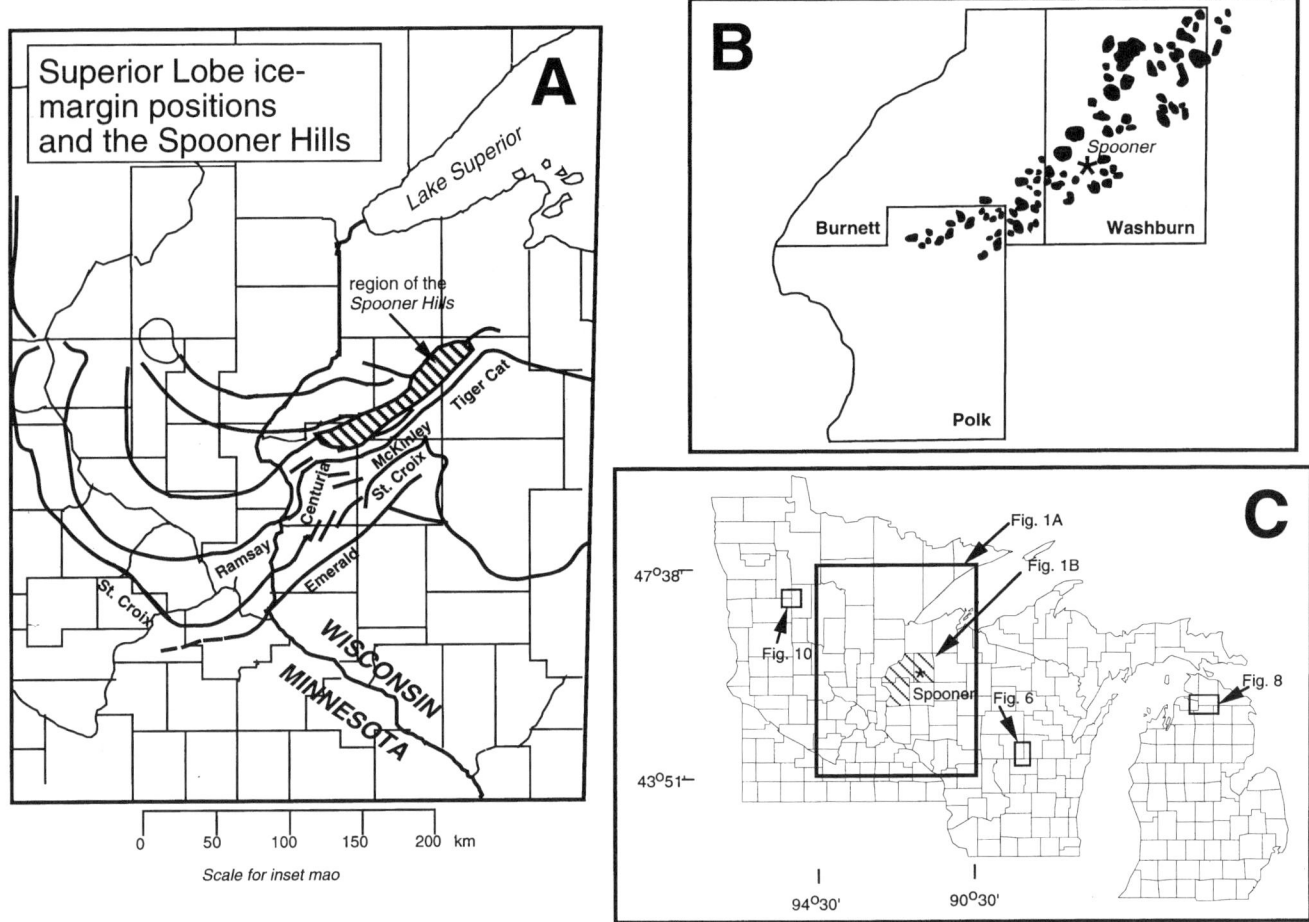

Figure 1. A, Map showing Superior Lobe ice-margin positions and the general location of the Spooner Hills. B, Polk, Burnett, and Washburn Counties showing the location of Spooner and individual Spooner Hills. Note that some of the Spooner Hills are aligned somewhat parallel to ice flow, which is to the southeast. C, State and county maps show location of Figures 1A, 1B, 6, 8, and 10.

monly described in glaciated regions. I will present a hypothesis that states these hills were made by the erosive action of subglacial meltwater. The Wisconsin hills are referred to as the Spooner Hills because the town of Spooner, Wisconsin, lies about in the middle of their distribution (Fig. 1B).

Additionally, I will describe similar hills that occur in several places in the Midwest. The Spooner Hills may represent a landform type more common than previously recognized.

Regional glacial geology

Surface features in western Wisconsin were formed primarily during the Wisconsinan Glaciation. During the last part of the Wisconsinan Glaciation, the Superior Lobe advanced through the study area and melted back, leaving several clear ice-margin positions (Fig. 1A) defined by tunnel-channel mouths, outwash heads, discontinuous ridges, and boundaries of hummock tracts (Chamberlin, 1905; Mathieson, 1940; Clayton, 1984; Johnson, 1986, 1998; Johnson et al., 1995; Johnson and Mooers, 1998). The remainder of the region behind the late Wisconsinan ice-margin limit is characterized by broad hummock tracts with ice-walled-lake plains, outwash plains, rolling till plains, isolated basalt bedrock knobs, as well as the broad valleys of the preglacial St. Croix, Wood, Clam, and Yellow Rivers that have become filled with fluvial sand and glaciolacustrine deposits of glacial Lakes Lind and Grantsburg (Johnson, 1986, 1998; Johnson and Hemstad, 1998).

The age of glacial advances in the study area is not well known. Few radiocarbon dates exist regionally, and this has led to different interpretations of the regional chronology (Wright et al., 1973; Clayton and Moran, 1982; Meyer and Knaeble, 1996; Mooers, 1997; Mooers and Lehr, 1997). Based on these chronologies, the Spooner Hills may have been formed sometime between 20,000 and 14,000 B.P.

THE SPOONER HILLS

The Spooner Hills occur in a band 5–15 km wide, parallel to and about 10 km behind a prominent ice-margin position of the Superior Lobe. This ice-margin position in Washburn and Sawyer Counties was named the Tiger Cat ice margin by Clayton (1984; Fig. 1A). It links with the McKinley and Centuria ice margins in Polk and Barron Counties (Johnson, 1998) and continues into the Twin Cities area, where it is called the Ramsey ice-margin position (Patterson, 1992). This ice margin occurs 15–25 km behind the St. Croix ice-margin position and about 35 km behind the terminal late Wisconsinan Emerald ice-margin position (Fig. 1).

Numerous tunnel channels are associated with the St. Croix and Tiger Cat–McKinley-Centuria-Ramsey ice-margin positions (Fig. 2; Attig et al., 1989; Johnson, 1986, 1998; Patterson, 1994). The up-ice ends of many tunnel channels occur within the region of the Spooner Hills and some tunnel channels clearly extend through the band of hills (Fig. 2, 3, 4). Outwash fans with broad outwash plains occur at the down-ice ends of tunnel channels where they meet former ice-margin positions (Johnson, 1998).

The Spooner Hills are generally 1–15 km² in area with as much as 60 m of relief. Some of the hills are elongate parallel to ice flow, whereas the geometry of others seem unrelated to ice flow. The Spooner Hills lie in a well-defined band regionally (Fig. 1A); they have a rather abrupt up-ice and down-ice extension, and they extend the entire length of the Tiger Cat–McKinley ice margin. Elevation rises from the proximal edge of this band of hills to the ice margin. Interhill valleys in the northwestern part of the hills show elevations of about 320 m, while the distal valleys in the band are at 450 m, an elevation gain of 130 m. Finally, the landscape in the region between the Spooner Hills and the former ice-margin position is characterized by extensive hummock tracts (Johnson et al., 1995). These hummocky areas obscure the relationship between the hills and the tunnel channels, outwash fans, and ice margins.

Drill holes, exposures, and well logs show the Spooner Hills to be composed of unlithified sediment, including diamicton as well as sorted sediment. Most surface exposures and shallow drill holes contain Superior Lobe till of the most recent advance. Well logs from northeastern Polk County located on Spooner Hills contain a variety of sequences of unlithified sediment. A drill hole in northeastern Polk County on the summit of a Spooner Hill contained 30 m of slightly pebbly sand (NE¼NE¼ sec. 14, T. 37 N., R. 15 W.). A gravel pit in southeastern Burnett County occurring on the flank of a Spooner Hill exposes several meters of meltwater stream sediment overlying an older Superior Lobe till (SE¼ sec. 4, T. 37 N., R. 14 W.). Because the outwash at this site overlies till and is in turn overlain by the most recent till, the site represents two ice advances. Although there are basalt hills in northern and western Polk County (Johnson, 1998), the basalt hills in western Wisconsin are distinct topographically from the Spooner Hills and outcrops of basalt are common among them. In summary, the Spooner Hills are made predominantly of unlithified sediment, and their stratigraphy is not uniform, including units that likely predate the till of the most recent advance.

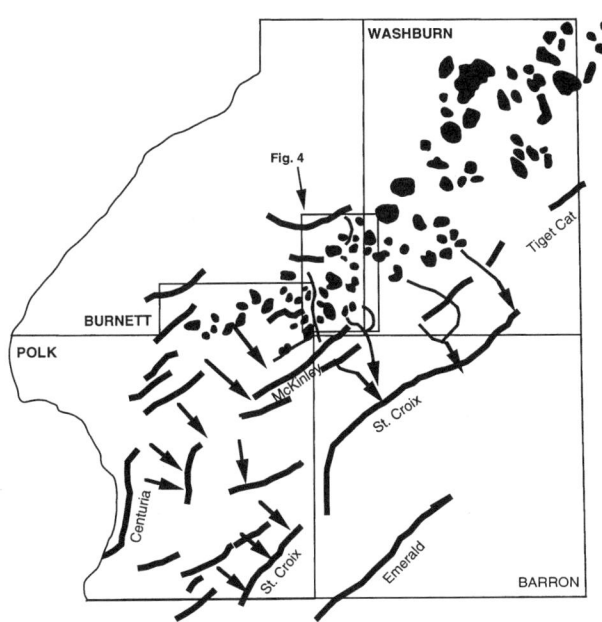

Figure 2. A part of Barron, Burnett, Polk, and Washburn Counties in western Wisconsin showing distribution of Spooner Hills (black forms), prominent tunnel channels (curved lines with arrowheads), prominent ice-margin positions (solid lines), some of which are named, and a location map for Figure 4.

Because the hills form a discrete band concentric to and behind the Superior Lobe ice-margin positions, and because the hills contain older till, I interpret the Spooner Hills to have been eroded subglacially. My hypothesis for the formation of the Spooner Hills is that the hills and the valleys surrounding them were excavated by subglacial meltwater. This erosion was accomplished by water flowing in subglacial conduits as opposed to subglacial sheet floods (Pair, 1997). The tunnel channels down-ice from the Spooner Hills represent the conduits for meltwater and eroded sediment. The water and sediment was released at the glacier margin building large outwash fans and plains (Fig. 5). A subglacial-meltwater hypothesis for the formation of the hills is preferred, rather than subglacial erosion by ice, for three reasons: (1) tunnel channels and eskers extend into and through the Spooner Hills regions; (2) the tunnel channels end at ice-margin positions that are characterized by large outwash fans and plains; and (3) the planforms of the interhill valleys are sinuous and anastomosing, similar to subglacial-meltwater features like eskers, some tunnel channels, and some subglacial bedrock channels (Walder and Hallet, 1979) and different from straight subglacial, ice-formed features like striations, flutes, and drumlins. It is likely that the interhill conduits operated at different times. The tunnel channels outlined in Figure 4 contain more collapsed sediment than other interhill valleys, which I interpret to have been cut by slightly earlier subglacial meltwater events during the most recent glaciation.

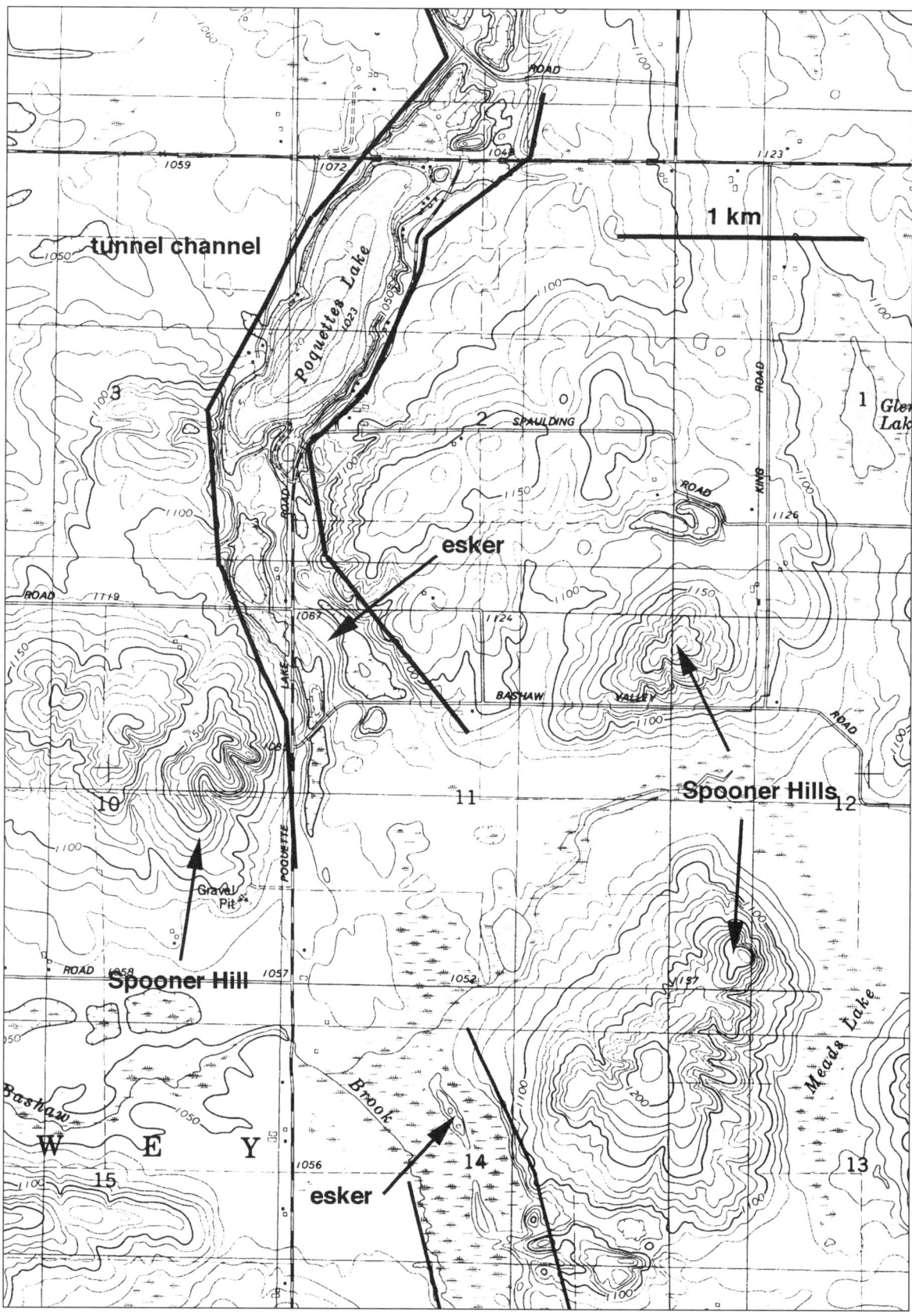

Figure 3. A portion of the Poquettes Lake Quadrangle, southeastern Burnett County, Wisconsin (U.S. Geological Survey, 7.5-minute series, topographic, 1983, contour interval 10 feet). Parts of several Spooner Hills occur in this figure. Poquettes Lake lies within a tunnel channel and separates two Spooner Hills. An esker that formed in this tunnel channels noted. The Poquettes Lake tunnel channel is identified as well in Figure 4.

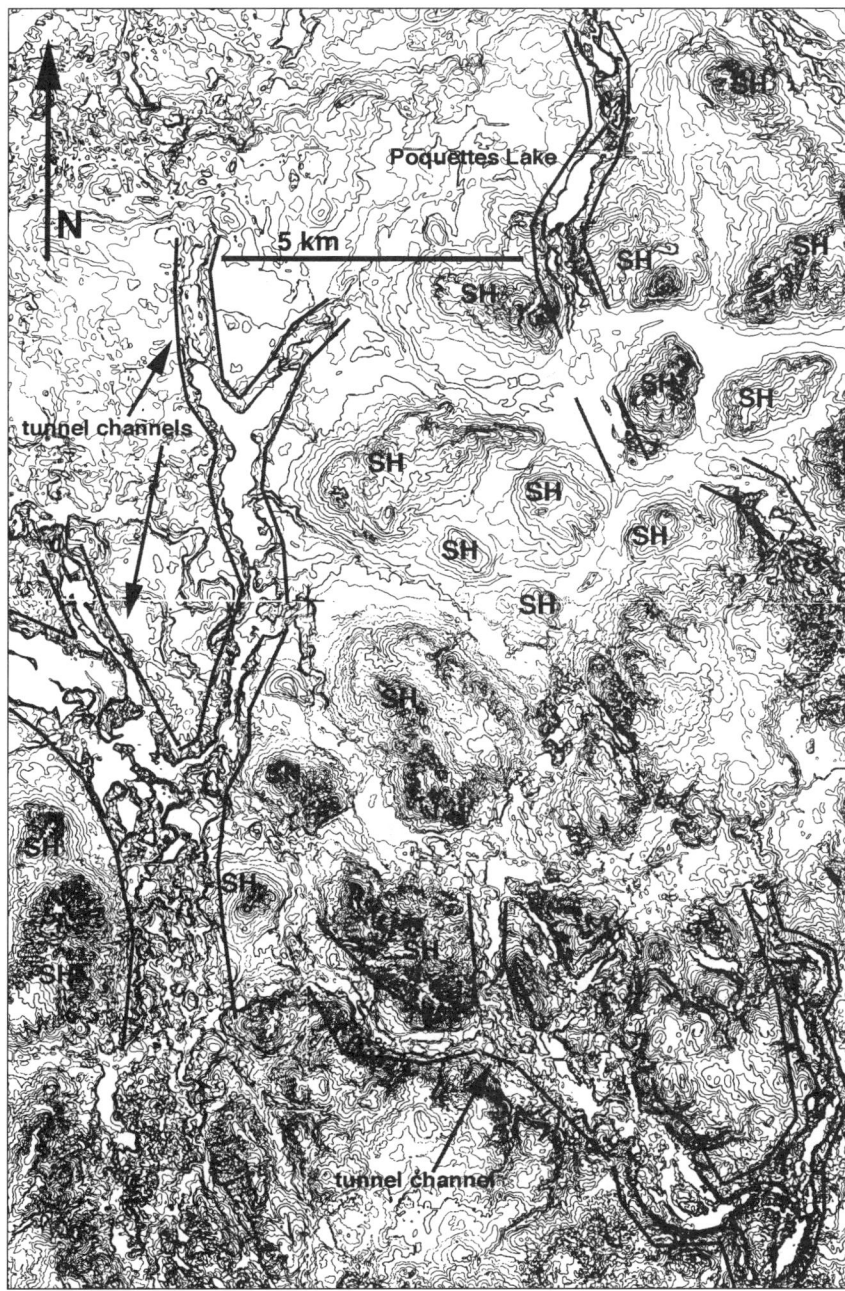

Figure 4. Mosaic of topographic-line separates from portions of the Poquettes Lake, Hertel, Indian Creek, and Timberland Quadrangles, Wisconsin (U.S. Geological Survey, 7.5-minute series, topographic). Poquettes Lake lies within the tunnel channel shown in Figure 3. Other prominent tunnel channels occur that traverse the Spooner Hill region. The tunnel channels outlined here contain more collapsed sediment and likely represent the last tunnel channels to form at the Superior Lobe ice margin. The other interhill valleys were also cut by tunnel channels but at earlier stages of the most recent glaciation, perhaps during earlier surge events. The position of this figure relative to the Spooner Hills is shown in Figure 2.

HILLS SIMILAR TO THE SPOONER HILLS IN THE MIDWEST

Landforms that resemble the Spooner Hills in morphology and that are interpreted to have formed by the erosion of subglacial meltwater have been reported from eastern Wisconsin by Clayton (1986), the Puget Lowland by Booth and Hallet (1993), and Antarctica (Denton, 1983). In this section, I would like to examine the landforms described by Clayton (1986) in eastern Wisconsin as well as hills in northern Michigan and central Minnesota that I interpret as having a origin like that proposed for the Spooner Hills.

Green Bay Lobe, eastern Wisconsin

Clayton (1986) described till ridges in Portage County, Wisconsin, that have similar characteristics to the Spooner Hills. He suggested that they "... may be partly the result of glacial erosion and deposition, but they are more likely till remnants resulting from erosion by subglacial rivers in the intervening areas"

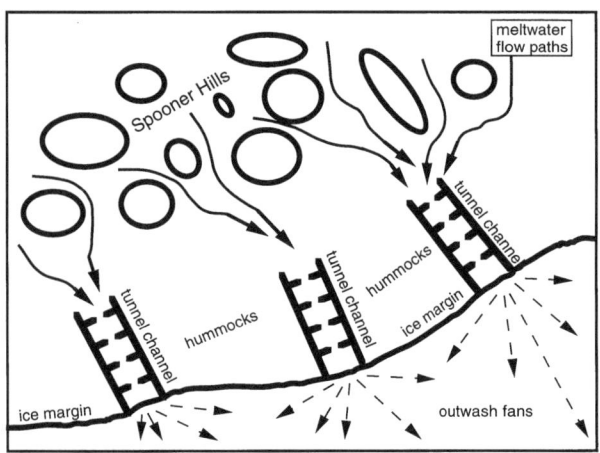

Figure 5. Cartoon diagram showing the relationship between the location of Spooner Hills, interhill meltwater flow paths, tunnel channels, outwash fans, and hummocks.

(Clayton, 1986, p. 8). These till ridges are similar in morphology to the Spooner Hills and occur 10–20 km behind the late Wisconsinan Hancock and Arnold ice-margin positions of the Green Bay Lobe. Well-developed end moraines with tunnel channels and outwash fans mark these ice-margin positions (Fig. 6). A 10-km-wide zone free from drumlins and till ridges occurs immediately behind the prominent ice-margin positions. The till ridges appear as elongate ovals, and many have drumlins on them (Fig. 7), although there are also numerous drumlins between the hills suggesting that the drumlins were formed after the hills. There is not a pronounced rise in elevation from east to west towards the former ice margin, but the valleys between the hills are 50 m lower than the level of the outwash plain. Cross sections drawn through these hills show that they are not made of bedrock (Clayton, 1986, Plate 2). I suggest that the till ridges described by Clayton (1986) are similar to the Spooner Hills and have the same origin.

Lake Michigan Lobe, northern Michigan

Landforms similar in size and scale to the Spooner Hills occur in the lower peninsula of Michigan around the town of Petosky. A prominent ice-margin position (Fig. 8) is fronted by extensive outwash plains. There are no clear tunnel channels, except for the fact that the interhill valleys are narrow and look like a maze of channels. There are a few drumlins. This area in Michigan shows the hills closer to a former ice margin than in the previous two examples. The relief from the hilltops to the edge of the outwash plain is 70 m; the relief from the valley bottoms is 200 m (Fig. 9). If the eroding agent was subglacial meltwater, the water flowed uphill towards the ice margin. The hills south of Petosky described here are similar to the Spooner Hills, have similar associated landforms, and likely have the same origin as the Spooner Hills.

Wadena Lobe, northern Minnesota

The Itasca ice-margin position of northern Minnesota (Fig. 10) is marked by a 15-km band of high-relief hummocks, behind which lie a band of hills etched by tunnel channels (Wright, 1972) and which are morphological similar to the Spooner Hills. These hills are not as widely spaced as at the other sites, but are separated by discrete tunnel channels. The hills occur in a narrow zone and are bounded sharply on the up-ice and down-ice sides. The exact relationship between the hills and the outwash to the south is obscured by the wide band of high-relief hummocky topography, however, these may be genetically related, as discussed below.

DISCUSSION AND CONCLUSIONS

The Spooner Hills are a band of hills of distinct morphology, parallel to and 10 km behind a known, prominent ice-margin position. The Spooner Hills are associated with tunnel channels

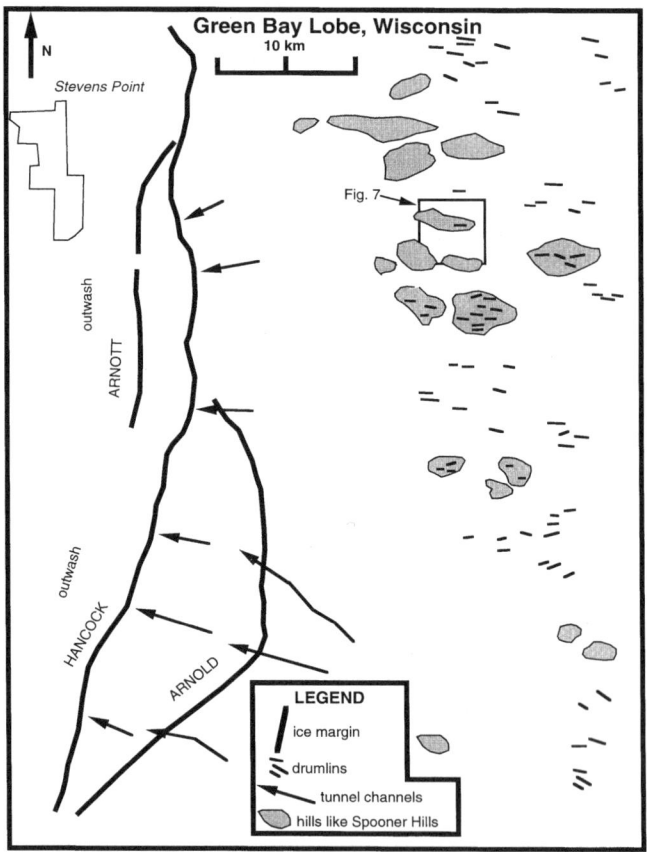

Figure 6. Sketch map of part of eastern Wisconsin (see Fig. 1 for location) showing hills similar in shape to the Spooner Hills (elongate ovals), drumlins (short lines), tunnel channels (arrows), and ice-margin positions (long lines), and location of Figure 7. Extensive outwash fans and plains occur to the west of the Hancock ice-margin position. The difference in elevation between the interhill valleys and the outwash plain is about 50 m.

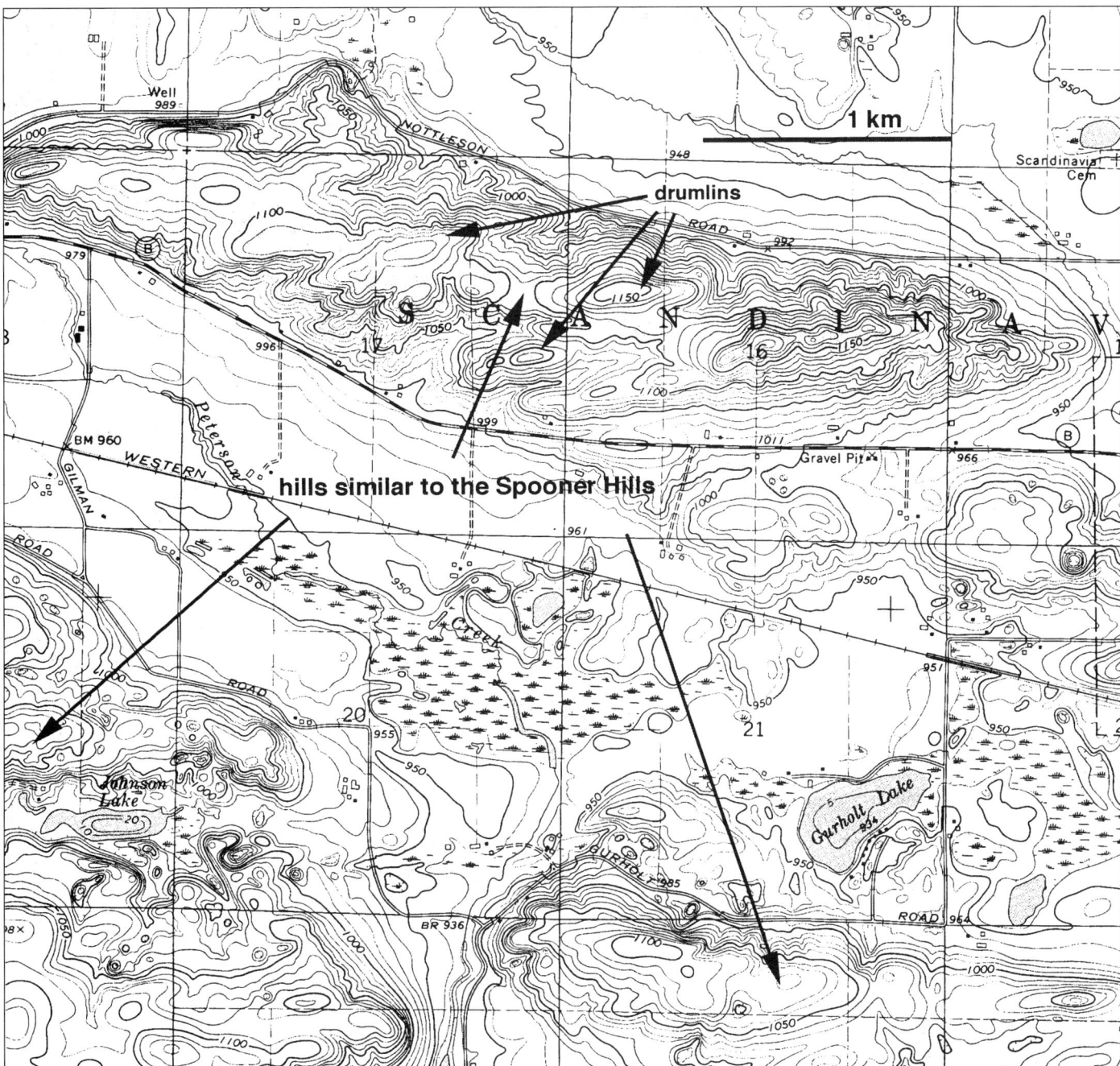

Figure 7. A portion of the Scandinavia Quadrangle, Wisconsin (U.S. Geological Survey, 7.5-minute series, topographic, 1969, contour interval 10 feet). The hill in the upper part of the map is about 4 km long and has about 70 m of relief and is interpreted as a hill similar to a Spooner Hill. Ice flow was from right to left. Location shown in Figure 6.

and extensive outwash fans, and the elevation of Spooner Hill landscape rises towards the ice margin. The hills are composed of unlithified sediment, some of which predates the most recent ice advance. These observations suggest that the Spooner Hills are erosional. The tunnel channels that dissect the hills and the fans at their mouths indicate erosion by subglacial meltwater. The water and sediment was released at the glacier margin and the sediment was deposited proglacially in large outwash fans and outwash plains. The morphology of the interhill valleys suggests that meltwater erosion was confined to discrete subglacial conduits rather than distributed as broad sheetfloods.

In northwestern Wisconsin and in the three other areas discussed in this paper, the hills occur in a band behind an ice margin. Ice-margin elevations are significantly higher than the valleys in between the hills, indicating that, if they were indeed cut by subglacial meltwater, the meltwater would have flowed up

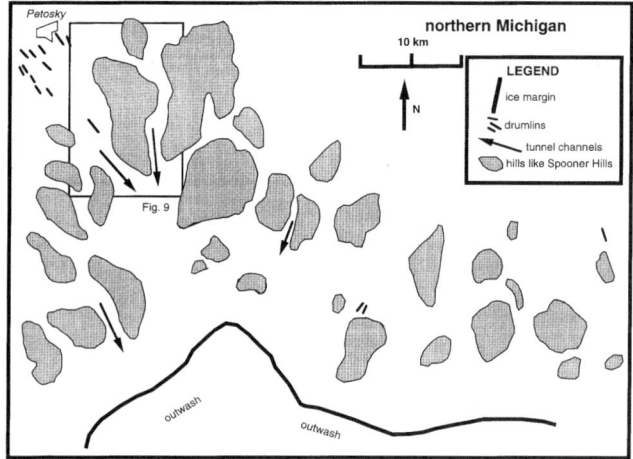

Figure 8. Sketch map of part of northern Michigan (see Fig. 1 for location) showing hills similar in shape to the Spooner Hills (elongate ovals), drumlins (short lines), ice-margin positions (long lines), and location map for Figure 9. Extensive outwash fans and plains occur to the south of the ice-margin position.

towards the ice margin. It may be that such a glaciohydraulic environment is necessary for the erosion of these hills.

In two of the examples (western Wisconsin and northern Minnesota), the hills are associated with broad tracts of hummocks that occur between the hills and the ice margin. Additionally, the end moraines of eastern Wisconsin are hummocky in places. The hummocks of western Wisconsin contain large amounts of till interpreted to be melt-out till or flow till derived from englacial debris (Johnson et al., 1995; Ham and Attig, 1997). Lawson et al. (1997) report that freezing-on of subglacial meltwater is a significant source of accretion of sediment to the bottom of glaciers, and they suggest that hummock tracts are composed of sediment acquired originally by the glacier by the freezing-on of subglacial meltwater near the margin. I suggest that the hummock tracts and the Spooner Hills may be genetically related, and may help answer the question of where the eroded interhill sediment was eventually deposited. I suggest that much of the sediment eroded from the Spooner Hill tracts ended up in outwash plains, but that a large amount may have been frozen on to the base of the ice sheet and is now present in the sediment of the hummock tracts. Additionally, the conditions necessary for extensive freezing-on of subglacial meltwater require a steep-bed slope that is 1.2–1.7 times greater than the ice-surface slope (Alley et al., 1997). The Spooner Hills and the three other areas discussed all exhibit bed slopes that slope upwards toward the ice margin.

Finally, I suggest that the Spooner Hills and hills similar to them from the three other sites show a variation in form that can be interpreted as evolutionary stages of development. A broad band of till cut by discrete, simple tunnel channels is illustrated by the Itasca example (Fig. 10). More extensive erosion and a greater number of channels, which are somewhat anastomosing,

is shown in the northern Michigan example (Fig. 8). The Spooner Hills themselves appear further eroded (Fig. 1). It is likely that not all of the interhill valleys were cut at the same time, but that different conduit systems established themselves at different times during glaciation. The Green Bay Lobe hills (Fig. 7) form an extremely dissected end member. The Green Bay Lobe hills were additionally molded by the ice at a slightly later time to form drumlins between and upon the hills.

ACKNOWLEDGMENTS

This project was supported by the Wisconsin Geological and Natural History Survey as part of a mapping program in Polk, Burnett, Barron, and St. Croix Counties, Wisconsin. I wish to thank John Attig, Tim Fisher, and Don Pair for their helpful and thoughtful reviews, and the Wisconsin Geological and Natural History Survey for help in the construction of Figure 4.

REFERENCES CITED

Alley, R. B., Evenson, E. B., Lawson, D. E., Strasser, J. C., and Larson, J. G., 1997, Conditions for Matanuska-type glaciohydraulic supercooling beneath other glaciers: Geological Society of America Abstracts with Programs, v. 29, p. 1.

Attig, J. W., Clayton, L., and Mickelson, D. M., 1989, Late Wisconsin landform distribution and glacier-bed conditions in Wisconsin: Sedimentary Geology, v. 62, p. 399–405.

Booth, D. B., 1994, Glaciofluvial infilling and scour of the Puget Lowland, Washington, during ice-sheet glaciation: Geology, v. 22, p. 695–698.

Booth, D. B., and Hallet, B., 1993, Channel networks carved by subglacial water: Observations and reconstruction in the eastern Puget Lowland of Washington: Geological Society of America Bulletin, v. 105, p. 671–683.

Chamberlin, R. T., 1905, The glacial features of the St. Croix Dalles region: Journal of Geology, v. 13, p. 238–256.

Clayton, L., 1984, Pleistocene geology of the Superior region, Wisconsin: Wisconsin Geological and Natural History Survey Information Circular 46, 40 p.

Clayton, L., 1986, Pleistocene geology of Portage County, Wisconsin: Wisconsin Geological and Natural History Survey Information Circular 56, 19 p.

Clayton, L., and Moran, S. R., 1982, Chronology of late Wisconsinan glaciation in middle North America: Quaternary Science Reviews, v. 1, no. 1, p. 55–82.

Cofaigh, C. Ó., 1996, Tunnel valley genesis: Progress in Physical Geography, v. 20, p. 1–19.

Denton, G. H., 1983, Giant labyrinth of subglacial stream channels cut beneath overriding ice, Transantarctic Mountains, Antarctica: Geology, v. 11, no. 12, cover photograph.

Fisher, T. G., and Shaw, J., 1992, A depositional model for Rogen moraine, with examples from the Avalon Peninsula, Newfoundland: Canadian Journal of Earth Science, v. 29, p. 669–686.

Fyfe, G. J., 1990, The effect of water depth on ice-proximal glaciolacustrine sedimentation: Salpausselkä I, southern Finland: Boreas, v. 19, p. 147–164.

Ham, N. R., and Attig, J. W., 1997, Pleistocene geology of Lincoln County, Wisconsin: Wisconsin Geological and Natural History Survey, Bulletin 93, 31 p.

Johnson, M. D., 1986, Pleistocene geology of Barron County, Wisconsin: Wisconsin Geological and Natural History Survey, Information Circular 55, 42 p.

Johnson, M. D., 1998, Pleistocene geology of Polk County, Wisconsin: Wisconsin Geological and Natural History Survey, Bulletin 92 (in press).

Johnson, M. D., and Hemstad, C., 1998, Glacial Lake Grantsburg: a short-lived lake recording the advance and retreat of the Grantsburg Sublobe, in

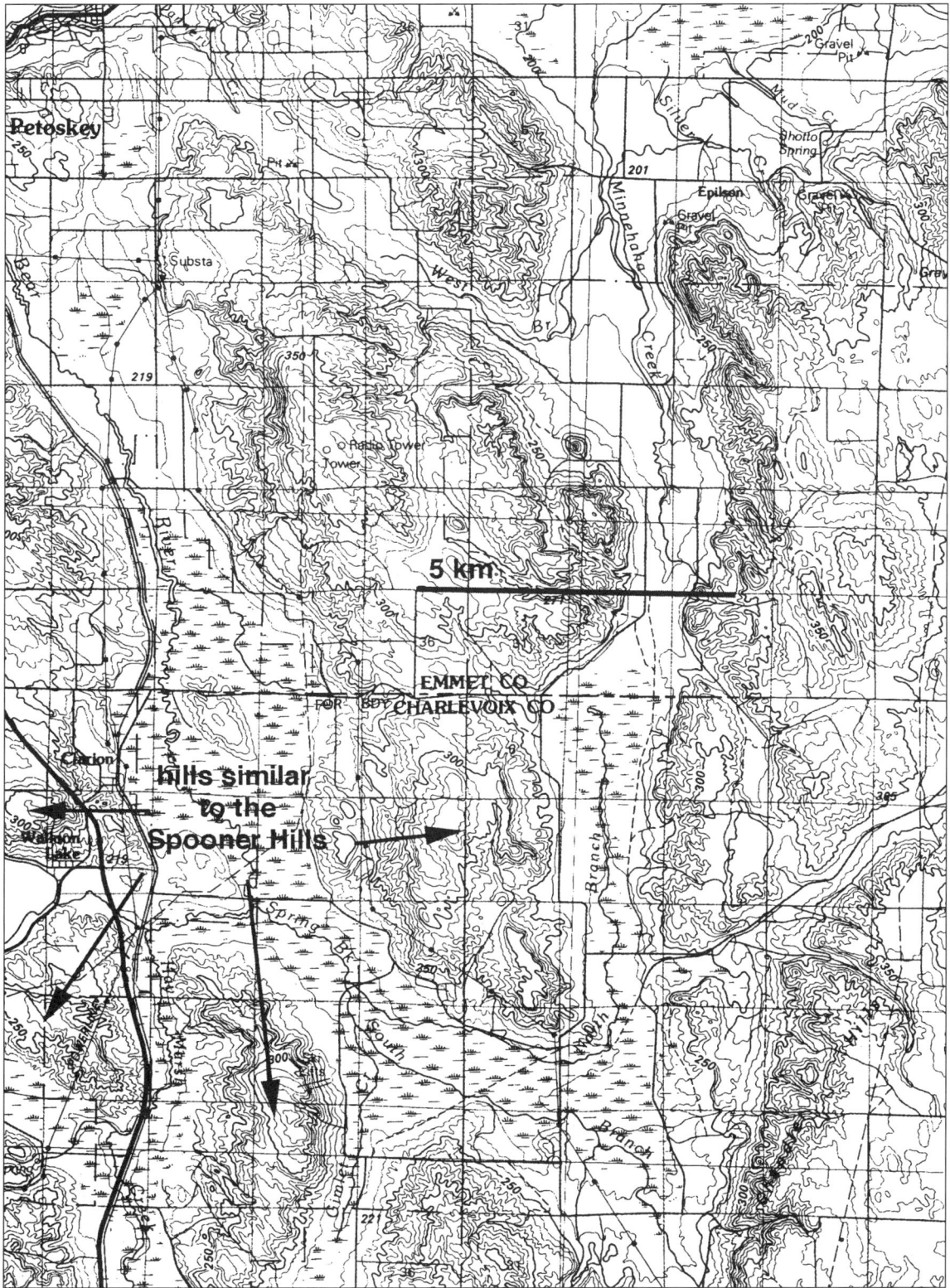

Figure 9. A portion of the Petosky Quadrangle, Wisconsin (U.S. Geological Survey, 30 × 60 minute series, topographic, 1982, contour interval 10 m). Ice flow to the southeast. These hills are more than 100 m in relief and occur about 10 km north of a prominent ice-margin position. Location can be found in Figure 8.

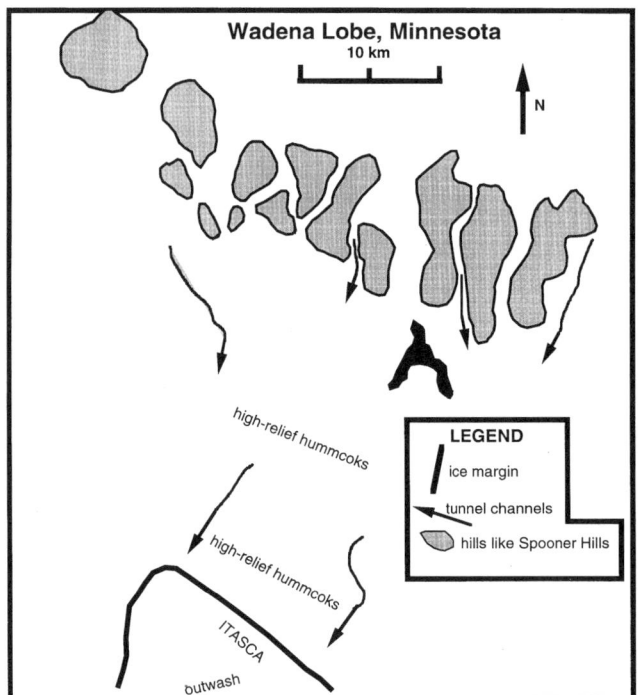

Figure 10. Sketch map of part of northern Minnesota (see Fig. 1 for location) showing hills similar in shape to the Spooner Hills (elongate ovals), tunnel channels (arrows), Lake Itasca, and the Itasca ice-margin position. Extensive outwash plains occur to the south of the ice-margin position. Note the extensive zone of high-relief hummocks between the ice-margin position and the hills.

Patterson, C. J., and Wright, H. E., Jr., eds., Contributions to Quaternary studies in Minnesota: Minnesota Geological Survey Report of Investigation 49, p. 49–60.

Johnson, M. D., and Mooers, H. D., 1998, Ice-margin positions of the Superior Lobe during late Wisconsinan deglaciation, in Patterson, C. J., and Wright, H. E., Jr., eds., Contributions to Quaternary studies in Minnesota: Minnesota Geological Survey Report of Investigation 49, p. 7–14.

Johnson, M. D., Mickelson, D. M., Clayton, L., and Attig, J. W., 1995, Composition and genesis of glacial hummocks, western Wisconsin: Boreas, v. 24, p. 97–116.

Lawson, D. E., Evenson, E. B., Larson, J. G., Alley, R. B., and Strasser, J. C., 1997, Net accretion model for sediment entraiment, transport, and deposition, and the formation of hummocky ground and end moraines by glaciers and ice sheets: Geological Society of America Abstracts with Programs, v. 29, p. 30.

Mathieson, J. T., 1940, The Pleistocene of part of northwestern Wisconsin: Wisconsin Academy of Sciences, Arts, and Letters, v. 32, p. 251–272.

Meyer, G. N., and Knaeble, A. R., 1996, Part C, Text Supplement, in Meyer, G. N., and Swanson, L., eds., Geologic atlas, Stearns County, Minnesota: Minnesota Geological Survey County Atlas Series C-10, 63 p.

Mooers, H. D., 1989, On the formation of tunnel valleys of the Superior Lobe: Quaternary Research, v. 32, p. 24–35.

Mooers, H. D., 1997, Numerical reconstruction of a soft-bedded Laurentide Ice Sheet during the last glacial maximum: Comment and Reply: Comment: Geology, v. 25, p. 379–380.

Mooers, H. D., and Lehr, J. D., 1997, Terrestrial record of Laurentide Ice Sheet reorganization during Heinrich events: Geology, v. 25, no. 11, p. 987–990.

Munro, M., and Shaw, J., 1997, Erosional origin of hummocky terrain in south-central Alberta, Canada: Geology, v. 25, p. 1027–1030.

Pair, D. L., 1997, Thin film, channelized drainage, or sheetfloods beneath a portion of the Laurentide Ice Sheet: an examination of glacial erosion forms, northern New York State, USA: Sedimentary Geology, v. 111, p. 199–215.

Patterson, C. J., 1992, Surficial geology, Plate 3, in Meyer, G. N., and Swanson, L., eds., Geologic atlas, Ramsey County, Minnesota: Minnesota Geological Survey County Atlas Series, Atlas C-7, scale 1:48,000.

Patterson, C. J., 1994, Tunnel-valley fans of the St. Croix moraine, east-central Minnesota, U.S.A., in Warren, W. P., and Croot, D. G., eds., Formation and deformation of glacial deposits: Rotterdam, A. A. Balkema, p. 69–87.

Price, R. J., 1973, Glacial and fluvioglacial landforms: London, Longman Group, Limited, 242 p.

Rains, B., Shaw, J., Skoye, R., Sjogren, D., and Kvill, D., 1993, Late Wisconsin subglacial megaflood paths in Alberta: Geology, v. 21, p. 323–326.

Shaw, J., and Kvill, D., 1984, A glaciofluvial origin for drumlins of the Livingstone Lake area, Saskatchewan: Canadian Journal of Earth Science, v. 21, p. 1442–1459.

Shaw, J., and Gilbert, R., 1990, Evidence for large-scale subglacial meltwater flood events in southern Ontario and northern New York State: Geology, 18, p. 1169–1172.

Shaw, J., Rains, B., Eyton, R., and Weissling, L., 1996, Laurentide subglacial outburst floods: landform evidence from digital elevation models: Canadian Journal of Earth Science, v. 33, p. 1154–1168.

Walder, J. S., and Hallet, B., 1979, Geometry of former subglacial water channels and cavities: Journal of Glaciology, v. 23, p. 335–346.

Wright, H. E., Jr., 1972, Physiography of Minnesota, in Sims, P. K., and Morey, G. B., eds., Geology of Minnesota: a centennial volume: St. Paul, Minnesota, Minnesota Geological Survey, p. 561–578.

Wright, H. E., Jr., 1973, Tunnel valleys, glacial surges and subglacial hydrology of the Superior Lobe, Minnesota, in Black, R. F., Goldthwait, R. P., and Willman, G. P., eds., The Wisconsinan Stage: Geological Society of America Memoir 136, p. 251–276.

Wright, H. E., Jr., Matsch, C. L., and Cushing, E. J., 1973, Superior and Des Moines Lobes, in Black, R. F., Goldthwait, R. P., and Willman, G. P., eds., The Wisconsinan Stage: Geological Society of America Memoir 136, p. 153–185.

MANUSCRIPT ACCEPTED BY THE SOCIETY OCTOBER 8, 1998

Origin of the Driftless Area by subglacial drainage—a new hypothesis

Howard Hobbs
Minnesota Geological Survey, 2624 University Avenue, St. Paul, Minnesota 55114

ABSTRACT

There is no evidence that the Driftless Area of southwestern Wisconsin was ever covered by continental ice sheets, even though they extended far to the south in the Midcontinent region. The Driftless Area occupies the eastern part of the Paleozoic Plateau, a relatively high area of Paleozoic sedimentary bedrock that is generally permeable, and is deeply dissected by the Mississippi River and its tributaries. Bedrock of the Paleozoic Plateau acted as a giant sieve that was able to dewater the base of advancing ice. The exposed bedrock created pinning points that inhibited ice advance across the Paleozoic Plateau. Ice therefore flowed around the eastern and western margins of the Driftless Area, and continued its advance as far as southern Illinois and central Missouri. The southern extent of Midcontinent glaciation has been attributed in part to low subglacial shear stress associated with deformable substrates, and high pore-water pressure. This giant sieve hypothesis provides a speculative explanation for the origin of the Driftless Area.

THE DRIFTLESS AREA

The Driftless Area (Fig. 1) of southwestern Wisconsin and small adjacent parts of southeastern Minnesota and northwestern Illinois lacks a cover of glacial till and erratics (Chamberlin and Salisbury, 1885). There is no positive evidence for it ever having been covered by continental ice sheets. Bedrock within the Driftless Area is now covered by Quaternary sediment; the uplands are mantled by loess, and the lowlands locally contain glacial lake sediment. Some valleys within the Driftless Area contain outwash derived from ice sheets. The stream-dissected, bedrock-dominated uplands of the Driftless Area contrast sharply with most of the surrounding terrain, which has been smoothed and mantled by glacial sediment during repeated glaciations, and in which bedrock is only rarely exposed at the surface. The topography and Quaternary sediments of the Driftless Area would be quite unremarkable if juxtaposed with the dissected topography south of the glacial limit in the Midcontinent region. According to Black (1960, 1970a, b) the entire Driftless Area was covered by ice at least several times, most recently in an Early Wisconsinan event that he termed the Rockian. Black's (1960, 1970a, b) ideas have been subsequently refuted (Knox, 1982; Knox and Attig, 1988).

The topographic boundary of the Driftless Area does not coincide perfectly with the absence of glacial sediment. There is a pseudo-driftless area (Fig. 1, dark stipple) west of the Mississippi River in Iowa and Minnesota; it is geomorphically indistinguishable from the Driftless Area, but contains erratics and patches of old till on the uplands (Chamberlin and Salisbury, 1885; Trowbridge, 1966; Hobbs, 1984). The pseudo-driftless area was covered by pre-Illinoian ice sheets but not by more recent ones. The pseudo-driftless area also has been dissected by tributaries of the Mississippi River, especially in Minnesota and Iowa. The boundary between areas of patchy till and those areas from which till is absent can only be determined by detailed fieldwork. The hypothesis presented here to explain the origin of the Driftless Area also explains why the pseudo-driftless area has not been affected by more recent glaciations.

The region underlain by topographically high Paleozoic sedimentary rock in the vicinity of the Driftless Area is the Paleozoic Plateau (Hobbs, 1992). Paleozoic Plateau was first used in a more restricted sense to mean the pseudo-driftless area of Iowa (Prior, 1976). Almost all of the Driftless Area is included within the area of the Paleozoic Plateau, but a considerable amount of the Paleozoic Plateau also has been glaciated. The extent of the Paleozoic Plateau is approximated on Figure 1 by a light stipple,

Hobbs, H., 1999, Origin of the Driftless Area by subglacial drainage—a new hypothesis, *in* Mickelson, D. M., and Attig, J. W., eds., Glacial Processes Past and Present: Boulder, Colorado, Geological Society of America Special Paper 337.

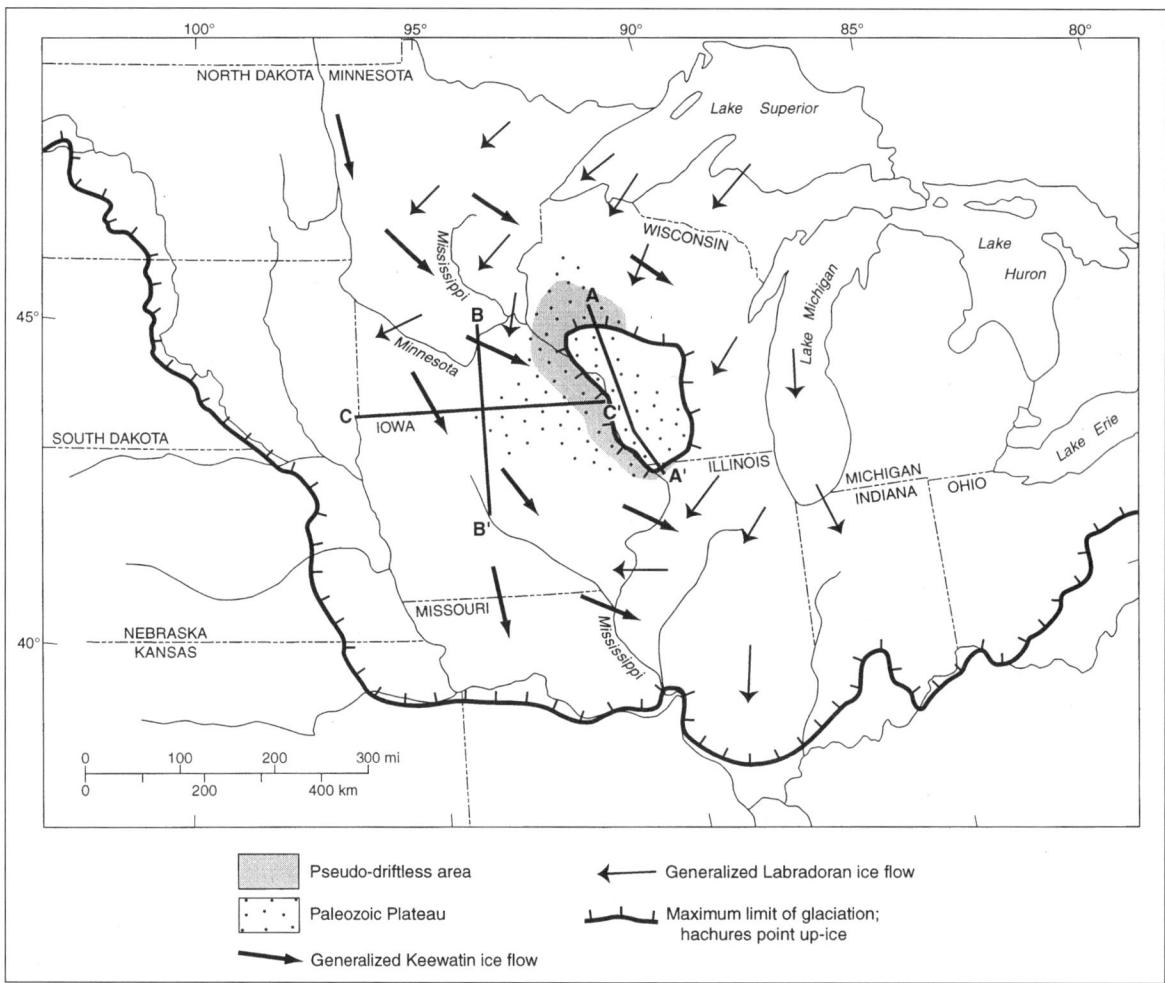

Figure 1. Map of the Midcontinent region showing maximum limit of Late Cenozoic glaciation and location of the Driftless Area and the Paleozoic Plateau. Ice margins were sketched from the Quaternary Geologic Atlas of the United States (U. S. Geological Survey Miscellaneous Investigations Series I-1420, sheets NJ-14, NJ-15, NJ-16, NJ-17, NK-14, NK-15, NK-16, NK-17, NL-14, NL-15) and modified from the author's experience in southeastern Minnesota. The extent of the Paleozoic Plateau was compiled from Morey et al. (1982), Mossler (1983), and the maps of bedrock topography in Iowa (U.S. Geological Survey Miscellaneous Geological Investigations Maps I-717, I-933 and I-1080). Generalized ice flowlines are from several glaciations. A–A' and B–B' are profiles shown in Figure 3. C–C' is profile shown in Figure 4.

which shows the area underlain by Paleozoic rocks that lie above 300 m in elevation. The type and extent of bedrock in Minnesota and Wisconsin is shown by Morey et al. (1982).

PREVIOUS MODELS FOR THE ORIGIN OF THE DRIFTLESS AREA

Models for the origin of the Driftless Area have in the past focused on the role of the Mississippi River as an ice-marginal stream that became entrenched. The timing of Mississippi River entrenchment, and the way in which the entrenchment may have influenced the spread of ice sheets and the development of the Driftless Area are key issues. Past models have also drawn on the role of Midcontinent uplift and preglacial topography as factors controlling the spread of ice sheets.

A remnant of preglacial topography

Early geologists interpreted the Driftless Area as a remnant of preglacial topography. They thought the bedrock topography of surrounding drift-covered areas was just like that of the Driftless Area, only buried (Chamberlin and Salisbury, 1885; McGee, 1891; Leverett, 1921). Subsurface data, mainly from water wells, showed that bedrock valleys are filled with more than a hundred meters of Quaternary sediment in the Minneapolis–St. Paul area (Sardeson, 1916). The bedrock valleys appear to be stream cut, at least in the area underlain by Paleozoic bedrock, because they form a dendritic stream pattern draining to the Mississippi (Olsen and Mossler, 1982).

Trowbridge (1921, 1954, 1966) argued that the deep dissection of the Mississippi and its tributaries postdated the earliest glaciation(s), but predated Kansan glaciation. Trowbridge suggested that in areas traversed by glacial ice during Kansan glaciation, glacial sediment mantles the uplands and fills deep bedrock valleys. In areas beyond the limits of Kansan glaciation (basically in the pseudo-driftless area), patchy pre-Kansan drift occupies the uplands but is absent from the valleys. He interpreted these relations to indicate general uplift and deep dissection of the region in early Pleistocene prior to Kansan glaciation. Willman and Frye (1969) also accepted the idea that the stream network had been greatly deepened since the earliest glaciation, on the basis of old outwash at a high level in the Driftless Area of Illinois, which was apparently derived from the region west of the Mississippi River. Willman and Frye (1969) concluded that the Mississippi River was not in its present position until after the Nebraskan glaciation.

Entrenchment of the Mississippi River

The origin and Pleistocene development of the Mississippi has been extensively debated, although it has not been recently summarized; some field trip guides provide useful minireviews (Knox, 1982; Lively, 1985; Hobbs, 1990). Two main interpretations exist: (1) the gross alignment and general drainage area of the preglacial Mississippi was similar to that of today, and (2) the Mississippi River adjacent to the Driftless Area originated as an ice-marginal stream associated with early glaciation. Both interpretations included variants which postulate that the Mississippi River north of LaCrosse originally flowed north.

Chamberlin and Salisbury (1885), Leverett (1921), and Trowbridge (1921) believed that the Mississippi existed prior to Plio-Pleistocene glaciation, although Trowbridge (1921) presented the idea that the preglacial course was the result of stream capture. He interpreted the original stream network as consequent on the bedrock structure; the ancestral Mississippi initially flowed northwest from a drainage divide at LaCrosse. After the drainage network reached grade at the Dodgeville Peneplain (the highest erosion surface that he recognized in the Driftless Area), the south-flowing drainage captured the north-flowing stream. Trowbridge later (1954) concluded that the Mississippi was an ice-marginal stream associated with Nebraskan glaciation, because its course approximates the southwestern edge of the Driftless Area, and because its present course is out of adjustment with bedrock structure. He suggested that its preglacial course was west of the current Mississippi River, and is now deeply buried by glacial sediment. Anderson (1988) agreed that where the Mississippi River borders the Driftless Area, it had its origins as an ice-marginal stream. But Anderson (1988) postulated that the Mississippi River had a preglacial origin along the axes of the Wisconsin and Kankakee Arches of the Driftless Area. He interpreted the ice-marginal Mississippi River to have become deeply entrenched as land was raised by a glacial forebulge during an early glacial advance, and that the Mississippi River remained in its new valley after ice retreated.

Alternatively, it is possible that the Paleozoic Plateau represents a preglacial continental divide; the northern part of which drained to the north (Hobbs, 1997). The divide between the north- and south-draining streams would have been in the area of La Crosse, Wisconsin. Drainage capture was accomplished by a south-flowing meltwater stream that demarcated the maximum extent of the early glaciation, or perhaps the lowest ice-free divide. The meltwater stream eroded headward such that it ultimately captured much of the formerly north-flowing drainage in Minnesota and Wisconsin. In this scenario, it is not necessary to postulate tectonic uplift to explain the great amount of fluvial downcutting accomplished by the Mississippi where it flows through the Paleozoic Plateau.

If the modern course of the Mississippi River across the Paleozoic Plateau resulted from early glacial diversion (regardless of where the preglacial drainage ran), the incision and development of the tributary system represents a youthful phase of dissection initiated in the late Pliocene. In many places the bedrock bluff tops overlooking the Mississippi Valley are nearly as high as the average bedrock elevations tens of kilometers from the river. This indicates that the upland topography has not yet fully adjusted to a major valley running through it. In this scenario the Driftless Area is not a remnant of unaltered preglacial topography, but a newly formed geomorphic region. The preglacial valleys would have been less deeply incised, but were taken over and deepened by the modern drainage net, so that no trace of their original rock walls remain, although their gross alignment may be the same.

FACTORS CONTROLLING THE ORIGIN OF THE DRIFTLESS AREA

The puzzle is not so much in what the Driftless Area *is*, but why it is so far north. Chamberlin and Salisbury (1885) were the first to map the Driftless Area and environs. They asked (p. 315), "What were the conditions that enabled the driftless area to escape the glaciation that repeatedly intruded itself upon the surrounding country?" Chamberlin and Salisbury (1885) argued that bedrock elevation was not sufficient to prevent glaciation, because higher bedrock to the north and west had been covered by glacial ice. They agreed with Winchell's (1877) observation that "this driftless tract" was in the lee of Precambrian rock highlands in northern Wisconsin and upper Michigan. This protective barrier or shield was enhanced by the troughs of Lake Superior and Lake Michigan, which diverted or channeled the ice past the Driftless Area to the south. The bedrock shielding and lowland channeling of glacial ice are valid mechanisms for the development of the Driftless Area; their explanation that the northern Great Plains were depressed before and during glaciation relative to the current elevation is no longer tenable.

The Driftless Area lies equidistant from the ice accumulation centers that controlled the southern margin of the Laurentide Ice Sheet. The Labradoran ice accumulation center lay to the northeast, and the Keewatin ice accumulation center lay to the northwest. The location would not have created the driftless area in and of itself, but it may have enhanced the effects of topography, shielding, and channeling.

The four major factors recognized to date as controlling the flow of the ice around the Driftless Area are (1) the elevation of bedrock surface of the Driftless Area is greater than in adjacent glaciated areas, (2) the Driftless Area is located at the maximum distance from both the Labradoran and Keewatin ice accumulation centers, (3) the bedrock highlands in northeastern Wisconsin and the Upper Peninsula of Michigan protect the Driftless Area on its northeast side, and (4) the basins of Lakes Michigan and Superior channeled Labradoran ice around the Driftless Area.

Elevated bedrock surface

Glaciated bedrock south and east of the Driftless Area is generally at a lower elevation than that of the Driftless Area; glaciated bedrock to the north, west, and southwest is as high on average, and higher in places. These relations imply that while elevation may have been a factor in limiting the spread of ice across the Driftless Area from the east, it cannot explain what held back the western ice. Most of the author's mapping experience has been in southeastern Minnesota, near the western margin of the Driftless Area (Hobbs, 1984, 1985, 1987, 1988, 1992, 1995); this paper focuses on why the Driftless Area was not covered by ice from the west.

Distance from ice-accumulation centers

Glacial ice moved from west to east, and from east to west, both south and north of the Driftless Area (Fig. 1), but did not cross the Driftless Area itself. Ice from the Labradoran (northeastern) ice accumulation center, from which the Superior and Rainy lobes were sourced, extended far west of the Driftless Area into Minnesota at many times (Meyer, 1997). Ice of the Lake Michigan lobe (also Labradoran) crossed what is now the Mississippi River into southeastern Iowa during the Illinoian (Richmond et al., 1991). On the other hand, pre-Illinoian gray tills of Keewatin provenance are known from northern Wisconsin (Attig and Muldoon, 1989). The pre-Illinoian Banner Formation of Illinois is of northwestern provenance as far east as the Illinois River (Richard Berg, Illinois State Geological Survey, oral communication, 1997). Thus, the existence of two ice-accumulation centers and their associated ice streams as well as their distance from the Driftless Area may have influenced the flow of ice around the Driftless Area, but these were not major factors controlling the development of the Driftless Area.

Bedrock shielding and lowland channeling

Before we attempt to evaluate the role of bedrock shielding and lowland channeling in the development of the Driftless Area, let us turn the question around. Rather than ask, "why does the driftless area extend so far north?," ask "why did ice extend so far south in the Midcontinent?" The maximum glacial boundary runs north-northwest to south-southeast in the Great Plains (Fig. 1), apparently reflects, and is possibly controlled by increased topographic elevation to the west and southwest. Elevation also seems to control the spread of ice onto the Appalachian Plateau. Extensive Midcontinent glaciation can be partially explained by topographic control of ice movement by the central lowlands. The Driftless Area, however, is topographically lower than the ice margins to the east and west at comparable latitudes. The roles of bedrock shielding and lowland channeling of ice in controlling the spread of ice across the Midcontinent region and the flow of ice around the Driftless Area are best explained in the context of mechanical models for ice flow that include subglacial pore-water pressure, deformable substrates, and basal sliding.

Mechanical models for ice flow

Fisher et al. (1985) reconstructed the Laurentide Ice Sheet at 18 ka B.P., using a mathematical model similar to the one developed by Reeh (1982) for the Greenland ice cap. The model included the ice margin, the present-day topography, and an assumed yield stress; no assumptions were made about ice divides or flow lines. The model was successfully tested by comparing the predicted ice margins with geological evidence. To generate a reasonable reconstruction, and to agree with Peltier's (1981) reconstruction of ice thickness based on rebound, Fisher et al. (1985) used a low value for yield stress in the Prairies and Great Lakes region (the region outside the Canadian Shield). They suggested that the low yield stress is due to deforming beds under ice that overrides tills derived from sedimentary rocks. The model indicates the dominance of thin ice with low slopes in the region outside the shield.

Clayton et al. (1985) attribute low yield stress in the Prairie and Great Lakes regions to high subglacial pore-water pressure that supported the overlying ice together with enhanced subglacial sliding. Much of this region is covered by thick, low-permeability tills, and bedrock under a large part of the Prairie region is shale. Clayton et al. (1985) noted that ice margins were extremely lobate and typically unstable in the region of fine-grained till, where they alternated between surging and stagnation. By contrast, in the sandy-till area (the Canadian Shield), ice margins were smoother and less influenced by bed topography. Ice in the Hudson Bay area was intermediate between smooth and lobate. Clayton et al. (1985) concluded that the main factor controlling subglacial friction was the ease with which water escapes from under the glacier.

Clark (1992) reconstructed several Late Wisconsinan ice lobes in the Midcontinent, and found them to be relatively thin and gently sloping, especially those that extended farthest south. Clark (1992) interpreted Midcontinent ice lobes as analogous to the distal ends of West Antarctic ice streams, whose low-gradient margins have been attributed to basal sliding or deforming substrate. Patterson (1997) postulated that the Des Moines lobe is the outlet glacier for a surging and stagnating ice stream. The Des Moines lobe was able to advance as long as adequate water pressure was maintained, but stagnated when subglacial water drained away through tunnel valleys. In summary, Late Wisconsinan ice lobes in the Midcontinent were probably thin and extensive because of low yield stress at their base. Pre-Wisconsinan ice lobes are more difficult to reconstruct. Aber (this volume, chapter 11) suggests that at least some pre-Wisconsinan Midcontinent ice lobes had low marginal gradients, low subglacial yield stresses, and were strongly controlled by topography.

The presence of a deformable substrate and subglacial water played an important part in keeping yield stresses low. Water could not easily escape into the substrate beneath the Midcontinent ice lobes. Water could escape from the edge of the ice through englacial and subglacial channels, but only very slowly through the substrate, which was mainly loam-textured till, derived for the most part from the underlying and up-ice sedimentary rocks (Richmond and Fullerton, 1983; Richmond et al., 1991). Drainage of water near the margins would have reduced the pore-water pressure and thus steepened the gradient at the ice margin, which would tend to enforce a lobate shape on the ice margin, in addition to any topographic constraints. In Minnesota, ice from the Keewatin center generally moved southward across the deformable substrate of older tills. The preglacial surface beneath the older tills was mainly Cretaceous shale and deeply weathered Precambrian rock. Local pinning points did exist, such as the Sioux Quartzite in southwestern Minnesota, but hard bedrock made up only a small proportion of the area. Ice advancing from the Labradoran center (Rainy and Superior lobes) was in contact with hard Precambrian bedrock over much of its extent in Minnesota during the Late Wisconsinan, and probably also during the Illinoian. Soller (1997) shows the total thickness of glacial sediment to be less than 15 m (50 ft) in much of the area covered by Labradoran ice; bedrock outcrops are locally extensive (Morey et al., 1982). Not surprisingly, the Labradoran-sourced ice lobes became pinned by bedrock, and extended only as far south as the northern border of the Driftless Area (Richmond and Fullerton, 1983).

The bedrock highlands northeast of the Driftless Area and the troughs of Lakes Superior and Michigan may have directed ice around the Driftless Area, but the impermeable soft substrate on either side of the Driftless Area allowed the ice to extend far southward. Was Late Wisconsinan ice prevented from extending across the Driftless Area due to the local reversal of the factors that brought it so far south elsewhere? Did the Paleozoic Plateau allow a reduction in subglacial pore-water pressure, resulting in higher shear stresses and the pinning of glacial ice?

THE GIANT SIEVE HYPOTHESIS

The Mississippi River and its tributaries are deeply incised into the Paleozoic Plateau. Ice that encroached onto the margins of the Driftless Area eroded weathered rock and older tills from the surface of the Paleozoic Plateau, exposing bedrock to the glacier sole. Although the exposed carbonate bedrock was relatively soft, and could be readily smoothed and striated by the encroaching ice, it was not deformable like till. At the margins of the Paleozoic Plateau, soft sediment at the base of the ice was trapped locally in the bedrock valleys (Fig. 2). Most of the bedrock forming the Paleozoic Plateau is highly permeable sandstone and karsted limestone. Water could infiltrate the bedrock wherever the base of the ice was in contact with bedrock; water could flow laterally through bedrock aquifers, and discharge into valleys outside the ice margin. Although these ice sheets were probably fringed by zones of permafrost, the depth of freezing would have needed to exceed at least 100 m to block flow in the bedrock aquifers. The net result was to dewater any wet-based ice lobe that advanced onto the Paleozoic Plateau, and to hold it back. The ice east and west of the plateau continued south until it reached a melting equilibrium. In short, the dissected highland acted as a giant sieve—it let through the water but held back the ice.

South of the Driftless Area the elevation of the bedrock surface decreases. The elevational difference between the upland surface and the valley bottoms also decreases; the sieve thus becomes less effective, and ice crosses from west to east, and east to west. There is no evidence for the Driftless Area ever being surrounded by ice at any one time. If it had been, the whole area would have flooded. McGee (1891) interpreted the loess mantling the Driftless Area as lake silt, and proposed that the area had been flooded by Lake Hennepin. The loess is now recognized to be an aeolian deposit; there is no evidence for widespread lacustrine deposition on the Paleozoic Plateau.

Profile reconstructions

In order to evaluate the effect of bedrock elevation on preventing the spread of glacial ice across the Driftless Area, the topographic profile of the Driftless Area (Fig. 3, profile a) was superimposed on the north-south profile inferred for the axis of the base of the Des Moines lobe (Fig. 3, profile b), which is derived using the current elevation of the Des Moines till as a proxy for the base of the ice. The highest points on the Driftless Area profile are only slightly higher than the most elevated parts of the base of the Des Moines lobe profile. In order to assess the effect of the topographic barrier presented to the Des Moines lobe ice by the Paleozoic Plateau, the thickness of the Des Moines lobe deposits must first be subtracted from the land surface. Des Moines lobe sediments are less than 50 m thick on average, and make little difference to the comparison. The most significant difference is in large-scale roughness of the Paleozoic Plateau compared to that of the Des Moines lobe.

A longitudinal ice profile of the Des Moines lobe based on Figure 5 of Clark (1992) is also shown (Fig. 3, profile c). The Des Moines lobe certainly appears capable of crossing the Driftless Area, if elevation alone was the sole consideration. The hypothetical elevation for ice sheets that extended farther south during earlier glaciations is also shown (Fig. 3, profile d). The elevation for the upper surface of these ice sheets was determined by translating Clark's profile for the Des Moines lobe (which

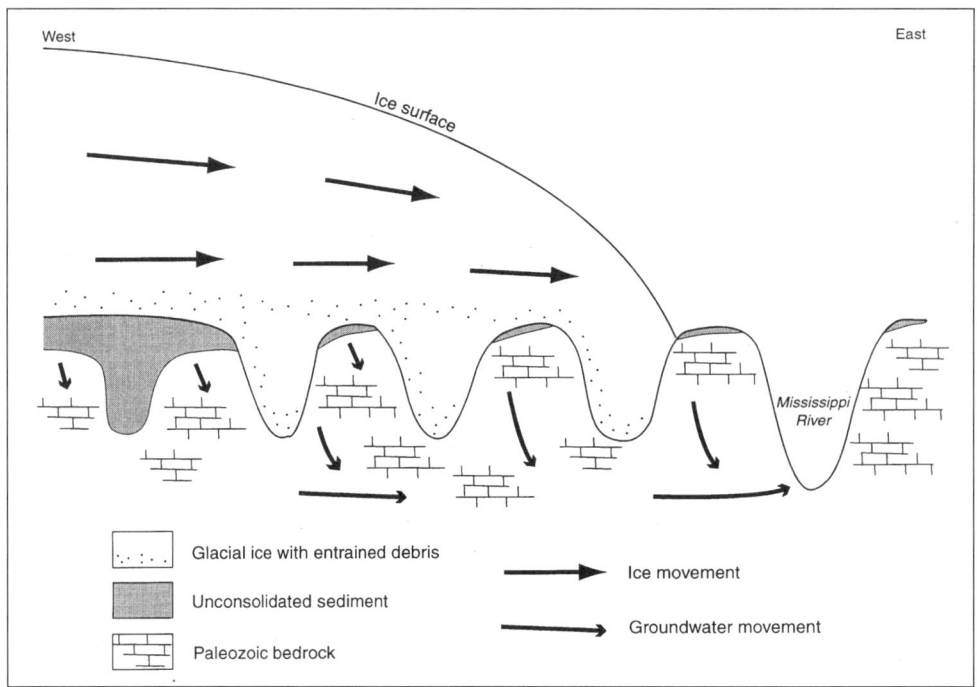

Figure 2. Schematic diagram showing proposed drainage of subglacial meltwater where ice sheet encroaches on the Paleozoic Plateau. Length of arrow on ground-water flowpaths indicates relative amount of subglacial meltwater draining into the bedrock.

terminated at Des Moines, Iowa) southward to the known limit of early glaciation in central Missouri. The upper surface of the ice was then extrapolated up-ice beyond the level indicated in Clark's (1992) curve. This ice sheet should have been even more capable of crossing the Driftless Area, but there is no evidence for it having done so. The elevation for the margin of the most extensive pre-Illinoian ice sheet (Fig. 3, profile e) is sketched on the profile at the north side of the Driftless Area for comparison. In order to attain the thickness required for the ice sheet to advance to its southern limit, the ice margin has to be steep adjacent to the Driftless Area. These hypothetical ice profiles are all relatively flat compared to profiles for most existing ice sheets.

The comparison between the gentle profile down the axis of a lobe with the elevation of an area off to the side of the main lines of advance is somewhat biased. An east-west cross section along the entire length of the Minnesota-Iowa boundary (Fig. 4) shows the elevation and distribution of bedrock and glacial sediments, and again shows Clark's (1992) profile for the Des Moines lobe (Fig. 4, profile a). The upper surface of the ice has a much higher elevation in the west than in the east; it also slopes rather sharply at its eastern margin. This asymmetry may result from the nonsymmetrical distribution of subglacial bedrock and till.

Note that the western portion of the cross section is composed of thick pre-Wisconsinan tills overlying clayey Cretaceous sedimentary rocks that are underlain by Precambrian crystalline rocks; water could not easily escape through this substrate. In contrast, the eastern part is underlain by thin pre-Wisconsinan tills that overlie permeable Paleozoic sedimentary rocks, which form an aquifer that allows escape of water to the Mississippi Valley.

Using the Des Moines lobe profile as a model, what would ice profiles for the more extensive earlier glaciations have looked like? Two hypothetical pre-Illinoian ice profiles (Fig. 4, profiles b and c) are sketched above the Des Moines lobe profile. They show (b) the maximum ice elevation (estimated using Clark's, 1992, Des Moines lobe profile, translated south), and (c) an intermediate position. The main constraint in developing the maximum profile was that the ice surface had to terminate near the Mississippi, and not enter the Driftless Area. The intermediate profile represents glaciations that exceeded the extent of the Des Moines lobe, but did not reach the glacial maximum.

These profiles show that the Midcontinent ice lobes had steep lateral slopes adjacent to the Driftless Area, particularly when compared to their longitudinal slopes. The features illustrated in Figures 2 and 4 are analogous to the end-member states presented by Boulton and Dobbie (1993, Fig. 8) in their rigorous treatment of subglacial ground-water flow and soft-bed glacier dynamics.

The western part of Figure 4 is comparable to Boulton and Dobbie's (1993) low-permeability and poor drainage scenario; (case D) the substrate is saturated and cannot transmit enough meltwater, so most of it escapes through subglacial channels. This results in deformation of the substrate, and maintenance of low slopes for the ice surface. To the east where the cover of glacial sediment is thinner and the Paleozoic rocks thicken, the situation comes to resemble case B (low permeability and good

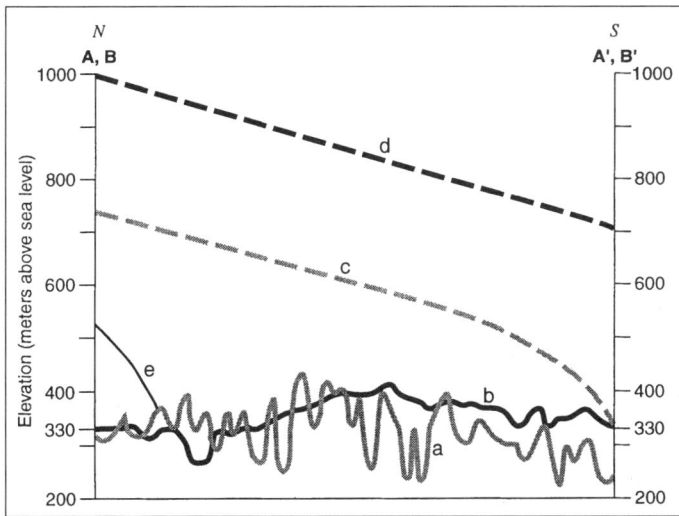

Figure 3. Profiles (vertical exaggeration ×3600) showing: a, topography of the Driftless area for section A–A′ on Figure 1; b, inferred elevation of the base of the Des Moines lobe as determined from the surface of the Des Moines till for B–B′ at approximately 94° longitude; c, Longitudinal ice profile for the upper surface of the Des Moines lobe ice, based on Clark (1992); d, Profile for the upper surface of the ice sheet(s) associated with the Midcontinent pre-Illinoian glacial maximum. See text for details of profile determination; e, Profile showing predicted slope of ice margin for the most extensive pre-Illinoian ice sheet at the northern margin of the Driftless Area.

drainage; Boulton and Dobbie, 1993). Most of the meltwater can be transmitted through the low-permeability till into the underlying aquifer, but the necessary pressure drop is great, and the upper part of the till deforms. Farther to the east the till is thin and patchy, and ice that extends this far will encounter case A (high permeability and good drainage; Boulton and Dobbie, 1993). Here, all the water is discharged into the aquifer and the thin till cover is not deformed. Ice is thus unable to advance eastward due to the low subglacial pore water pressure, and the consequent development of high yield stress and the lack of a deformable substrate; the ice becomes pinned, and a steep eastern ice margin develops.

Glacial recharge of aquifers

Can recharge of aquifers by former ice sheets be demonstrated? Research has necessarily concentrated on the characteristics of water in the aquifer. For example, Carlson (1994) identified an isotopic and hydrochemical anomaly in what is now the discharge area of the Fox Hills aquifer in North Dakota, which she interpreted as evidence for subglacial recharge. Low-chloride water in the discharge area is anomalously fresh compared to water upgradient in the aquifer, which is of the sodium-bicarbonate-chloride type. Moreover, stable-isotope data from the anomalous water suggests that it was precipitated in a cooler climate. However, Carlson (1994, p. 83) cautions that, "Modeling the response of an aquifer to Pleistocene glaciation is a speculative endeavor because of our limited knowledge of both the pre-glacial aquifer system and the ice sheet configuration and hydraulics. Assumptions must be made that will *directly* influence the outcome of the simulation. Thus, attempts at simulating an aquifer's response to glaciation can only indicate the feasibility of various postulated ways in which the aquifer could have behaved under the specified conditions."

If the sieve mechanism proposed herein for the origin of the Driftless Area is valid, the composition of the ground water in the bedrock aquifers should provide geochemical evidence compatible with glacial recharge. Siegel (1984, 1989) reports evidence for major recharge of lower Paleozoic aquifers of the Midcontinent region during glaciation. Distribution of "fresh" ground water (low total dissolved solids), and oxygen-isotope ratios suggest flowpaths much different from those of today, but consistent with recharge from thick ice sheets. The relative contribution of the most recent ice sheets versus that of earlier ice sheets could not be determined.

More recently, Siegel (1991) identified anomalous water (low dissolved solids) in the Cambrian-Ordovician aquifer of Iowa, which he interpreted to result from recharge during the last glaciation. The anomaly extends across a geographic area equivalent to that of the Des Moines lobe at its maximum extent, but is displaced 100 km southeast from the Des Moines lobe, down the regional flowpath. Siegel (1991) calculated that the anomaly could have formed by vertical recharge over 600 years, even though it had to pass through hundreds of meters of confining beds; it is significant that such a large amount of glacial meltwater can be transferred to a bedrock aquifer despite the existence of an aquitard. Based on the dynamic behavior of the Des Moines lobe, and on the amount of water required to completely displace preglacial water, only a small proportion of the total amount of meltwater from the Des Moines lobe would have been able to infiltrate the aquifer through the aquitard.

In contrast, the proposed sieve mechanism would drain nearly all of the meltwater into the bedrock aquifers; most of the water would ultimately discharge into the Mississippi River system. The giant sieve may have been most active during the most extensive pre-Wisconsinan glaciations, because the farther the ice rode onto the Paleozoic Plateau, the more effectively it would drain.

Earliest glaciations: thin extensive ice sheets

The deep-sea oxygen-isotope record for the late Pliocene–early Pleistocene implies that global ice volume was only about half to two-thirds that of the late Pleistocene (Shackleton et al., 1984; Morley, 1991). If late Pliocene–early Pleistocene ice sheets were more extensive than the Late Wisconsinan ice sheet, something about them must have been different. Field evidence for the earlier glaciations having extended farther towards the Driftless Area does not appear to be compatible with lower ice volumes. In a very general way, late Pleistocene ice sheets were thick over the hard rock and thin, sandy tills of the Canadian Shield, but thin and gently sloping over the sedimentary rocks and thicker, finer-

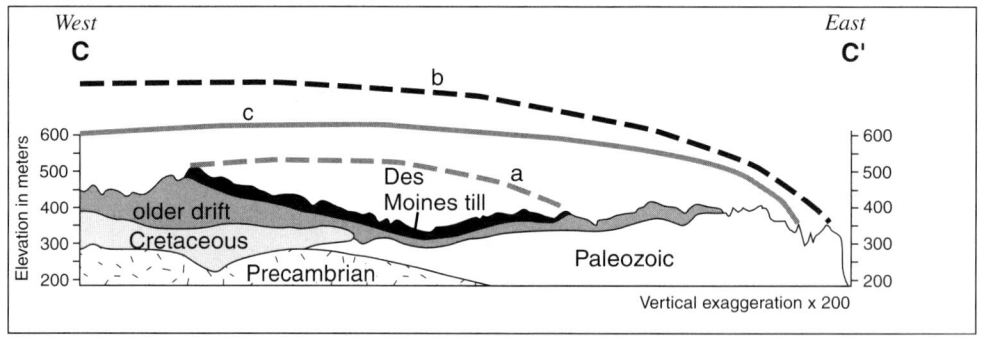

Figure 4. Topographic and generalized stratigraphic profile along the Iowa-Minnesota border (C–C′ on Fig. 1) at latitude 43°30′N, and hypothetical ice profiles. Vertical exaggeration ×200. a, Upper surface of Des Moines lobe ice from Clark (1992); b, Maximum elevation of upper surface of ice during pre-Illinoian glacial maximum; c, "Intermediate" elevation of the upper surface of the ice during pre-Illinoian glacial maximum. See text for discussion of curves b and c.

grained tills. The decreased global ice volume indicates that the early continental ice sheets were thin throughout.

Boellsdorf (1978) reports that a till in southwestern Iowa underlies volcanic ash that is dated at 2.2 Ma. The site is inside the all-time glacial limit, but lies to the south of the Driftless Area, implying that the southern margin already was showing the differential southern limit which has characterized the later glaciations. The tentative correlation of the old till described by Boellsdorf (1978) with old tills fringing the Driftless Area on its west side suggests that the Paleozoic Plateau was acting as a barrier during early stages of Plio-Pleistocene glaciation, otherwise pre-2.2 Ma glacial sediments should be found in the Driftless Area.

Prior to the Late Cretaceous a thick weathering saprolith had developed over the Canadian Shield in Minnesota (Parham, 1970). It is still fairly thick and extensive in Minnesota where protected by Cretaceous sedimentary rocks and thick glacial sediment (Setterholm et al., 1989) but is thin and extremely patchy in areas of the state where glacial sediments are thin. Presumably it has been stripped away by repeated glaciation. Was saprolith thick and widespread over the Canadian Shield before the first glaciation? If so, there would have been essentially no hard substrate anywhere to provide pinning points, and subglacial meltwater could not easily percolate into the underlying jointed bedrock. Then the whole ice sheet would have behaved as the extra-shield part of the ice sheet did in later glaciations; it would have been thin, gently sloping, and unstable.

Clark and Pollard (1998) tested this idea with an ice sheet and bedrock model that included transport of sediment and ice by subglacial sediment deformation. They demonstrated that a widespread deforming layer maintains thin ice sheets that respond to the dominant orbital forcing of 23 and 41 k.y. Moreover, repeated runs of the model progressively removed the sediment layer, and eventually caused a transition to thicker ice sheets with a dominant timescale of 100 k.y. Thus it is quite plausible that the patchy tills of the pseudo-driftless area were deposited by one or more very early, very extensive ice sheets. More recent pre-Illinoian glaciers, though they contained more ice, may not have been able to extend as far east onto the Paleozoic Plateau.

Testing the giant sieve hypothesis

Possible avenues for testing the hypothesis fall into three main groups: (1) karst geology, (2) aquifer geochemistry, and (3) numerical modeling of ice sheets. Each method has its limitations, and a definitive test is not likely in the near future.

The karst geology of the Paleozoic Plateau could provide some clues to the glacial recharge of aquifers, because much of the recharge took place through sinkholes and solution cavities that presumably still exist. Many bedrock sinkholes in the area are filled with sediment, and have no surface expression; they can only be seen in roadcuts and quarry walls. However, unweathered till is scarcely ever observed in them. Most of the fills present in the karst areas of southeastern Minnesota consist of weathering products of bedrock mixed with weathered pre-Illinoian drift. They were emplaced long after the most recent local glaciation, and do not hold a record of glacial events. The shape of the solution cavities themselves might contain evidence of rapid recharge under pressure, but it is not certain what this shape should be. Even if it can be established that large amounts of glacial meltwater were discharged through the karst system, it would not prove that this discharge held back the ice from the Driftless Area.

Geochemical studies of water in bedrock aquifers are unlikely to provide definitive answers regarding the role of the giant sieve in controlling aquifer recharge or in controlling the spread of ice sheets across the Paleozoic Plateau, because most of the glacial recharge was quickly discharged into the Mississippi River. It is possible that deep, seldom-used aquifers contain some record that has been erased in the shallower aquifers, although sampling opportunities in deep aquifers are limited.

Numerical modeling of former ice sheets offers the best prospect for testing the giant sieve hypothesis, but the pitfalls of such an approach should not be underestimated. The elevation and location of a considerable length of a correlative ice margin needs to be specified, and at least an approximation of the subglacial topography of the entire portion of the ice sheet modeled. This is most easily done for more recent glacial periods; Late Wisconsinan ice margins are recognized by their morphology, and the subglacial topography can be approximated by the modern topography. But even in this case, errors are not hard to find. In many places, the modern topography is much higher than the base of the last ice sheet, because Late Wisconsinan deposits are 50 m to more than 100 m thick. Temporal variability also induces errors; the deposition of terminal Late Wisconsinan deposits is not synchronous.

Numerical modeling of pre-Wisconsinan ice sheets is even less certain. To test the giant sieve hypothesis, one would need (at minimum) to specify correlative Labradoran and Keewatin ice margins along the border of the Driftless Area, and include the area at least as far north as the International Boundary (area covered by Soller, 1997). Although individual ice margins are recognized within this area, they cannot be correlated, because (1) subsequent erosion has removed most of the geomorphic evidence, and (2) sediments associated with these old margins are beyond the range of radiocarbon dating.

Any viable modeling efforts depend on advances in pre-Illinoian glacial stratigraphy. Ice margins can theoretically be correlated by their till sheets, which would also provide rough elevations of the base of the glacier throughout its extent. The accessibility, complexity, and poorly understood nature of the pre-Wisconsinan glacial records renders it difficult to attempt this type of modeling. However, a rough reconstruction could be attempted, using the maximum position of ice, and the modern topography as a proxy for the subglacial surface.

CONCLUSIONS

The roles of substrate and subglacial drainage in controlling the extent of glacial ice and the location of the ice margin are crucial; they control the amount of subglacial yield stress, and the likelihood of the ice being pinned to the bedrock. The dissected limestones and sandstones of the Paleozoic Plateau prevented the spread of glacial ice onto the Driftless Area, because they allowed dewatering at the base of the ice sheet.

REFERENCES CITED

Anderson, R. C., 1988, Reconstruction of preglacial drainage and its diversion by earliest glacial forebulge in the Upper Mississippi Valley region: Geology, v. 16, p. 254–257.

Attig, J. W., and Muldoon, M. A., 1989, Pleistocene geology of Marathon County, Wisconsin: Wisconsin Geological and Natural History Survey Information Circular 65, 27 p., scale 1:100,000, 1 sheet.

Boellsdorf, J., 1978, North American Pleistocene stages reconsidered in light of probable Pliocene-Pliestocene continental glaciation: Science, v. 202, p. 305–307.

Black, R. F., 1960, "Driftless Area" of Wisconsin was glaciated [abs.]: Geological Society of America Bulletin, v. 71, no. 12, pt. 2, p. 1827.

Black, R. F., 1970a, Blue Mounds and the erosional history of southwestern Wisconsin, in Pleistocene geology of southern Wisconsin: Wisconsin Geological and Natural History Survey Information Circular No. 15, p. H1–H11.

Black, R. F., 1970b, Residuum and ancient soils of the Driftless Area of southwestern Wisconsin, in Pleistocene geology of southern Wisconsin: Wisconsin Geological and Natural History Survey Information Circular No. 15, p. I1–I12.

Boulton, G. S., and Dobbie, K. E., 1993, Consolidation of sediments by glaciers: relations between sediment tectonics, soft-bed glacier dynamics and subglacial ground-water flow: Journal of Glaciology, v. 39, p. 26–44.

Carlson, C. A., 1994, Identification of an isotopic and hydrochemical anomaly in the discharge area of the Fox Hills Aquifer, south-central North Dakota: Evidence for Pleistocene subglacial recharge? [Ph.D. thesis]: East Lansing, Michigan State University, 215 p.

Chamberlin, T. C., and Salisbury, R. D., 1885, Preliminary paper on the Driftless Area of the Upper Mississippi Valley: U.S. Geological Survey 6th Annual Report, p. 199–322.

Clark, P. U., 1992, Surface form of the Laurentide Ice Sheet and its implications to ice-sheet dynamics: Geological Society of America Bulletin v. 104, p. 595–605.

Clark, P. U., and Pollard, D., 1998, Origin of the middle Pleistocene transition by ice sheet erosion of regolith: Paleoceanography, v. 13, p. 1–9.

Clayton, L., Teller, J. T., and Attig, J. W., 1985, Surging of the southwest part of the Laurentide Ice Sheet: Boreas, v. 14, p. 235–242.

Fisher, D. A., Reeh, N., and Langley, K., 1985, Objective reconstructions of the Late Wisconsinan Laurentide Ice Sheet and the significance of deformable beds: Geographie Physique et Quarternaire, v. 39, no. 3, p. 229–238.

Fullerton, D. S., 1995, Quaternary geologic map of the Dakotas Quadrangle, United States: U.S. Geological Survey Miscellaneous Investigations Series I-1420(NL-14), Scale 1:1,000,000.

Fullerton, D. L., and Richmond, G. M., 1991, Quaternary geologic map of the Lake Erie Quadrangle, United States and Canada: U.S. Geological Survey Miscellaneous Investigations Series I-1420(NK-17), scale 1:1,000,000.

Hobbs, H. C., 1984, Surficial geology, Plate 3, in Balaban, N. H., and Olsen, B. M., eds., Geologic atlas of Winona County, Minnesota: Minnesota Geological Survey County Atlas Series C-2, scale 1:100,000.

Hobbs, H. C., 1985, Quaternary history of southeastern Minnesota, in Lively, R. S., ed., 1985, Pleistocene geology and evolution of the Upper Mississippi Valley, Winona, Minnesota, August 13–16, 1985, Abstracts and Field Trip Guide: Minnesota Geological Survey, 105 p., 1 pl.

Hobbs, H. C., 1987, Quaternary geology of southeastern Minnesota, in Balaban, N. H., ed., Field Trip Guidebook for the upper Mississippi Valley Minnesota, Iowa, and Wisconsin: Geological Society of America, North-Central Section 21st Annual Meeting: Minnesota Geological Survey Guidebook Series no. 15, 185 p.

Hobbs, H. C., 1988, Surficial geology, Plate 3, in Balaban, N. H., ed., Geologic atlas, Olmsted County, Minnesota: Minnesota Geological Survey County Atlas Series C-3, scale 1:100,000.

Hobbs, H. C., 1990, Geologic history and development of the Upper Mississippi River, in Hammer, W. R., and Hess, D. F., eds., Geology Field Guidebook: Current perspectives on Illinois Basin and Mississippi Arch geology: Geological Society of America, North-Central Section 24th Annual Meeting, April 1990, Macomb, Illinois, p. F1–F52.

Hobbs, H. C., 1992, Paleozoic Plateau of southeastern Minnesota, in Nater, E. A., ed., Soils-geomorphology pre-conference tour guidebook, Oct. 29–Nov. 1: Annual Meeting of the Soil Science Society of America, Nov. 1–6, 1992, Minneapolis, Minnesota, unpaginated.

Hobbs, H. C., 1995, Surficial geology, Plate 3, in Mossler, J. H., project manager, Geologic atlas of Fillmore County, Minnesota: Minnesota Geological Survey County Atlas Series C-8, Part A, scale 1:100,000.

Hobbs, H. C., 1997, The preglacial Upper Mississippi River flowed north and west into the Arctic Ocean: Geological Society of America Abstracts with Programs, v. 29, p. 20.

Knox, J. C., 1982, Quaternary history of the Driftless Area: Wisconsin Geological and Natural History Survey Field Trip Guidebook 5, 169 p.

Knox, J. C., and Attig, J. W., 1988, Geology of the pre-Illinoian sediment in the Bridgeport Terrace, lower Wisconsin River Valley, Wisconsin: Journal of Geology, v. 96, p. 505–513.

Leverett, F., 1921, Outline of the Pleistocene history of the Mississippi Valley: Journal of Geology, v. 29, p. 615–626.

Lively, R. S., editor, 1985, Pleistocene geology and evolution of the Upper Mississippi Valley, Winona, Minnesota, August 13–16, 1985, Abstracts and Field Trip Guide: Minnesota Geological Survey, 105 p., 1 pl.

Meyer, G. N., 1997, Pre-late Wisconsinan till stratigraphy of north-central Minnesota: Minnesota Geological Survey Report of Investigations 48, 67 p.

McGee, W. J., 1891, The Pleistocene history of northeastern Iowa: U.S. Geological Survey 11th Annual Report, p. 199–586.

Morey, G. B., Sims, P. K., Cannon, W. F., Mudrey, M. G., Jr., and Southwick, D. L., 1982, Geologic map of the Lake Superior region; Minnesota, Wisconsin, and northern Michigan: Minnesota Geological Survey State Map Series S-13, scale 1:1,000,000.

Morley, J. J., 1991, Evolving Pliocene-Pleistocene climate: A North Pacific perspective: Quaternary Science Reviews, v. 10, p. 225–237.

Mossler, J. H., 1983, Bedrock topography and isopachs of Cretaceous and Quaternary strata, east-central and southeastern Minnesota: Minnesota Geological Survey Miscellaneous Map Series M-52, scale 1:500,000.

Olsen, B. M., and Mossler, J. H., 1982, Geologic map of Minnesota, bedrock topography: Minnesota Geological Survey State Map Series S-15, scale 1:1,000,000.

Parham, W. E., 1970, Clay mineralogy and geology of Minnesota's kaolin clays: Minnesota Geological Survey Special Publication Series SP-10, 142 p.

Patterson, C. J., 1997, Southern Laurentide ice lobes were created by ice streams: Des Moines Lobe in Minnesota, USA: Sedimentary Geology, v. 111, p. 249–261.

Peltier, W. R., 1981, Ice age geodynamics: Annual Review of Earth and Planetary Sciences, v. 9, p. 199–226.

Prior, J. C., 1976, A regional guide to Iowa landforms: Iowa Geological Survey Educational Series 3, 71 p.

Reeh, N., 1982, A plasticity theory approach to the steady-state shape of a three-dimensional ice sheet: Journal of Glaciology, v. 28, no. 100, p. 431–455.

Richmond, G. M., 1994, Quaternary geologic map of the Platte River Quadrangle, United States: U.S. Geological Survey Miscellaneous Investigations Series I-1420(NK-14), scale 1:1,000,000.

Richmond, G. M., and Christiansen, A. C., 1993, Quaternary geologic map of the Wichita Quadrangle, United States: U.S. Geological Survey Miscellaneous Investigations Series I-1420(NJ-14), scale 1:1,000,000.

Richmond, G. M., and Fullerton, D. S., 1983, Quaternary geologic map of the Minneapolis 4 × 6 quadrangle, United States: U.S. Geological Survey Miscellaneous Investigations Series I-1420(NL-15), scale 1:1,000,000.

Richmond, G. M., Fullerton, D. S., and Christiansen, A. C., editors, 1991, Quaternary geologic map of the Des Moines 4 × 6 Quadrangle, United States: U.S. Geological Survey Miscellaneous Investigations Series I-1420(NK-15), scale 1:1,000,000.

Richmond, G. M., and Fullerton, G. S., 1983, Quaternary geologic map of the Chicago Quadrangle, United States: U.S. Geological Survey Miscellaneous Investigations Series I-1420(NK-16), scale 1:1,000,000.

Richmond, G. M., and Fulerton, G. S., 1991, Quaternary geologic map of the Louisville Quadrangle, United States: U.S. Geological Survey Miscellaneous Investigations Series I-1420(NK-16), scale 1:1,000,000.

Richmond, G. M., and Weide, D. L., 1993, Quaternary geologic map of the Ozark Plateau Quadrangle, United States: U.S. Geological Survey Miscellaneous Investigations Series I-1420(NJ-15), scale 1:1,000,000.

Richmond, G. M., Fullerton, D. S., and Christiansen, A. C., 1991, Quaternary geologic map of the Blue Ridge Quadrangle, United States: U.S. Geological Survey Miscellaneous Investigations Series I-1420(NJ-17),

Sardeson, F. W., 1916, Minneapolis–St. Paul folio: U.S. Geological Survey Geologic Atlas of the United States, 14 pl., maps, photos.

Setterholm, D. R., Morey, G. B., Boerboom, T. J., and Lamons, R. C., 1989, Minnesota kaolin clay deposits: A subsurface study in selected areas of southwestern and east-central Minnesota: Minnesota Geological Survey Information Circular 27, 99 p.

Siegel, D. I., 1984, Isotopic evidence for glacial meltwater recharge to the Cambrian-Ordovician aquifer, north-central United States: Journal of Quaternary Research, v. 22, p. 328–335.

Siegel, D. I., 1989, Geochemistry of the Cambrian-Ordovician aquifer system in the northern Midwest, United States: U.S. Geological Survey Professional Paper 1405-D, 76 p.

Siegel, D. I., 1991, Evidence for dilution of deep confined ground water by vertical recharge of isotopically heavy Pleistocene water: Geology, v. 19, no. 5, p. 433–436.

Shackleton, N. J., Blackman, J., Zimmerman, H., Kent, D. V., Hall, M. A., Roberts, D. G., Schnitker, D., Baldauf, J. G., Desprairies, A., Homrighausen, R., Huddlestun, P., Keene, J. B., Kaltenback, A. J., Krumsiek, K. A. O., Morton, A. C., Murray, J. W., and Westberg-Smith, J., 1984, Oxygen isotope calibration of the onset of ice-rafting and history of glaciation in the North Atlantic region: Nature, v. 307, p. 620–623.

Soller, D. R., 1997, Map showing the thickness and character of Quaternary sediments in the glaciated United States east of the Rocky Mountains; northern and central plains states (90 to 102 degrees West longitude): U.S. Geological Survey Map I-1970-C, scale 1:1,000,000.

Trowbridge, A. C., 1921, The erosional history of the Driftless Area: University of Iowa Studies in Natural History, v. 9, p. 55–127.

Trowbridge, A. C., 1954, Mississippi River and Gulf Coast terraces and sediments as related to Pleistocene history—a problem: Geological Society of America Bulletin, v. 65, p. 793–812.

Trowbridge, A. C., 1966, Glacial drift of the "Driftless Area" of northeast Iowa: Iowa Geological Survey Report of Investigations 2, 28 p.

Willman, H. B., and Frye, J. C., 1969, High-level glacial outwash in the Driftless Area of northwestern Illinois: Illinois State Geological Survey Circular 440, 23 p.

Winchell, N. H., 1877, The geological and natural history of Minnesota: The Fifth Annual Report, submitted to the President of the University [of Minnesota]: Saint Paul Pioneer Press, 205 p.

MANUSCRIPT ACCEPTED BY THE SOCIETY OCTOBER 8, 1998

Paleoglacier reconstruction and late Pleistocene equilibrium-line altitudes, southern Sawatch Range, Colorado

Keith A. Brugger
Department of Geology, University of Minnesota-Morris, Morris, Minnesota 56267
Barry S. Goldstein
Department of Geology, University of Puget Sound, Tacoma, Washington 98416

ABSTRACT

Equilibrium-line altitudes (ELAs) for six reconstructed, late Pleistocene glaciers in the southern Sawatch Range were calculated using the accumulation-area ratio (AAR) method. An AAR of 0.65 ± 0.05 was used for five paleoglaciers having typical area-altitude distributions at the late glacial maximum. An AAR of 0.55 was used for one glacier with a piedmont-type geometry. The ELAs thus determined varied little from an average of 3,375 m in a nonsystematic way over the study area. Additional estimates and checks of paleo-ELAs were made using toe-to-headwall altitude ratios (THARs) and the maximum elevations of lateral moraines. Both methods (using THAR = 0.45) suggest an elevation of 3,375 m for regional equilibrium line, which is remarkably consistent with that determined by the AAR method.

The paleoclimatic significance of ELA depression in the study area is interpreted within the range of temperature and precipitation conditions found at equilibrium lines on modern glaciers. Mean summer temperatures were on the order of 7–9 °C cooler during the last glacial maximum if winter precipitation was not significantly different than it is today. Lesser reductions in late Pleistocene temperatures would have required substantial increases in precipitation.

INTRODUCTION

Paleosnowline elevations, or specifically equilibrium-line altitudes (ELAs), are commonly estimated from reconstructions of former valley glaciers, and form a basis for inferring late Pleistocene climatic change in the Rocky Mountain region (Pierce, 1979; Meierding, 1982; Porter et al., 1983; Leonard, 1984; Murray and Locke, 1989; Brugger, 1996). The ELA is by definition the elevation where the net mass-balance of a glacier is zero, and as such divides a glacier's accumulation area from its ablation area. In brief, an estimate of the ELA is obtained by determining that altitude which yields some specified ratio of the reconstructed glacier's accumulation area to its total area. This ratio is referred to as the accumulation-area ratio or AAR. Although ELAs can be established using other criteria, recent comparisons of methodologies (Meierding, 1982; Hawkins, 1985; Torsnes et al., 1993) have concluded that the AAR method has a more sound physical basis and renders the most accurate determination.

In this paper, we present detailed reconstructions of paleoglaciers in the Taylor River basin (Fig. 1) of the southern Sawatch Range, Colorado, and late Pleistocene ELAs derived from them using the AAR method. A preliminary analysis of late Pleistocene ELA depression in terms of climate change in this region is also discussed.

GLACIAL HISTORY OF THE TAYLOR RIVER BASIN

Glacial deposits

Deposits of two late Quaternary glaciations are present within the Taylor River basin and can be distinguished on the

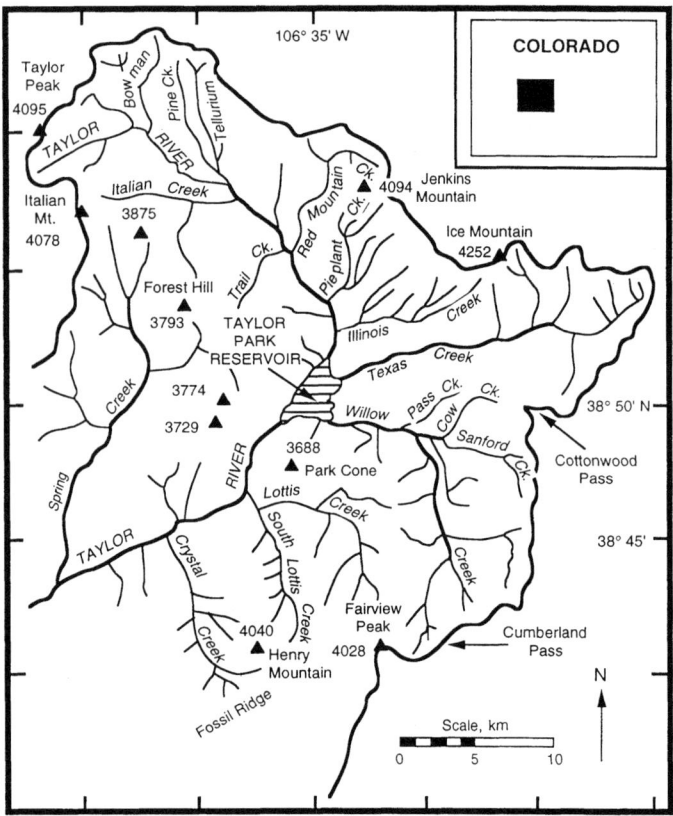

Figure 1. Location and selected geographic features of the study area.

basis of relative age-dating criteria (Goldstein et al., 1988). Moraines of the older Cow Creek glaciation tend to be massive with subdued surface morphologies and they are extensively dissected by streams. Few boulders are exposed on moraine surfaces, and those that are exposed are highly weathered. In contrast, moraines assigned to the younger Taylor River glaciation are sharp crested with hummocky surface morphologies owing to preservation of ice-disintegration features. These younger moraine surfaces are undissected and exhibit a relative abundance of unweathered boulders.

No absolute ages have been obtained for glaciations in the southern Sawatch Range, although a minimum age of 15.6 ka (thousands of years before present) is suggested for the last deglaciation. Age estimates for the Cow Creek and Taylor River glacial maximums can, therefore, only be made through morphostratigraphic correlations to the Bull Lake (older) and Pinedale (younger) glaciations of the Wind River Mountains of Wyoming. Recent cosmogenic dating and stratigraphic relationships of glacial deposits in the Wind River Mountains (including the Bull Lake type locality) suggests that the Bull Lake glaciation consisted of four separate ice advances, ranging in age from 95 to greater than 130 ka (Chadwick et al., 1997; Phillips et al., 1997). Cosmogenic dates from the type Pinedale terminal moraines (Gosse et al., 1995) and elsewhere in the Wind River Mountains (Chadwick et al., 1997; Phillips et al., 1997) indicate

that Pinedale glaciers attained their maximum extent about 23 ka with major ice recession beginning at 16 ka. Regionally, however, there is evidence suggesting earlier advances of Pinedale glaciers (e.g., Nelson et al., 1979; Sturchio et al., 1994). Thus, assuming regional synchroneity of glacial advances, these ages constrain those of the Cow Creek and Taylor River glaciations in the southern Sawatch Range.

Extent and style of the last glaciation

Figure 2 illustrates the maximum extents of glaciers during the Taylor River glaciation determined by the distribution of glacial erosional and depositional features. Glacial systems at this time existed as small cirque glaciers, larger valley and piedmont glaciers, and upland ice fields. The largest of these systems, the Taylor River glacier complex, covered about 250 km^2 of the headwaters of the Taylor River. This complex consisted of coalescing valley glaciers sourced from the Taylor River proper, Italian, South Italian, Bowman, Pine, Tellurium, and Red Mountain Creeks, and thin upland ice fields. The latter were centered on the upland plateau between American Flag Mountain and Forest Hill and over the head of the Bowman Creek drainage (Fig. 1). Two major ice lobes were nourished by the complex: one extending about 20 km down the Taylor River valley, the other flowing about 10 km down the Spring Creek drainage. Smaller ice lobes supported solely by the American Flag–Forest Hill ice field flowed into the valley of Trail Creek. Maximum ice thickness in the Taylor River valley was in excess of 500 m while that in the ice fields did not greatly exceed 100 m.

A second large glacier flowed westward out of Texas Creek to form a piedmont lobe terminating at what is now the Taylor Park Reservoir. At its maximum, the area of the Texas Creek glacier was approximately 125 km^2 with ice thicknesses as much as 300 m. Larger composite glaciers also occupied the valleys of the Willow Creek, Lottis Creek, and South Lottis Creek drainages.

We envision the growth and decay of glacial systems in the Taylor River basin during the Taylor River glaciation as follows: 1. Early in the glaciation, small cirque glaciers in the headwaters of major drainages began to advance in response to a lowering of snowline. In some valleys the cirque glaciers coalesced to form small composite glaciers. 2. While these glaciers thickened and advanced, thin ice began to form in protected areas on the American Flag–Forest Hill and Bowman Creek uplands. Bedrock exposures in several small, shallow basins in this area are polished and striated, and streamlined features, including roches moutonées, are common. These basins were the "nuclei" of the developing ice field. 3. Presumably by 20–25 ka, glaciers in the Taylor River basin were at their maximum extents and volumes, as depicted in Figure 2. At this point, a substantial part of American Flag–Forest Hill upland was above the lowered snowline and thus served as a center of accumulation and outflow for large glaciers advancing down the Taylor River, Italian Creek, South Italian Creek, and Spring Creek valleys. Similarly, ice from the Bowman Creek uplands supported large tributary glaciers augmenting the Taylor River lobe. Else-

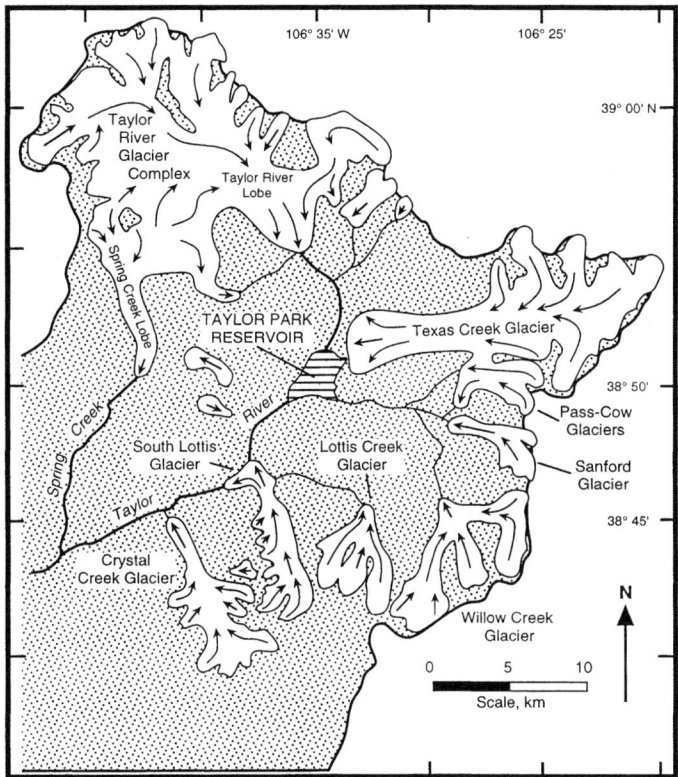

Figure 2. Extent of glaciers in the Taylor River basin during the Taylor River glaciation. (Note that several small ice masses are not shown in this figure.)

where within the basin, the Texas Creek and South Lottis Creek glaciers were expanding into Taylor Park and the Taylor valley, respectively. Glaciers in the tributary basins of both Lottis and Willow Creeks merged at common termini (Fig. 2). In several valleys (e.g., South Lottis and Lottis Creeks), maximum ice extent exceeded that during the earlier Cow Creek glaciation. In others (e.g., Taylor River, Texas Creek), Cow Creek–aged moraines acted to constrain later ice flow. 4. A date of 15,620 ± 250 ^{14}C yr B.P. (Pitt-0120) obtained from basal sediments in a bog in the upland area between Italian Creek and American Flag Mountain (Fig. 1) provides a minimum age for the disintegration of the ice field. This date is consistent with those recognized for deglaciation of similar climatically sensitive ice fields in Yellowstone and the nearby San Juan Mountains (Porter et al., 1983; Maher, 1972; Carrara et al., 1984; Barnowski et al., 1987). Two points need to be emphasized here. First, the dates reported by Maher (1972) and Carrara et al. (1984) for deglaciation of upland ice in the San Juan Mountains are not universally accepted (cf. Elias et al., 1991). Secondly, we believe deglaciation of the upland ice fields in the Taylor River basin is quite distinct from that which occurred in the major valleys. Tentative support for this view comes from a presumed ice-dammed pond along the right-lateral moraine of the Taylor River lobe (Fig. 2). This pond is thought to have formed prior to approximately 14 ka (Brugger et al., 1990), indicating that ice still occupied the lower reach of Taylor valley while the upland ice thinned and stagnated. 5. Ice recession in the main valleys was punctuated by several stillstands or readvances before the glaciers retreated into their respective cirque basins. Ice retreat in the Taylor River valley, where as many as twelve spatially distinct recessional moraines exist, may have been particularly sluggish.

METHODS

Reconstruction of the paleoglaciers

The Texas, Pass, Cow, Sanford, Lottis, and South Lottis Creek paleoglaciers (Fig. 2) were selected for reconstruction because: (1) the glaciers' areal extents are known from a combination of detailed field mapping and analyses of aerial photos and topographic maps; (2) each glacier is of sufficient area so that errors introduced in the AAR method by uncertainties in determining its areal extent are minimized; and (3) each glacier exhibits an area-altitude distribution that is somewhat typical of simple valley glaciers. The significance of each of these requirements, if not immediately evident, will become more clear in subsequent discussions.

Ice-surface contours of the six paleoglaciers were reconstructed for the Taylor River glaciation (Fig. 3) following the procedure given by Pierce (1979) and Porter (1975, 1981). The areal extent of each glacier and its tributaries was determined using the geometry of moraine segments, the upper limits of erratic boulders, and the upper limits of glacial erosion. Elevations at glacier margins were used to establish initial ice-surface contours by assuming convergent and divergent flow in accumulation and ablation areas respectively, and with due consideration of the constraints on local ice flow imposed by bedrock topography. Longitudinal profiles drawn from the reconstructed ice surfaces provided estimates of ice thicknesses and surface slopes, which were then used to calculate basal shear stress along the length of the glacier. The shear stress τ is given by

$$\tau = f\rho g h \sin \alpha \qquad (1)$$

where f is a "shape factor" to account for drag from the valley sides, ρ is the density of ice, g is the acceleration due to gravity, h is the ice thickness, and α is the surface slope averaged over a distance of one order of magnitude larger than the local ice thickness to account for longitudinal stress gradients (Paterson, 1994). Ice-surface contours were then "corrected" in an iterative manner until the computed shear stresses were within, or close to, the accepted range of 50–150 kPa (Paterson, 1994).

Determination of ELAs

Area-altitude distributions for the paleoglaciers (Fig. 4) were quantified by planimetering areas between successive contours on the glacier's surface. From these distributions the ELAs of the glaciers could be found using an appropriate value for the accumulation-area ratio. Modern glaciers in quasi equilibrium with existing climates have AAR values between 0.50 and 0.80 (Meier and Post, 1962; Grosval'd and Kotlyakov, 1969; Andrews and

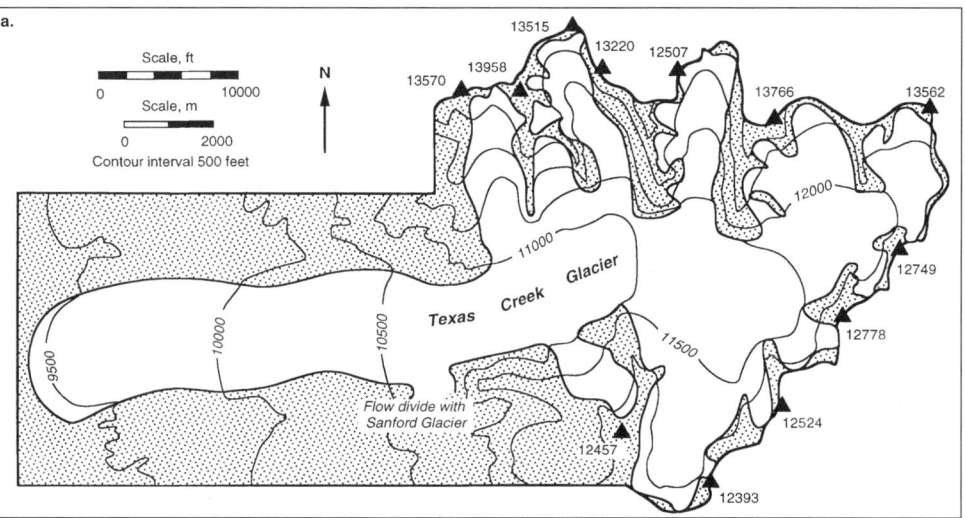

Figure 3 (this page). Reconstructions of the six paleoglaciers in the Taylor River basin. a., Texas Creek glacier; b., Pass and Cow glaciers; c., Sanford glacier; d., Lottis Creek glacier; e., South Lottis Creek glacier. (Elevations are in feet to conform with U.S.G.S. topographic maps.) Bedrock contours are omitted in places for clarity. Selected summit elevations are also shown.

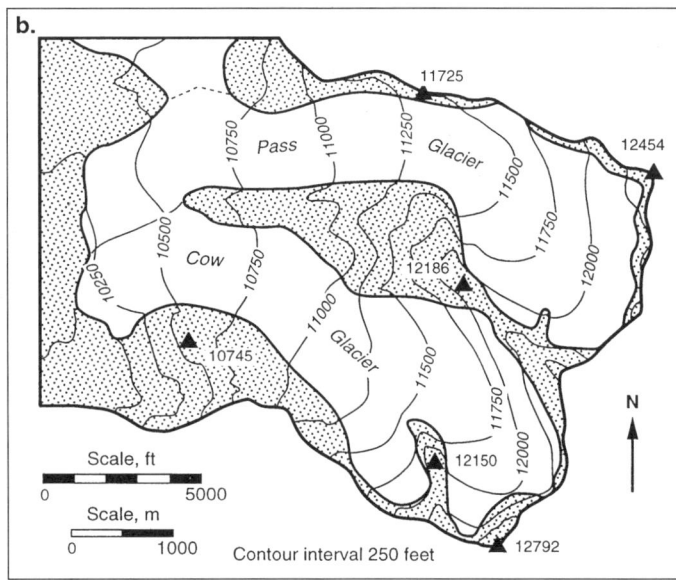

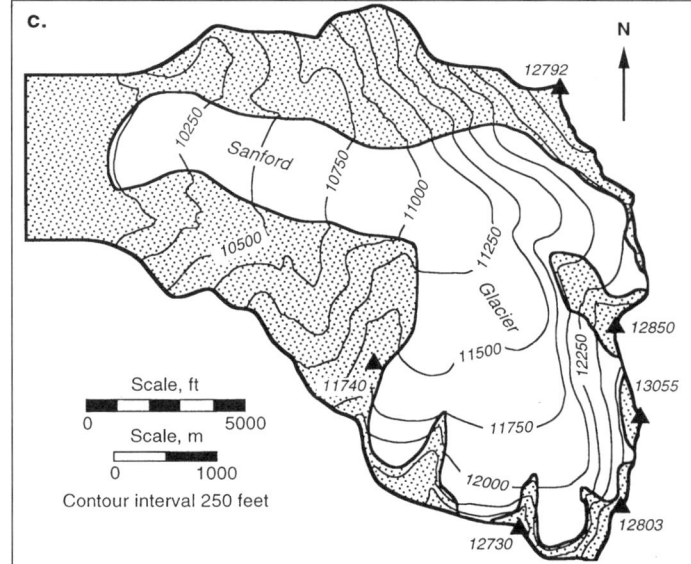

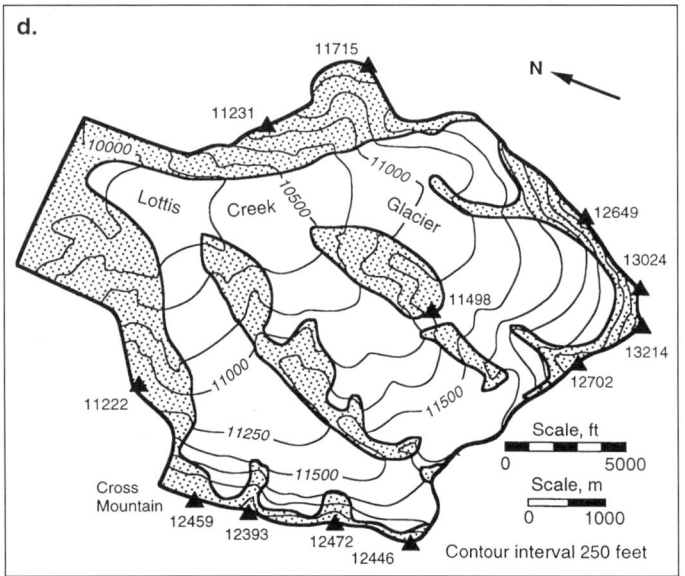

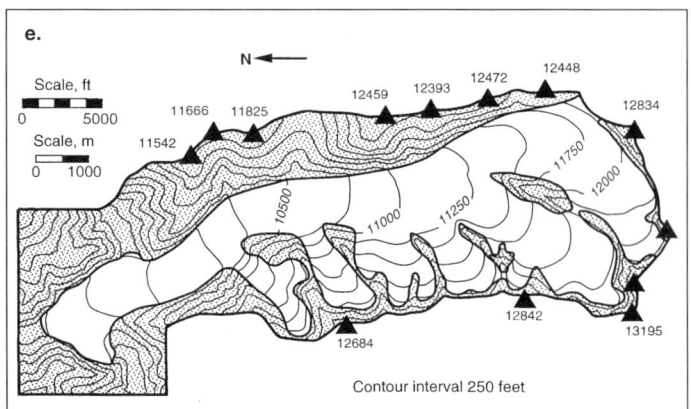

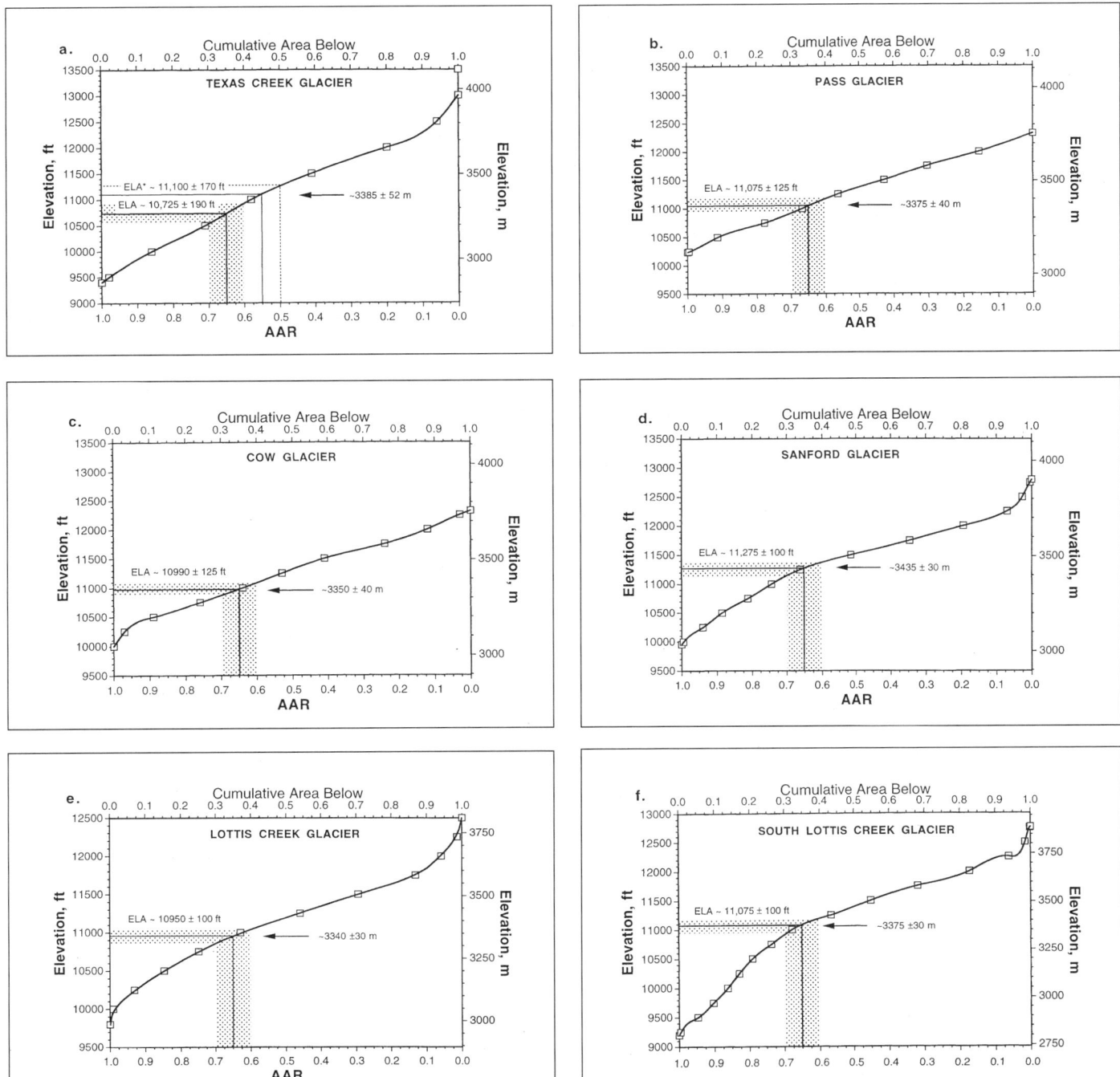

Figure 4. Area-altitude distributions and equilibrium-line altitudes (ELAs) of the six paleoglaciers. ELA* in (a) is that determined with AAR = 0.55 (accumulation-area ratio).

Miller, 1972; Porter, 1975). AARs for piedmont-type glaciers, having an "excess" of area at lower elevations, might fall at the lower end of this range; glaciers emanating from ice fields, having large accumulation areas, might fall at the higher end (Meier and Post, 1962; Pierce, 1979; Leonard, 1984; Hawkins, 1985). Thus we used the median value of 0.65 to determine steady-state ELAs of the paleoglaciers, as have other investigators (Porter, 1975, 1981; Meierding, 1982; Leonard, 1984; Hawkins, 1985; Murray and Locke, 1989; Burbank, 1991; Torsnes et al., 1993).

Uncertainties in the ELA estimates

Details and analyses of the errors inherent in this method of ELA determination can be found elsewhere (Porter, 1975;

Meierding, 1982; Leonard, 1984; Hawkins, 1985), but some discussion is warranted here. Errors that might be introduced by the inability to *precisely* define the ice margins are relatively insignificant. Even where the potential uncertainties in delimiting ice margins are large, as in former accumulation areas, the effect on glacier area is small (a few percent) because valley sides are rather steep and topographic contours are closely spaced (Porter, 1975). Error associated with planimetry was also small (<1%). Both errors become smaller with increasing surface area of the glacier. To account for these quantifiable sources of error we allowed for ±0.05 variation in the AAR (equivalent to ±5% in area). For paleoglaciers of the Taylor River basin, this corresponds to a mean uncertainty in ELA of about ±40 m (~125 ft).

Perhaps the most significant (and least quantifiable) error arises from the uncertainties in reconstructing the ice-surface contours (Leonard, 1984; Hawkins, 1985) because a unique glacier surface cannot be obtained. However, calculation of basal shear stresses allows only glaciologically reasonable reconstructions to be considered (Pierce, 1979), which are often a small, convergent set of ice surfaces. Sensitivity analysis of such a set for the Sanford paleoglacier suggests that resulting uncertainties in ELA are comparable to those given above. Moreover, Hawkins (1985) noted that errors in the positions of ice-surface contours should be randomly distributed and therefore should not introduce any systematic error in determining the ELA.

Finally, Meier and Post (1962) cautioned against using the AAR method for glaciers having complex geometries and/or highly skewed area-altitude distributions (e.g., ice caps, outlet and large piedmont glaciers). The single equilibrium line afforded by the AAR method implies that mass balance over a glacier's surface is uniform, that is, unaffected by local nonaltitudinal variations in precipitation, energy balance, and so forth. If a single equilibrium line can be problematic for a composite valley glacier of moderate size (cf. Brugger, 1996), it might be more so for the Taylor River glacier complex. Accumulation and ablation processes would have varied considerably over its broad open icefields and large valley systems. It is principally for this reason the Taylor River glacier complex was not used in the present analysis.

RESULTS AND DISCUSSION

Late Pleistocene snowline elevation

The reconstructed glaciers and their area-altitude distributions (or hypsometries) are shown in Figures 3 and 4. Table 1 presents a summary of the relevant parameters obtained from the reconstructions. ELAs determined using an AAR of 0.65 ranged from a high of 3,435 ± 40 m (11,275 ± 125 ft) for the Sanford Creek glacier to a low of 3,275 ± 45 m (10,750 ± 150 ft) for the much larger Texas Creek glacier. Although trend-surface analyses of cirque floor elevations suggest a slight rise in paleosnowline to the northeast (Brugger, unpublished), no such systematic variation is evident in ELA values computed here (Fig. 5). We thus interpret the reconstructed ELAs to be independent measurements of snowline during the Taylor River glaciation, the mean of which is 3,360 m. The standard error of the mean yields an uncertainty of ±20 m.

Estimates of paleo-ELAs also can be determined using a "toe-to-headwall" altitude ratio (THAR). As has been noted (Meierding, 1982; Hawkins, 1985; Torsnes et al., 1993), this method is problematic because it does not consider the area-altitude distribution of the glacier and it is sensitive to minor variations in estimating headwall elevation. This is reflected in the variation in THAR values used by different investigators. Porter (1981), Leonard (1984), and Hawkins (1985) used THAR values of 0.50 (the median elevation of the glacier), while Meierding (1982) and Murray and Locke (1989) concluded that values as low as 0.35–0.40 gave the best estimates of paleo-ELAs.

For the paleoglaciers of the Taylor River basin, THARs corresponding to the *calculated* ELAs generally fall well within this range with an average of 0.44 ± 0.03 (Table 1). Therefore it is not surprising that ELAs calculated directly from a THAR value of 0.45 show the best agreement with those determined by the AAR method. The two exceptions are the Texas Creek and South Lottis Creek glaciers. The ELA of the Texas Creek glacier also shows *both* the greatest deviation from the mean ELA (Fig. 5) and from the median elevation (THAR = 0.5; Table 1). The latter is significant (Leonard, 1984) in that it substantiates the piedmont-type hypsometry of the Texas Creek glacier. Thus an AAR of 0.65 might be too high for this glacier so an alternative value of 0.55 was used, again with an uncertainty of ±0.05. The recalculated ELA (Fig. 4a) is 3,385 ± 50 m (11,100 ± 170 ft), corresponding to a THAR of 0.42 (Table 1). Although perhaps a better estimate, the resulting regional paleosnowline elevation of 3,375 ± 15 m is not considerably different from that given earlier. Again no statistically significant trend is seen in ELAs along a southwest-northeast transect (Fig. 5).

In theory, the upvalley extent of a lateral moraine, being a depositional feature, should approximate the ELA since the equilibrium line divides the zones of glacial erosion and deposition (Andrews, 1975). In practice, however, ELA estimates based on maximum moraine elevations can be ambiguous for several reasons, including: nondeposition of moraine near the equilibrium line, lack of postglacial preservation, difficulties in moraine recognition or in their distinction from other nonglacial deposits, proximal moraine growth during successive glacial advances, and miscorrelations made between moraine segments of different ages (Meierding, 1982; Hawkins, 1985). In view of this it may seem fortuitous that our estimate of 3,370 ± 10 m (Table 1) for paleosnowline elevation based on this method is remarkably similar to that derived using the AAR method. Nevertheless, it should be emphasized that elsewhere in the Taylor River basin, the lateral moraines of the last glacial maximum consistently terminate at elevations close to 3,350 m (Goldstein et al., 1988).

Regional and local comparisons of paleosnowlines

Our estimate of late Pleistocene ELA in the southern Sawatch Range differs significantly from earlier estimates

TABLE 1. RECONSTRUCTION PARAMETERS AND DERIVED EQUILIBRIUM-LINE ALTITUDES

Glacier and moraine data, and derived ELAs

Glacier	Area (10^6 m^2)	Minimum elevation (m)	Maximum elevation (m)	ELA* (m)	THAR	Maximum elevations of lateral moraines Range (m)	Mean (m)
Texas Creek	89.9	2865	4115	3275 ± 45	0.33	3377 - 3389	3383
Pass Creek	6.4	3048	3755	3375 ± 40	0.45	3414 (2)	3414
Cow Creek	6.1	3048	3755	3350 ± 40	0.43	3334 - 3402	3368
Sanford Creek	9.7	3036	3901	3435 ± 40	0.46	3322 - 3338	3330
Lottis Creek	20.9	2987	3810	3340 ± 30	0.43	3328 - 3389	3359
South Lottis Creek	21.3	2804	3889	3375 ± 30	0.53	3377 - 3383	3380
Mean ± Standard Error				3360 ± 20	0.44 ± 0.03		3370 ± 10
Texas Creek recalculated with AAR = 0.55				3385 ± 50	0.41		
Revised means				3375 ± 15	0.45 ± 0.02		

ELAs derived from THAR values

Glacier	THAR = 0.40 (m)	|ELA$_{THAR}$ − ELA$_{AAR}$| (m)	THAR = 0.45 (m)	|ELA$_{THAR}$ − ELA$_{AAR}$| (m)	THAR = 0.50 (m)	|ELA$_{THAR}$ − ELA$_{AAR}$| (m)
Texas Creek	3365	90	3425	150	3490	215
Pass Creek	3330	45	3365	10	3400	25
Cow Creek	3330	20	3365	15	3400	50
Sanford Creek	3380	55	3425	10	3470	35
Lottis Creek	3315	25	3355	15	3400	60
South Lottis Creek	3240	135	3290	85	3345	30
Mean ± Standard Error	3335 ± 20	60 ± 20	3370 ± 20	44 ± 25	3420 ± 20	69 ± 30
Texas Creek recalculated with AAR = 0.55		20		40		105
Revised means		50 ± 20		30 ± 10		50 ± 10

Note: Measured elevations (e.g. glacier and moraine elevations) are reported to the nearest meter; derived elevations (e.g. ELAs) are reported to the nearest 5 m.
*AAR = 0.65 unless noted.

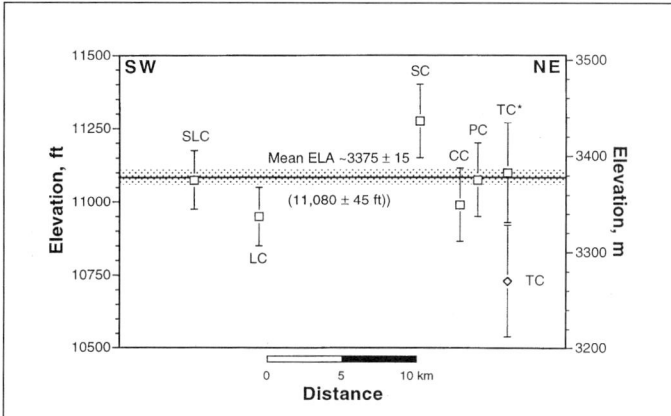

Figure 5. Elevation of paleoequilibrium lines along a southwest-northeast transect within the study area. Data points are: CC, Cow Creek; LC, Lottis Creek; PC, Pass Creek; SC, Sanford Creek; SLC, South Lottis Creek; and TC, Texas Creek. TC* is the equilibrium-line altitude (ELA) for the Texas Creek glacier calculated with AAR = 0.55 (accumulation-area ratio).

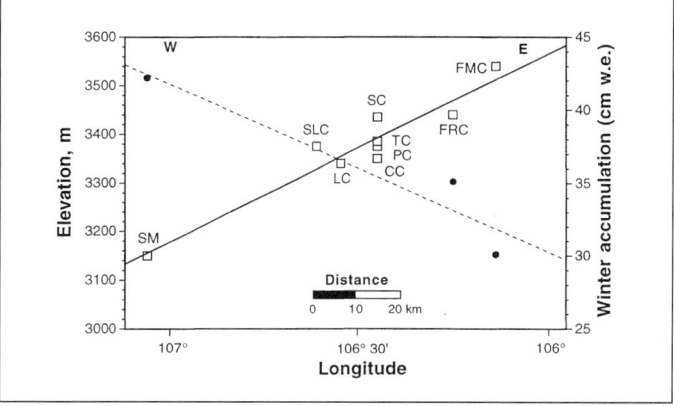

Figure 6. Regional trend of paleoequilibrium lines (solid line; squares), and modern snowpack at paleo-ELAs (equilibrium-line altitudes) (dashed line; circles), along a 100-km, east-west transect at 39° N latitude. Data points are: CC, Cow Creek; LC, Lottis Creek; PC, Pass Creek; SC, Sanford Creek; SLC, South Lottis Creek; TC, Texas Creek; FMC, Four Mile Creek (Mosquito Range); FRC, Frenchman Creek (eastern Sawatch Range); and SM, Snowmass (Elk Mountains).

based on regional trends of paleosnowline in the Rocky Mountains (Richmond, 1965; Flint, 1971). The latter suggest ELAs for central Colorado between 3,500 and 3,650 m. However, a more recent compilation of regional trends for the western United States (Porter et al., 1983) places the ELA for this same area slightly higher than 3,300 m, which is in excellent agreement with our value. For the Front Range of north-central Colorado, Meierding (1982) suggests a late Pleistocene ELA of 3,160 m. Leonard (1984) calculated a range of paleo-ELAs for the San Juan Mountains in southwestern Colorado from 3,000 to 3,700 m, the variation being attributed to local orographic effects.

Leonard (1989) also determined late Pleistocene ELAs for the eastern Sawatch Range and nearby Elk Mountains and Mosquito Range. Figure 6 shows these elevations, together with those determined in this study, projected onto a 100-km, east-west transect at 39° N latitude. The data define a statistically significant (p-value ≤ 0.01) rise in ELAs of about 4.5 m/km to the east, which is very similar to those shown by Porter et al. (1983, their Fig. 4-30) for the interior ranges of Idaho and Wyoming (about 4.0 and 4.5 m/km, respectively).

Paleoclimatic implications

The difference between modern and late Pleistocene ELAs has often been interpreted in terms of temperature change by applying a "standard" atmospheric lapse rate. This concept is rather simplistic as it fails to consider any changes in the altitudinal distribution of precipitation, overall precipitation, atmospheric lapse rates, and radiation balances that might have occurred since the last glacial maximum (Porter et al., 1983; Bradley, 1985; Leonard, 1989; Seltzer, 1994). In addition, no glaciers now exist in the Taylor River basin, so it is difficult to know what elevation is representative of modern regional snowline. Although the mean summer (or mean July) 0°C isotherm has been used as an approximation for modern snowline, their respective elevations are often quite different (Andrews and Miller, 1972; Porter, 1977, Bradley, 1985). Therefore, differences in paleo- and modern snowline elevations cannot be the basis for inferring late Pleistocene climate in the Taylor River basin.

The analysis used here follows that of Leonard (1989) who extended the earlier work of Loewe (1971). Leonard found that climates at the equilibrium line on 32 modern glaciers fall within a well-defined envelope of temperature-precipitation (T-P) requirements. A more recent study by Ohmura et al. (1992) reached a similar conclusion. The envelopes of both Leonard and Ohmura et al. are shown on Figure 7, but the latter is preferred here because it was developed for a larger sampling of modern glaciers (70). Therefore, by defining modern climate in terms of mean summer temperature and annual precipitation at former ELAs, one can determine the changes necessary in these parameters to meet the T-P conditions *required* at equilibrium lines. We emphasize that a unique paleoclimatic interpretation is not possible using this approach, but rather one of plausible temperature-precipitation combinations that might have existed in the Taylor River basin during the last glacial maximum.

Lacking instrumental records, modern climate at the elevation corresponding to paleo-ELA in the Taylor River basin was obtained using linear regressions of climate data versus altitude for Colorado (Leonard, 1989; his Tables 1 and 2). The equations for mean summer temperature, both statistically significant (p ≤ 0.001), are

East-central Colorado:
$T = -0.00721A + 33.2$ ($n = 20$, $r^2 = 0.906$), and (2)
West-central Colorado:
$T = -0.00844A + 35.3$ ($n = 26$, $r^2 = 0.956$),

where T is in °C and the altitude A is in meters. Both regressions were used as the study area lies along their boundary. Similarly, mean winter snowpack is calculated using regressions (p ≤ 0.05) developed for the adjacent eastern Sawatch and Elk Ranges:

Eastern Sawatch Range:
$P = 0.0379A - 95.1$ ($n = 6$, $r^2 = 0.807$), and (3)
Elk Range:
$P = 0.0179A - 14.4$ ($n = 7$, $r^2 = 0.937$),

where P is cm w.e. (water equivalent). Again, both regressions were used. For a paleosnowline elevation of 3,375 m, equations (2) predicts a mean summer temperature of 7.9 ± 1.5 °C. Winter accumulation at this altitude is predicted to be 40 ± 9 cm w.e. These values are supported by Fall's (1997) recent analyses of regional climate data that indicate a mean *July* temperature of 9.3 °C and a mean *annual* precipitation of 69 cm w.e. at this elevation.

The predicted modern T-P conditions are shown on Figure 7.

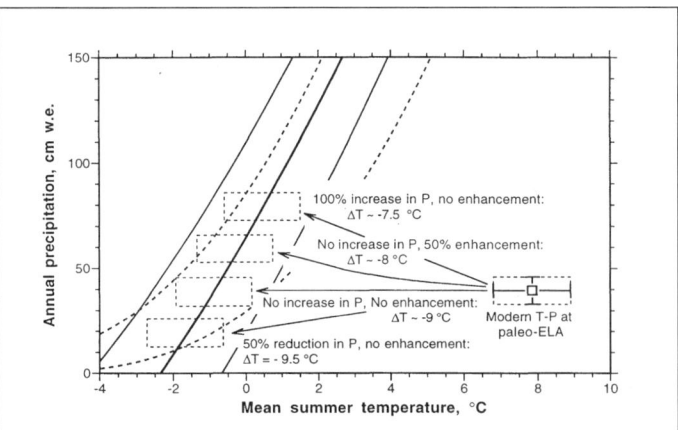

Figure 7. Climate parameters at modern equilibrium lines, and modern climate at the late Pleistocene equilibrium-line altitude (ELA) in the Taylor River basin. Solid bold line is derived from Ohmura et al. (1992) with the associated error delineated by the lighter lines. Dashed curves defines the Leonard's (1989) envelope. (See text for discussion.) P, precipitation; T, temperature; ΔT, change in temperature.

If it is assumed (1) that winter accumulation accounted for the bulk of annual precipitation, and (2) that local winter accumulation during the last glacial maximum was the same as it is today, then mean summer temperature would have to have been about 9 °C cooler. The first assumption arises from the fact that annual precipitation, and not winter accumulation, at the paleo-ELA must be known. At modern equilibrium lines, annual precipitation often exceeds winter accumulation by as much as 20–50% (Ohmura et al., 1992), and therefore equation (3) might underestimate annual precipitation. Using Fall's (1997) value of 69 cm w.e. as an upper limit for annual precipitation in the Taylor River basin only reduces the estimated temperature change by about 1 °C.

With regard to the second assumption, Leonard (1989) showed that even in the absence of an increase in winter precipitation, winter *accumulation* might increase as a result of a decrease in temperature. Presumably this would occur because of an increase in the duration of the accumulation season and the accompanying effect on the seasonal distribution of precipitation types (snow versus rain). Leonard (1989) attempted to quantify, at least in part, such temperature-enhanced accumulation. His work suggests that for modern climates and late Pleistocene temperature changes comparable to those described here, precipitation was effectively increased by about 50%. The effect of this enhancement is shown schematically in Figure 7 and implies a cooling of about 8 °C.

Late glacial climates in some regions of the Rocky Mountains were apparently wetter, and in others regions drier than today (Porter et al., 1983; Barnosky et al., 1987; Locke, 1990; Thompson et al., 1993; Hostetler and Clark, 1997). Synoptic summaries of paleoclimate data for the western United States (Barnosky et al., 1987; Thompson et al., 1993) indicate that central Colorado may have been drier during the last glacial maximum. (Thompson et al., 1993, did point out, however, that "effective moisture," a term that reflects decreased evaporation rates due to reduced temperatures and increased cloud cover, was greater.) Moreover, recent climate modeling of late Pleistocene glacial environments in the western United States by Hostetler and Clark (1997) suggests that paleoglaciers of the central Rockies existed under precipitation amounts similar to, or slightly less than, modern amounts.

Within the region of the study area, there is a close inverse relationship between modern snowpack at late Pleistocene equilibrium lines and paleo-ELAs (Fig. 6). A comparable relationship led Porter et al. (1983) to conclude that moisture sources and transport during the last glacial maximum were similar to that of the present. Leonard (1984) reached a similar conclusion, and furthermore argued that this precluded any increase in precipitation in southwestern Colorado.

It is thus reasonable to assume that because of lower sea-surface temperatures during the last glacial maximum (Kutzbach et al., 1993), precipitation in the Taylor River basin would have been reduced. However, Figure 7 shows that the magnitude of inferred temperature change is not particularly sensitive to changes in precipitation. For example, a 50% reduction in annual precipitation (assuming no temperature-enhanced accumulation) would imply a depression in mean summer temperature of about 9.5 °C. Likewise a doubling of precipitation yields a temperature depression on the order of 7.5 °C (about 6.5 °C with enhancement). These values are not significantly different from the 9 °C obtained under the assumption of no change in precipitation. Conversely, changes in paleoprecipitation are very sensitive to assumed temperature changes. Without invoking temperature change, a ninefold increase in annual precipitation would have been necessary to maintain the equilibrium state of the glaciers. A modest reduction in mean-summer temperature of about 4 °C would still require a concomitant increase in precipitation of nearly 500%. This sensitivity was first noted by Leonard (1989) and is also apparent in the climate modeling of Hostetler and Clark (1997).

It is likely then, that mean summer temperatures in the Taylor River basin during the last glacial maximum were about 7–9 °C cooler than they are at present. This range agrees well with other quantitative estimates of late Pleistocene climate change in the central Rocky Mountains. In his study of paleo-ELAs in Colorado, Leonard (1989) determined that mean summer temperatures were depressed about 8.5 °C assuming no changes in the amount and seasonal distribution of precipitation. Temperature depression could have been as much as 10–13 °C under drier conditions. Paleoclimate reconstructions based on fossil beetle assemblages (Elias, 1996) indicate mean July temperatures 10–11 °C colder than present. The modeling efforts of Hostetler and Clark (1997) suggest that paleoglaciers of the central and southern Rocky Mountains were sustained largely through decreases in mean July temperatures in the range of 9–12 °C with little or no change in precipitation.

ACKNOWLEDGMENTS

We are grateful to William Locke, Montana State University, and William Manley, Institute for Arctic and Alpine Research, for their thoughtful comments on an earlier draft of this paper. We also thank the Enrichment Committee of the University of Puget Sound, the Department of Geology and Geophysics, University of Minnesota, and The Trimble Corporation.

REFERENCES CITED

Andrew, J. T., 1975, Glacial systems: North Scituate, Massachusetts, Duxberry, 191 p.

Andrews, J. T., and Miller, G. H., 1972, Quaternary history of the northern Cumberland Peninsula, Baffin Island, N. W. T., Canada: Part IV. Maps of the present glaciation limits and lowest equilibrium line altitudes for north and south Baffin Island: Arctic and Alpine Research, v. 4, p. 45–59.

Barnowski, C. W., Anderson, P. M., and Bartlein, P. J., 1987, The northwestern U.S. during deglaciation: vegetational history and paleoclimatic implications, *in* Ruddiman, W. F., and Wright, H. E., eds., North America and adjacent oceans during the last deglaciation: Boulder, Colorado, Geological Society of America, The Geology of North America, v. K-3, p. 289–321.

Bradley, R. S., 1985, Quaternary paleoclimatology: Methods of paleoclimatic reconstruction: Boston, Unwin Hyman, 472 p.

Brugger, K. A., 1996, Implications of till-provenance studies for glaciological reconstructions of the paleoglaciers of Wildhorse Canyon, Idaho, U.S.A.: Annals of Glaciology, v. 22, p. 93–101.

Brugger, K. A., Goldstein, B. S., and Cotter, J. F. P., 1990, Growth, decay, and dynamics of late Quaternary glacial systems in the southern Sawatch Range, Colorado: a working hypothesis: Geological Society of America Abstracts with Programs, v. 22, no. 6, p. 4.

Burbank, D. W., 1991, Late Quaternary snowline reconstructions for the southern and central Sierra Nevada, California, and a reassessment of the "Recess Peak Glaciation": Quaternary Research, v. 36, p. 294–306.

Cararra, P. E., Mode, W. N., Meyer, R. L., and Robinson, S. W., 1984, Deglaciation and postglacial timberline in the San Juan Mountains: Quaternary Research, v. 21, p. 42–56.

Chadwick, O. A., Hall, R. D., and Phillips, F. M., 1997, Chronology of Pleistocene glacial advances in the central Rocky Mountains: Geological Society of America Bulletin, v. 109, p. 1443–1452.

Elias, S. A., 1996, Late Pleistocene and Holocene seasonal temperatures reconstructed from fossil beetle assemblages in the Rocky Mountains: Quaternary Research, v. 46, p. 311–318.

Elias, S. A., Carrara, P. E., Toolin, L. J., and Jull, J. T., 1991, Revised age of deglaciation of Lake Emma based on new radiocarbon and macrofossil analyses: Quaternary Research, v. 36, p. 307–321.

Fall, P. L., 1997, Timberline fluctuations and late-Quaternary paleoclimates in the southern Rocky Mountains, Colorado: Geological Society of America Bulletin, v. 109, p. 1306–1320.

Flint, R. F., 1971, Glacial and Quaternary geology: New York, John Wiley, 892 p.

Goldstein, B. S., Brugger, K. A., and Cotter, J. F. P., 1988, The chronology and style of glaciation in the southern Sawatch Range, Colorado: Geological Society of America Abstracts with Programs, v. 20, p. 416.

Gosse, J. C., Klein, J., Evenson, E. B., Lawn, B., and Middleton, R., 1995, Beryllium-10 dating of the duration and retreat of the last Pinedale glacial sequence: Science, v. 268, p. 1329–1333.

Grosval'd, M. G., and Kotlyakov, V. M., 1969, Present-day glaciers in the U.S.S.R. and some data on their mass balance: Journal of Glaciology, v. 8, p. 9–22.

Hawkins, F. F., 1985, Equilibrium-line altitudes and paleoenvironment in the Merchants Bay area, Baffin Island, N. W. T., Canada: Journal of Glaciology, v. 31, p. 205–213.

Hostetler, S. W., and Clark, P. U., 1997, Climatic controls of western U.S. glaciers at the last glacial maximum: Quaternary Science Reviews, v. 16, p. 505–511.

Kutzbach, J. E., Guetter, P. J., Behling, P. J., and Selin, R., 1993, Simulated climate changes: results of the COHMAP climate-model experiments, in Wright, H. E., Kutzbach, J. E., Webb, T., III, Ruddiman, W. F., Street-Perrott, F. A., and Bartlein, P. J., eds., Global climates since the last glacial maximum: Minneapolis, Minnesota, University of Minnesota Press, p. 24–93.

Leonard, E. M., 1984, Late Pleistocene equilibrium-line altitudes and modern snow accumulation patterns, San Juan Mountains, Colorado, U.S.A.: Arctic and Alpine Research, v. 16, p. 65–76.

Leonard, E. M., 1989, Climatic change in the Colorado Rocky Mountains: estimates based on modern climate at late Pleistocene equilibrium-lines: Arctic and Alpine Research, v. 21, p. 245–255.

Locke, W. W., 1990, Late Pleistocene glaciers and climate of western Montana, U.S.A.: Arctic and Alpine Research, v. 22, p. 1–13.

Loewe, F., 1971, Considerations on the origin of the Quaternary ice sheet of North America: Arctic and Alpine Research, v. 3, p. 331–344.

Maher, L. J., Jr., 1972, Absolute pollen diagram of Redrock Lake, Boulder County, Colorado: Quaternary Research, v. 2, p. 531–553.

Meier, M. F., and Post, A., 1962, Recent variations in mass net budgets of glaciers in western North America: International Association of Hydrological Sciences Publication, v. 58, p. 63–77.

Meierding, T. C., 1982, Late Pleistocene glacial equilibrium-line altitudes in the Colorado Front Range: a comparison of methods: Quaternary Research, v. 18, p. 289–310.

Murray, D. R., and Locke, W. W., III, 1989, Dynamics of the late Pleistocene Big Timber glacier, Crazy Mountains, Montana, U.S.A.: Journal of Glaciology, v. 35, p. 183–190.

Nelson, A. R., Millington, A. C., Andrews, J. T., and Nichols, H., 1979, Radiocarbon-dated upper Pleistocene glacial sequence, Fraser Valley, Colorado Front Range: Geology, v. 7, p. 410–414.

Ohmura, A., Kasser, P., and Funk, M., 1992, Climate at the equilibrium line of glaciers: Journal of Glaciology, v. 38, p. 397–411.

Paterson, W. S. B., 1994, The physics of glaciers: Oxford, Pergamon Press, 480 p.

Phillips, F. M., Zreda, M. G., Gosse, J. C., Klein, J., Evenson, E. B., Hall, R. D., Chadwick, O. A., and Sharma, P., 1997, Cosmogenic ^{36}Cl and ^{10}Be ages of Quaternary glacial and fluvial deposits of the Wind River Range, Wyoming: Geological Society of America Bulletin, v. 109, p. 1453–1463.

Pierce, K. L., 1979, History and dynamics of glaciation in the northern Yellowstone Park area: U.S. Geological Survey Professional Paper 729-F, 90 p.

Porter, S. C., 1975, Equilibrium-line altitudes of late Quaternary glaciers in the Southern Alps, New Zealand: Quaternary Research, v. 5, p. 27–47.

Porter, S. C., 1977, Present and past glaciation thresholds in the Cascade Range, Washington, U.S.A.: topographic and climatic controls, and paleoclimatic implications: Journal of Glaciology, v. 18, p. 101–116.

Porter, S. C., 1981, Glaciological evidence of Holocene climatic change, in Wigley, T. M. L., Ingram, M. J., and Farmer, G., eds., Climate and history: Studies of past climates and their impact on man: Cambridge, United Kingdom, Cambridge University Press, p. 82–110.

Porter, S. C., Pierce, K. L., and Hamilton, T. D., 1983, Late Wisconsin mountain glaciation in the western United States, in Wright, H. E., ed., Late-Quaternary environments of the United States, Volume 1, The late Pleistocene: Minneapolis, Minnesota, University of Minnesota Press, p. 71–111.

Richmond, G. M., 1965, Glaciation in the Rocky Mountains, in Wright, H. E., and Frey, D. G., eds., The Quaternary of the United States: Princeton, New Jersey, Princeton University Press, p. 217–230.

Seltzer, G. O., 1994, Climatic interpretation of alpine snowline variations on millennial time scales: Quaternary Research, v. 41, p. 154–159.

Sturchio, N. C., Pierce, K. L., Murrell, M. T., and Sorey, M. L., 1994, Uranium-series ages of travertines and timing of the last glaciation in the northern Yellowstone area, Wyoming-Montana: Quaternary Research, v. 41, p. 265–277.

Thompson, R. S., Whitlock, C., Bartlein, P. J., Harrison, S. P., and Spaulding, W. G., 1993, Climatic changes in the western United States since 18,000 yr B.P., in Wright, H. E., Kutzbach, J. E., Webb, T., III, Ruddiman, W. F., Street-Perrott, F. A., and Bartlein, P. J., eds., Global climates since the last glacial maximum: Minneapolis, Minnesota, University of Minnesota Press, p. 468–513.

Torsnes, I., Rye, N., and Nesje, A., 1993, Modern and Little Ice Age equilibrium-line altitudes on outlet valley glaciers from Josterdalbreen, western Norway: an evaluation of different approaches to their calculation: Arctic and Alpine Research, v. 25, p. 106–116.

MANUSCRIPT ACCEPTED BY THE SOCIETY OCTOBER 8, 1998

Pre-Illinoian glacial geomorphology and dynamics in the central United States, west of the Mississippi

James S. Aber
Earth Science Department, Emporia State University, Emporia, Kansas 66801-5087; aberjame@emporia.edu

ABSTRACT

A geographic information system (GIS) database has been compiled for pre-Illinoian glacial margins and related melt-water drainage features in the region west of the Mississippi. The GIS database was assembled from various published and unpublished sources; it includes glacial margins of different ages, melt-water channels, preglacial (buried) valleys, preglacial drainage divides, and related features, which are portrayed on a background constructed from a digital elevation model.

From this database, several observations have emerged. (1) Many more ice-margin positions are mapped than ever described before, especially in eastern Nebraska and southern Iowa. (2) Melt-water drainage valleys are well preserved in bedrock and represent the most important long-term geomorphic legacy of glaciation. (3) Melt-water drainage routes are situated in two positions—parallel and normal to ice margins. The parallel routes are arranged in concentric, nested patterns that mark positions of glacier advance and retreat. The normal routes are either parallel or slightly diverging in the downstream direction. These drainage routes, at right angles to ice margins, presumably formed as either subglacial tunnel valleys or proglacial spillways, in places following preglacial valleys. They probably served repeatedly in both roles during multiple glacier advances and retreats. The overall pattern of ice margins and drainage routes strongly suggests a lobate style for pre-Illinoian glaciations, in which ice-lobe surging may have taken place.

BACKGROUND

The Wisconsin and Illinoian glacial stages are accepted and widely used for glacial stratigraphy and correlation in North America and elsewhere. However, the "Kansan" and "Nebraskan" are no longer considered valid glacial stages in light of much evidence for more complicated pre-Illinoian glaciation of the central United States (Richmond and Fullerton, 1986). Hallberg's (1986) early and middle Pleistocene stratigraphy for Iowa and Nebraska now serves as the standard for the region. More recently, Aber (1991) proposed a revision for glaciation in northeastern Kansas, which is now called the Independence Glaciation, dated approximately 600 to 700 k.a.

The Glacial Map of the United States East of the Rocky Mountains (Flint et al., 1959) displayed the distribution of glacial deposits and related features in the central United States, as they were known and interpreted in the 1950s. Systematic mapping for the *Quaternary Geologic Atlas of the United States* included much detailed information on Wisconsin glaciation (e.g., Fullerton et al., 1995). However, interpretation of pre-Wisconsin glaciation was necessarily more limited in scope with little or no chronologic control available, especially for pre-Illinoian glaciations.

The work group on Geospatial Analysis of Glaciated Environments (GAGE) of the INQUA Commission on Glaciation undertook a project in 1995 to compile a database and produce maps of the glaciated region south of the Wisconsin ice limit, in the central United States. This project represents application of new methods—geographic information systems (GIS)—for compiling and producing synthesis maps of glacial features in central

North America. The overall goal of this project is to better document the geomorphic expression and limits of pre-Wisconsin glaciations and thereby to gain insight to possible glacier dynamics. This article presents initial project results for pre-Illinoian glaciation in the region west of the Mississippi River.

METHODOLOGY

GIS data entry was completed for the region from Illinois to North Dakota: latitude 37°–49° N, longitude 87°–104° W. Glacial features were derived from various published sources (Table 1), and much additional unpublished information was provided by contributors (Table 2). Among the most important features included in the database are limits of glaciation. These limits were identified following the method of Wayne (1985), in which drainage anomalies are key indicators for blockage and diversion of streams at ice margins. Other important indicators for glacial limits include:

- Heads or margins of drainage features, such as outwash fans and plains, eskers, spillway valleys, proglacial lakes, etc.
- Extents of mapped till sheets and distribution of large erratic boulders bearing striae or other glacial markings.
- Presence of other traces of direct glacier action, such as striations, grooves, and glaciotectonic structures in bedrock or unconsolidated sediment.

Sedimentary and geomorphic features of uncertain or ambiguous origin were excluded as evidence for ice margins. For example, exotic cobbles and boulders are scattered beyond the western limit of glaciation in the Dakotas, Nebraska, and Kansas. Such erratics may have been transported by melt-water floods (Aber, 1991), or they could be remnants of preglacial Tertiary gravel deposits of Rocky Mountain provenance, but they do not document glacial limits.

Glacial features were plotted on U.S. Geological Survey 1° by 2° topographic map sheets (scale 1:250,000), and then manually digitized in the latitude-longitude coordinate system.

TABLE 1. REFERENCES FOR PRE-ILLINOIAN MAP WEST OF MISSISSIPPI RIVER

State	References
Iowa	Dreeszen and Burchett (1971)
	Wright and Ruhe (1965)
Kansas	Aber (1991)
	Chelikowski (1976)
	Colgan (1992)
	Denne et al. (1982)
	Dort (1987)
	Dreeszen and Burchett (1971)
	Jewett (1964)
Minnesota	Hobbs and Goebel (1982)
	Patterson (1995, 1996)
Missouri	Colgan (1992)
	Dreeszen and Burchett (1971)
	Flint et al. (1959)
	Guccione (1983, 1985)
	Heim (1961)
	McCourt (1917)
Nebraska	Dreeszen and Burchett (1971)
	Wayne (1981, 1985)
North Dakota	Clayton et al. (1980)
	Fullerton et al. (1995)
South Dakota	Flint (1955)
	Fullerton et al. (1995)
	Lemke et al. (1965)

TABLE 2. CONTRIBUTORS TO THE PRE-WISCONSIN MAPPING PROJECT

Contributor	Region
Patrick M. Colgan	NW Missouri & Wisconsin
B. Brandon Curry	Southern Illinois
Margaret Guccione	Central Missouri
Howard Hobbs	Southeastern Minnesota
Mark D. Johnson	Central Minnesota
Robert Krumm	Illinois
Carrie Patterson	Southwestern Minnesota

Glacial features were digitized and stored as vector line objects. The minimum size criterion for inclusion of objects was features whose extents are at least 1 km wide or long. Objects <1 km in length or width generally were not included; however, a few smaller features of special interest were entered as point objects. *Idrisi for Windows* and *Tosca* were the GIS software packages used for data entry and manipulation.

The glacial features were combined with digital line graphs (basic geography) and digital elevation models (basic topography). A digital elevation model was utilized to prepare shaded-relief images that serve as backgrounds for displays of glacial features, which are portrayed as colored line symbols. This approach gives an impression of landscape terrain on which glacial features are superimposed without obscuring regional geomorphology. A reduced and simplified display portrays the overall map region from Illinois to North Dakota (Fig. 1).

PRE-ILLINOIAN GLACIAL FEATURES WEST OF THE MISSISSIPPI

The region of extensive pre-Illinoian glaciation is located west of the Mississippi River in the Central Lowlands and Great Plains of the north-central United States. The compilation of glacial features for this region has led to several observations about pre-Illinoian glaciations.

Ice-margin positions

Many more glacial margins are compiled here than have been mapped or recognized on a regional basis before. Only the more conspicuous or well-documented ice margins have been included in the database, as provided by contributors. Many smaller or less continuous ice margins exist locally and can be identified readily on the shaded-relief images and topographic maps. The ice margins form sets comprised of arcuate, parallel glacial borders. This is especially evident in eastern Nebraska and southern Iowa (Fig. 2), where the pattern of ice margins resembles the concentric "festooned moraines" of late Wisconsin glaciation in Indiana (Wayne, 1966, 1967).

Most of the glacial borders probably were created during recession of the ice margin (Wayne, 1985). Glacier retreat may have been in an active mode rather than by regional stagnation and downwasting, as suggested by the closely spaced, parallel pattern of the ice borders. However, it is also feasible that some ice borders were formed during separate glacial advances or retreats following periods of ice stagnation. It is possible that some ice margins were reoccupied more than once in different glacial episodes, during either advancing or retreating phases.

The ages and stratigraphy of some pre-Illinoian ice margins are known, for example the Independence glaciation in northeastern Kansas (Aber, 1991) and the Cedar Bluffs glaciation in eastern Nebraska (Reed et al., 1965), but most of the ice margins are of uncertain or unknown ages. The lack of detailed chronologic control makes problematic any attempt to work out the number, duration, or extent of various glacial advances across the region. A minimum age for all pre-Illinoian glaciations of the region is provided by cosmogenic dating of Sioux Quartzite bedrock knobs in southwestern Minnesota, beyond the limit of late Wisconsin glaciation (Fig. 3). The bedrock has been exposed continuously since last glaciation for at least half a million years (575 ± 57 k.a.; Bierman et al., 1999). This age is consistent with the presence of Pearlette "O" volcanic ash (610 k.a.) below the youngest till of the region (Hallberg, 1986).

Glacial geomorphology

Constructional glacial geomorphology is largely lacking for the region of pre-Illinoian glaciation, as has been well known since the late 1800s. Moraines, eskers, kames, and other landforms of glacial deposition are scarce, because of prolonged postglacial erosion and/or covering by younger sediment (loess). In contrast, bedrock valleys and channels attributed to glacial meltwater erosion are abundant and well preserved at all scales of observation. Many, if not most, of these valleys presumably came into being as a result of glacial diversion of drainage. In several portions of the region, modern drainage systems cut across preglacial buried valleys and divides. Asymmetric and barbed drainage patterns bear witness to the effects of glacial stream diversions (Wayne, 1985). The network of bedrock valleys, thus, represents the most important long-term geomorphic legacy of pre-Illinoian glaciation in the region.

Many of these valleys contain modern rivers. However, other

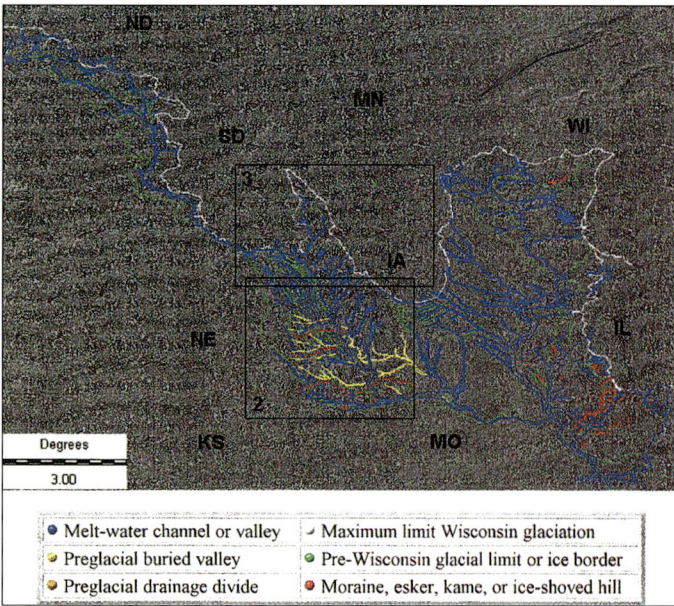

Figure 1. Overview of pre-Wisconsin glacial features in the central United States from Illinois (southeast) to North Dakota (northwest). Locations for Figures 2 and 3 indicated by boxes. Preglacial buried valleys and drainage divides mapped only for the region of Figure 2. Moraines and other constructional glacial landforms are restricted mostly to the region east of the Mississippi. Image processing by N. Wilkins and J. S. Aber.

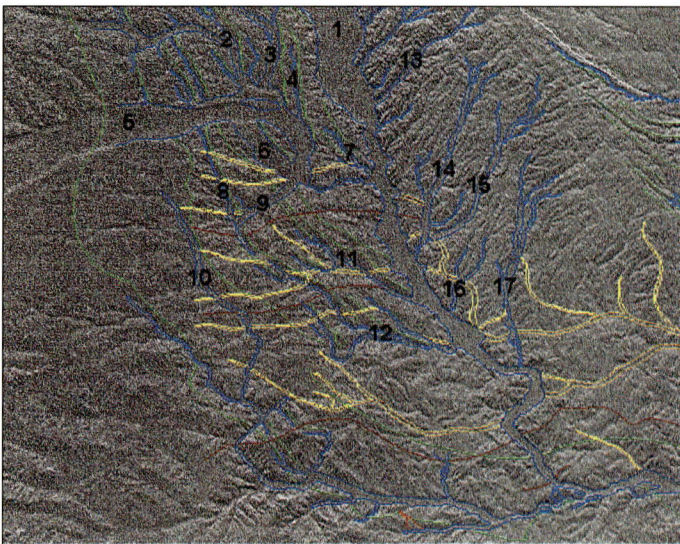

Figure 2. Detailed display of pre-Illinoian glacial features in southeastern Nebraska, northeastern Kansas, southwestern Iowa, and northwestern Missouri. Numbered valleys: 1, Missouri; 2, Elkhorn; 3, Logan; 4, Bell; 5, Platte; 6, Todd; 7, Papillion; 8, Oak; 9, Salt; 10, Big Blue; 11, Little Nemaha; 12, Big Nemaha; 13, Boyer; 14, West Nishnabotna; 15, East Nishnabotna; 16, Tarkio; and 17, Nodaway. Color coding of glacial features same as Figure 1. Image processing by N. Wilkins and J. S. Aber.

valleys are largely abandoned or hold greatly underfit streams, such as Todd Valley (Fig. 2). The bedrock valleys in general have been little modified by postglacial events. They are similar in many respects to melt-water spillways formed during the late Wisconsin glaciation in the northern Great Plains (Kehew and Lord, 1986) and southern Great Lakes (Kehew, 1993). Obvious exceptions are inferred spillways in northwestern Missouri. Downstream portions of the Nodaway and other tributaries to the Missouri River are quite narrow and tortuous, presumably because of downcutting

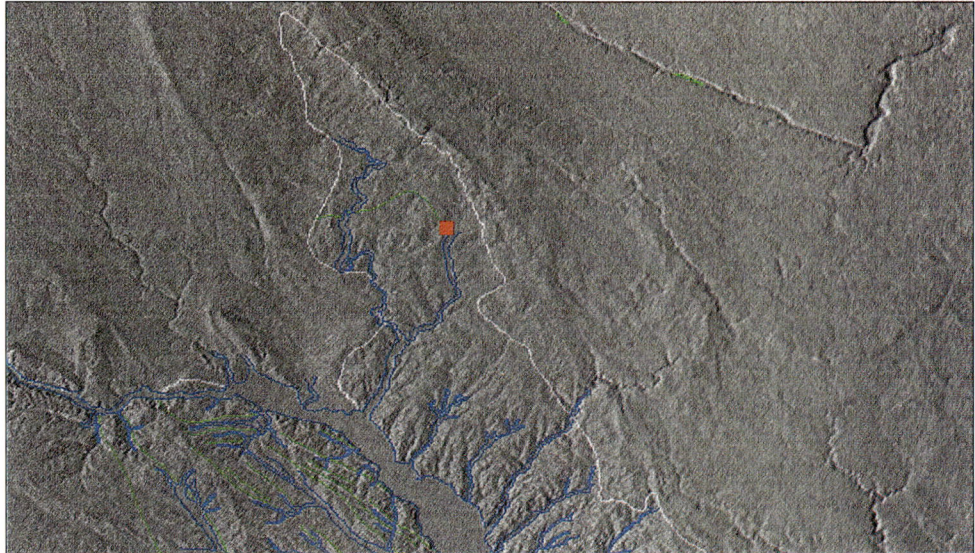

Figure 3. Detailed display of pre-Illinoian glacial features in southeastern South Dakota, southwestern Minnesota, northeastern Nebraska, and northwestern Iowa. Red square indicates location of Blue Mound, Minnesota, where cosmogenic dating was performed on Sioux Quartzite exposures. Blue Mound is located in the interlobate zone of pre-Illinoian glaciation. Color coding of glacial features same as Figure 1. Image processing by N. Wilkins and J. S. Aber.

into resistant Paleozoic limestone. Similar presumed spillway valleys are found in southeastern Minnesota and parts of Iowa.

Drainage patterns

Melt-water drainage routes are situated in two positions relative to ice margins—parallel and normal. The parallel routes are arranged in concentric, nested patterns that mark drainage paths along glacial borders. The normal routes are at right angles to inferred glacial borders and are either parallel or slightly diverging in the downstream direction. Eastern Nebraska serves as a type region for these drainage patterns (Fig. 2). The parallel drainage pattern is demonstrated by the Big Blue, Big and Little Nemaha, Oak, Papillion, Bell, Elkhorn, Logan, Todd, and other valleys (Wayne, 1985). Salt Creek follows a normal valley that extends from the lower Platte Valley toward the southwest to Lincoln. In northwestern Missouri and western Iowa, stream valleys are oriented normal to the ice margins of eastern Nebraska and northeastern Kansas. Representative examples include the Nodaway, Tarkio, East and West Nishnabotna, and Boyer Valleys.

The pattern of parallel and normal valleys bears a striking resemblance to glacial drainage patterns associated with late Wisconsin glaciation in the northern Great Plains and southern Great Lakes, where normal drainage routes are interpreted as tunnel valleys eroded by subglacial streams (Wayne, 1967; Patterson, 1993, 1997). The pre-Illinoian normal drainage routes presumably formed as either subglacial tunnel valleys or proglacial spillways, in some cases following preglacial valleys. Once formed, they probably served repeatedly in both roles during multiple glacial advances and retreats across the region. Valleys have survived because they are formed in well-consolidated Paleozoic bedrock. However, distinct variations can be seen in valley widths, which correspond to differences in bedrock lithology. The pre-Illinoian drainage network, thus, represents a mature product of multiple glaciations in which bedrock valleys were eroded both parallel and normal to glacial margins.

PRE-ILLINOIAN GLACIAL DYNAMICS WEST OF THE MISSISSIPPI

Ice lobes

The lack of end moraines, eskers, drumlins, and other constructional glacial landforms has long hampered efforts to interpret glacial dynamics for pre-Illinoian glaciations west of the Mississippi. I earlier proposed that pre-Illinoian glaciation took place in lobate fashion similar to the late Wisconsin glaciation (Aber, 1982, 1991). The late Wisconsin glaciation of the Great Plains is characterized by two major ice lobes—Des Moines and James—that followed lowlands on either side of the Coteau des Prairies upland (Fig. 4). Much of the northern Coteau des Prairies is built of thick glacial deposits, and so probably was not such a prominent topographic feature in pre-Wisconsin time (Patterson, 1993). However, the resistant Sioux Quartzite in the southern Coteau region must have formed a conspicuous upland, which led to lobate-style pre-Illinoian glaciation.

The overall pattern of ice margins and drainage routes strongly confirms a lobate style for pre-Illinoian glaciations in the region south of the Coteau. In southern Iowa, glacial limits and marginal drainage routes form a series of arcuate loops extending south of the maximum limit of the Des Moines lobe (Fig. 1). These ice margins represent stillstand positions of an expanded ice lobe, the Minnesota lobe, that can be traced clearly to near the Iowa-Missouri boundary. Most ice margins and drainage routes in eastern Nebraska trend northwest-southeast and would appear to relate to the western lateral margin of the Minnesota lobe (Fig. 2). In like manner, pre-Illinoian ice margins in southeastern Minnesota trend roughly north-south, parallel to each other and to the eastern edge of the Des Moines lobe (Fig. 1).

Evidence for another pre-Illinoian lobe is found in northeasternmost Nebraska (Fig. 3), where a series of ice margins and drainage routes correspond with an expanded ice lobe, the

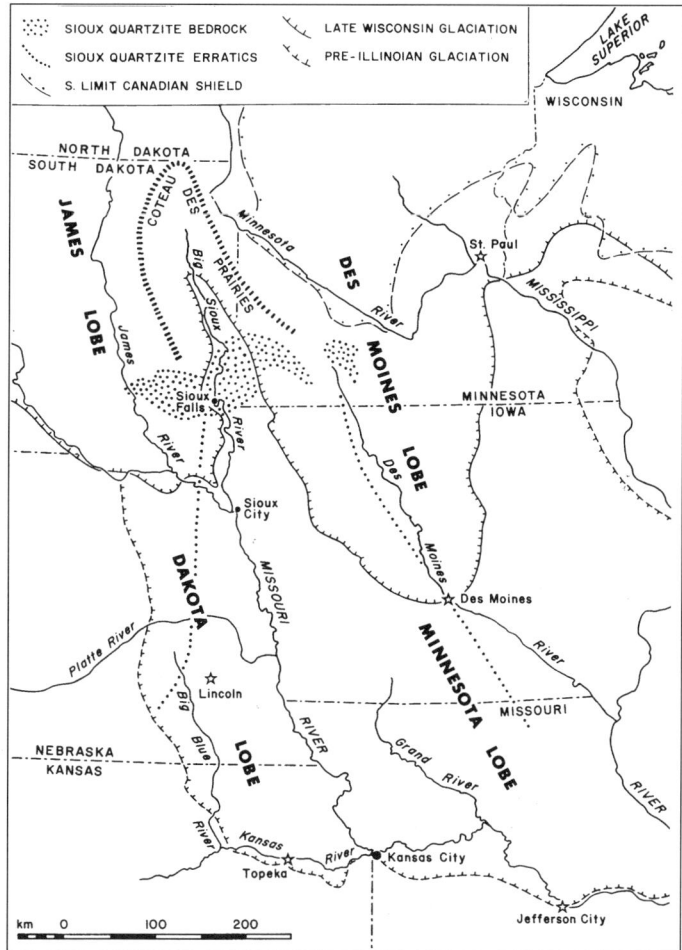

Figure 4. Major ice lobes of the Great Plains region of the north-central United States. Des Moines and James were late Wisconsin lobes; Minnesota and Dakota were presumed equivalent pre-Illinoian lobes. Limits of Sioux Quartzite erratic fan from Willard (1980); adapted from Aber (1991, Fig. 10).

Dakota (Wayne, 1985). The Dakota lobe may also have been responsible for well-documented regional ice movements from the north-northwest during the maximum Independence glaciation across northeastern Kansas (Dellwig and Baldwin, 1965; Aber, 1991). The existence of the Minnesota and Dakota lobes is confirmed also in the pre-Illinoian interlobate zone of the southern Coteau des Prairies (Fig. 3), where striations, grooves, and various fractures record ice movements from both the northeast and northwest. Similar crossing directions of ice movement are marked by striations and glaciotectonic structures in Nebraska, Iowa, and Missouri (Dort, 1987, Fig. 3).

On this basis, it seems that most pre-Illinoian glacial features of the region may be explained by a Minnesota ice lobe that occupied an expanded position similar to the late Wisconsin Des Moines lobe. The distribution of Sioux Quartzite erratics forms a fan (Fig. 4), which is consistent with transport by a Minnesota lobe (Willard, 1980). Evidence for the Dakota lobe is confined to more restricted areas mainly along and west of the Missouri Valley. Given the widespread geomorphic patterns, it appears the Minnesota lobe was probably the dominant glacier body for most pre-Illinoian glaciations of the region (Wayne, 1985).

Ice-lobe gradients

The Des Moines lobe serves as a well-known model for late Wisconsin glacier dynamics in the Great Plains. The western lateral margin, as marked by the Bemis moraine, may be taken as representative of low-gradient ice margins in a low-relief terrain. Table 3 presents the gradient of the Des Moines lobe in comparison to estimated gradients for assumed Minnesota and Dakota lobe margins. These gradients represent the adaption of ice-lobe dynamics to the regional topography, which is assumed to have been approximately similar for each glaciation. The Des Moines gradient equals a Minnesota margin following the Big Sioux and Missouri Valleys southward to the vicinity of Kansas City. This relationship, along with the general pattern of ice margins and drainage routes, supports the notion of Todd (1914) that the Missouri Valley trench likely came into existence as an ice-marginal drainage along the western side of the Minnesota lobe. The western maximum of pre-Illinoian glaciation represents a much lower gradient, however, the significance of which is unknown at this time. The western maximum ice limit could be a composite of multiple pre-Illinoian glacial advances by both Dakota and Minnesota lobes.

Many features of the late Wisconsin Des Moines and James lobes resemble those created by surging glaciers (Clayton et al., 1985). It is now generally accepted that thin, low-gradient ice lobes of the Great Plains may have advanced suddenly and repeatedly over water-lubricated or deformable beds (Clark, 1992), and Patterson (1997) extended this idea to include the "Winnipeg ice stream" that fed into the lobes from the north. By analogy, the pre-Illinoian Minnesota and Dakota lobes may also have been subject to glacier surging or ice streaming, as suggested earlier (Aber, 1991), given the equally low gradients and similar sedimentary substrata throughout the region.

CONCLUSIONS

1. Many more pre-Illinoian glacial margins and melt-water drainage features exist in the region west of the Mississippi than have been mapped or described before.

2. Proglacial and subglacial melt-water drainage valleys are well preserved in bedrock and represent the most important long-term geomorphic legacy of glaciation.

3. Glacial margins and drainage features strongly support the model of lobate-style glaciation for pre-Illinoian ice advances of the region. The Dakota and Minnesota lobes were analogous respectively to the late Wisconsin James and Des Moines lobes.

4. Pre-Illinoian ice lobes had low lateral gradients similar

TABLE 3. APPROXIMATE GRADIENTS ALONG ICE-LOBE LATERAL MARGINS

Ice Lobe	Vertical	Horizontal	Gradient
DES MOINES[a]	300 meters	440 km	0.7 m/km
DAKOTA[b]	165 meters	400 km	0.4 m/km
MINNESOTA[c]	370 meters	520 km	0.7 m/km

Note: a. Western lateral margin, as marked by the Bemis moraine crest, from near Watertown, South Dakota to Des Moines, Iowa.

b. Western maximum limit of pre-Illinoian glaciation from Creighton, Nebraska, to Manhattan, Kansas (on bedrock upland).

c. Assumed minimum ice thickness of 200 m on bedrock at Sioux Falls, South Dakota to maximum limit of glaciation at Kansas City, Missouri (on bedrock upland).

to or less than the late Wisconsin Des Moines lobe. The latter is thought to have advanced by surging, which suggests by analogy that pre-Illinoian lobes also may have advanced by surging.

ACKNOWLEDGMENTS

This project was made possible by the contributors who supplied published and unpublished information (Table 2). GIS data entry and editing were carried out by John Morettini and Jaap Jan Zeeberg under the direction of Naomi Wilkins, who was responsible for creation of map images. Financial support was provided by grants from NASA and from Emporia State University. H. C. Hobbs, C. J. Patterson, and W. J. Wayne reviewed an early draft of this article and offered valuable suggestions for improvement. A. Bettis provided additional comments for a later draft. I thank all these individuals and institutions.

REFERENCES CITED

Aber, J. S., 1982, Two-ice lobe model for Kansan glaciation: Nebraska Academy Sciences, Transactions, v. 10, p. 25–29.

Aber, J. S., 1991, The glaciation of northeastern Kansas: Boreas, v. 20, p. 297–314.

Bierman, P. R., Marsella, K. A., Patterson, C., Davis, P. T., and Caffee, M., 1999, Mid-Pleistocene cosmogenic minimum-age limits for pre-Wisconsinan glacial surfaces in southwestern Minnesota and southern Baffin Island: A multiple nuclide approach: Geomorphology, v. 27, p. 25–39.

Chelikowski, J. R., 1976, Pleistocene drainage reversal in the Upper Tuttle Creek Reservoir area of Kansas: Kansas Geological Survey, Bulletin, v. 211, part 1, 10 p.

Clark, P. U., 1992, Surface form of the southern Laurentide Ice Sheet and its implications to ice-sheet dynamics: Geological Society America Bulletin, v. 104, p. 595–605.

Clayton, L., Moran, S. R., Bluemle, J. P., and Carlson, C. G., 1980, Geological map of North Dakota: North Dakota Geological Survey, scale 1:500,000.

Clayton, L., Teller, J. T., and Attig, J. W., 1985, Surging of the southwestern part of the Laurentide Ice Sheet: Boreas, v. 14, p. 235–241.

Colgan, P. M., 1992, Stratigraphy, sedimentology, and paleomagnetism of pre-Illinoian glacial deposits near Kansas City, Kansas and Kansas City, Missouri [M.A. thesis]: University of Kansas, 201 p.

Dellwig, L. F., and Baldwin, A. D., 1965, Ice-push deformation in northeastern Kansas: Kansas Geological Survey Bulletin 175, part 2, 16 p.

Denne, J. E., Steeples, D. W., Sophocleous, M. A., Severini, A. F., and Lucas, J. R., 1982, An integrated approach for locating glacial buried valleys: Kansas Geological Survey, Groundwater Series, v. 5, 22 p.

Dort, W., Jr., 1987, Salient aspects of the terminal zone of continental glaciation in Kansas: Kansas Geological Survey, Guidebook Series, v. 5, p. 55–66.

Dreeszen, V. H., and Burchett, R. R., 1971, Buried valleys in the lower part of the Missouri River Basin: Kansas Geological Survey, Special Distribution Publication, v. 53, p. 21–25.

Flint, R. F., 1955, Pleistocene geology of eastern South Dakota: U.S. Geological Survey, Professional Paper, v. 262, 173 p.

Flint, R. F., Colton, R. B., Goldthwait, R. P., and Willman, H. B., 1959, Glacial map of the United States east of the Rocky Mountains: Geological Society of America, scale 1:1,750,000.

Fullerton, D. S., Bluemle, J. P., Clayton, L., Steece, F. V., Tipton, M. J., Bretz, R., and Goebel, J. E., 1995, Quaternary geologic map of the Dakotas 4° × 6° quadrangle, United States, in Quaternary Geological Atlas of the United States: U.S. Geological Survey, Miscellaneous Investigations Series, Map I-1420 (NL-14), scale 1:1,000,000.

Guccione, M., 1983, Quaternary sediments and their weathering history in north-central Missouri: Boreas, v. 12, p. 217–226.

Guccione, M. J., 1985, Quantitative estimates of clay-mineral alteration in a soil chronosequence in Missouri, U.S.A.: Soils and Geomorphology, Catena Supplement, v. 6, p. 137–150.

Hallberg, G. R., 1986, Pre-Wisconsin glacial stratigraphy of the Central Plains region in Iowa, Nebraska, Kansas, and Missouri: Quaternary Science Reviews, v. 5, p. 11–15.

Heim, G. E., Jr., 1961, Quaternary System, in Howe, W. B., and Koenig, J. W., eds., The stratigraphic succession in Missouri: Missouri Geological Survey and Water Resources, v. 40, 2nd Series, p. 130–136.

Hobbs, H. C., and Goebel, J. E., 1982, Geologic map of Minnesota, Quaternary geology: Minnesota Geological Survey, Map S-1, scale 1:500,000.

Jewett, J. M., 1964, Geologic map of Kansas: Kansas Geological Survey, Map M-1, scale 1:500,000.

Kehew, A. E., 1993, Glacial-lake outburst erosion of the Grand Valley, Michigan, and impacts on glacial lakes in the Lake Michigan basin: Quaternary Research, v. 39, p. 36–44.

Kehew, A. E., and Lord, M. L., 1986, Origin and large-scale erosional features of glacial-lake spillways in the northern Great Plains: Geological Society America Bulletin, v. 97, p. 162–177.

Lemke, R. W., Laird, W. M., Tipton, M. J., and Lindvall, R. M., 1965, Quaternary geology of northern Great Plains, in Wright, H. E., Jr., and Frey, D. G., eds., The Quaternary of the United States: Princeton, New Jersey, Princeton University Press, p. 15–27.

McCourt, W. E., 1917, The geology of Jackson County: Missouri Bureau of Geology and Mines, v. 14, 2nd Series, 158 p.

Patterson, C. J., 1993, Mapping glacial terrain in southwestern Minnesota, in Aber, J. S., ed., Glaciotectonics and mapping glacial deposits: Regina, Canada, Canadian Plains Research Center Proceedings 25, v. 1, p. 155–176.

Patterson, C. J., 1995, Regional hydrologic assessment, Quaternary geology—southwestern Minnesota: Minnesota Geological Survey, Regional Hydrologic Assessment, RHA-2, part A, pl. 1—Surficial geologic map, scale 1:200,000.

Patterson, C. J., 1996, The glacial geology of southwestern Minnesota with emphasis on the deposits and dynamics of the Des Moines lobe [Ph.D. thesis]: University of Minnesota, 143 p.

Patterson, C. J., 1997, Surficial geology of southwestern Minnesota: Minnesota Geological Survey, Report of Investigations, v. 47, p. 1–45.

Reed, E. C., Dreeszen, V. H., Bayne, C. K., and Schultz, C. B., 1965, The Pleistocene in Nebraska and Kansas, in Wright, H. E., Jr., and Frey, D. C., eds., The Quaternary of the United States: Princeton, New Jersey, Princeton University Press, p. 187–202.

Richmond, G. M., and Fullerton, D. S., 1986, Summation of Quaternary glaciations in the United States of America: Quaternary Science Reviews, v. 5, p. 183–196.

Todd, J. E., 1914, The Pleistocene history of the Missouri River: Science, New Series, v. 39, p. 263–274.

Wayne, W. J., 1966, Ice and land: A review of the Tertiary and Pleistocene history of Indiana, in Natural Features of Indiana, Indiana Sesquicentennial Volume: Indiana Academy of Science, p. 21–39.

Wayne, W. J., 1967, The Erie lobe margin in east-central Indiana during the Wisconsin glaciation: Indiana Academy Science, Proceedings for 1967, v. 77, p. 279–291.

Wayne, W. J., 1981, Kansan proglacial environment, east-central Nebraska: American Journal Science, v. 281, p. 375–389.

Wayne, W. J., 1985, Drainage patterns and glaciations in eastern Nebraska: Ter-Qua Symposium Series, v. 1, p. 111–117.

Willard, J. E., 1980, Regional directions of ice flow along the southwestern margin of the Laurentide Ice Sheet as indicated by distribution of Sioux Quartzite erratics [M.A. thesis]: University of Kansas, Lawrence, 87 p.

Wright, H. E., Jr., and Ruhe, R. V., 1965, Glaciation of Minnesota and Iowa, in Wright, H. E., Jr., and Frey, D. G., eds., The Quaternary of the United States: Princeton, New Jersey, Princeton University Press, p. 29–41.

MANUSCRIPT ACCEPTED BY THE SOCIETY OCTOBER 8, 1998

Wisconsin Episode glacial landscape of central Illinois: A product of subglacial deformation processes?

W. Hilton Johnson
Department of Geology, University of Illinois, Urbana, Illinois 61820
Ardith K. Hansel
Illinois State Geological Survey, 615 East Peabody Drive, Champaign, Illinois 61801

ABSTRACT

The Wisconsin Episode glacial landscape of central Illinois is characterized by a series of subparallel end moraines separated by low-relief till plain and lake plain. The end moraines are asymmetric with steeper and more prominent distal slopes and long, gentle proximal slopes that merge with till plain. Some end moraines are broadly arcuate; others are more lobate in map view. Notably lacking or poorly developed in the moraines are such typical features as kettles and hummocky topography, outwash plains, and kames. Supraglacial sediment is thin or absent; end moraines and till plain are composed primarily of uniform till.

Characteristics of the central Illinois landscape and sediment sequences are consistent with subglacial deposition from basal ice and/or a deforming bed. We explain moraine formation at stationary ice margins in central Illinois using a deforming-bed model. The model is based in part on processes inferred to operate at present at the margin of Fláajökull in Iceland. In our deforming bed model, debris is transported to and deposited on a proximal ramp beneath the glacier, which results in asymmetric end moraines. The moraines are composed of till and merge in the up-ice direction with till plain formed as the ice margin retreated.

Broadly arcuate end moraines and associated outwash and overridden proglacial sediment represent major ice-margin advances of the Lake Michigan Lobe; lobate moraines, which tend to lack outwash, represent recessional ice-margin positions within a sublobe. Although the landscape and sediment record reflect predominantly active-ice deposition from a wet-based glacier, the subglacial hydrology must have varied between times of instability and ice margin fluctuation and times of stability and moraine formation.

INTRODUCTION

Background

The landscape of central Illinois is largely the result of glaciation during the Wisconsin Episode. It contains a large number of end moraines (Fig. 1) and, as a result of their exceptional development, maps of the moraines are often included in Quaternary geology texts (e.g., Sugden and John, 1976; Bowen, 1978; Ehlers, 1996; Menzies, 1995). Overall, however, the landscape, as well as the sediment record, lacks many of the characteristics typical of a glaciated landscape. Kettles and hummocky topography are notably lacking, even in the end moraines. Outwash plains with their associated glaciofluvial deposits are either absent or discontinuous, small, and poorly developed on the distal sides of many end moraines, particularly recessional ones. Kames and ice-walled lake plains are uncommon, whereas drumlins, eskers, and tunnel valleys/channels are lacking. Supraglacial

---

Johnson, W. H., and Hansel, A. K., 1999, Wisconsin Episode glacial landscape of central Illinois: A product of subglacial deformation processes?, in Mickelson, D. M., and Attig, J. W., eds., Glacial Processes Past and Present: Boulder, Colorado, Geological Society of America Special Paper 337.

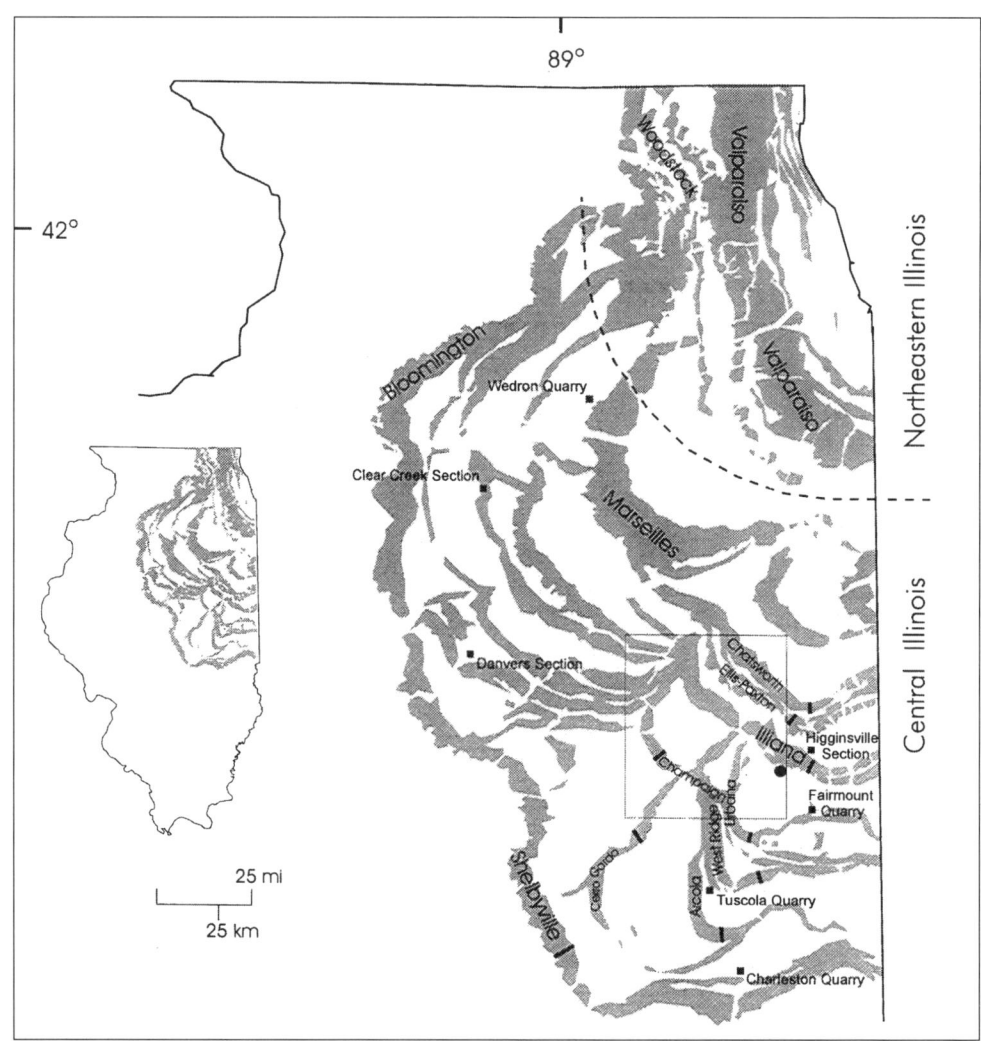

Figure 1. Map of Illinois showing location of Wisconsin Episode end moraines (gray) with respect to northeastern and central Illinois (separated by dashed line), moraine profiles in Figure 2 (shown by dark lines), satellite view in Figure 3 (shown by rectangle), key sections studied in central Illinois (shown by black squares), and site for which sediment flux was calculated (shown by black dot).

sediment is thin to absent, and the end moraines and till plain are composed predominantly of uniform till (Mickelson et al., 1983; Johnson and Hansel, 1995).

The low-relief landscape of the Wisconsin Episode Lake Michigan Lobe in central Illinois has been attributed to wet-based, fast-moving ice that had a low ice surface profile and deposited thick, widespread, uniform till sheets in a relatively short time (Willman and Frye, 1970; Mickelson et al., 1981, 1983; Clayton et al., 1985; Begét, 1986; Clark, 1992, 1994; Johnson and Hansel, 1990; Hansel and Johnson, 1992; Alley, 1991). Yet, how the glacier eroded, transported, and deposited sediment to form the landscape of end moraines and till plain is not fully understood.

Modern glacier studies in Iceland and Antarctica and theoretical modeling have led to speculation by some glaciologists and geologists that subglacial deformation of unlithified substrate materials may have contributed to or even dominated ice movement and sediment flux along the southern margin of the Laurentide Ice Sheet. Some previous studies have concluded that the character of the widespread, uniform till sheets that occur in the Lake Michigan Lobe area of Illinois is (most) consistent with deposition from the base of a deforming bed (e.g., Boulton, 1996a, b; Alley, 1991; Clark, 1992, 1994; Hansel and Johnson, 1997; Hansel et al., 1993; Johnson et al., 1991), although the evidence remains equivocal. In recent papers, Boulton (1996a, b) has used theoretical models for the processes of erosion, transport, and deposition by subglacial sediment deformation to explain the origin of till sequences in midlatitude glaciated areas. However, the formation of end moraines, such as the series that formed on the Illinois landscape as the ice retreated from the glacial maximum between about 20,000 and 14,000 ^{14}C years ago, has not been previously addressed in the context of the deforming bed model. In this paper, we examine the landscape and the sediment record of central Illinois to see if they are consistent with a deforming bed origin as speculated by Boulton and others (Boulton, 1996a, b; Boulton and Jones, 1979; Boulton and Hindmarsh, 1987), Clark (1997), and Alley (1991), and we offer a model of moraine formation.

Deforming-bed model

The conceptual description of the deforming-bed model presented here is summarized from theoretical explanations by Boulton and others (Boulton and Jones, 1979; Boulton and Hindmarsh, 1987; Boulton and Dobbie, 1993; Boulton, 1996a, b) and Alley (1991). In the deforming-bed model the glacier is moving over a soft substrate of unfrozen, unlithified sediment. Ice velocities are high (>≈100 m/a) and basal shear stresses low (≈10 kPa), which requires (1) ice at the pressure melting point, (2) saturated substrate materials, and (3) high basal water pressures. According to the model, the soft, unlithified materials fail and begin to deform when their shear strength is less than that of the ice. As these materials dilate and take in pore water, their bulk density decreases, and a zone of pervasive deformation develops beneath the glacier. The effect of bed deformation added to ice motion by basal sliding and ploughing increases ice velocity, causing the ice to undergo extensive flow and thinning, which lowers the ice surface profile. Fine-grained substrate materials of low permeability are more likely to deform because high pore water pressures and low effective pressures are more likely to develop in them than in coarse-grained materials, which are more readily drained.

Boulton and Hindmarsh (1987) referred to a zone of pervasive deformation, immediately beneath the glacier, as the A horizon of the deforming bed. Underlying the A horizon is a more stable zone, which they called the B horizon. Erosion occurs when there is a net loss of material from the stable B horizon, whereas deposition results in a net gain to the B horizon. Whether or not any given site beneath the glacier is undergoing erosion or deposition is related to ice flux and the location of the zones of accumulation and ablation (Boulton, 1996a). In the accumulation zone, shear stress and velocity increase down-ice resulting in erosion (lowering of the A/B interface) and an increase in discharge to the A horizon. In the ablation zone, shear stress and velocity decrease down-ice resulting in deposition (raising of the A/B interface) and a decrease in discharge to the A horizon. Sediment can be added to the deforming bed both by meltout of debris from basal ice and by erosion of sediment from the B horizon; the sediment is advected toward the ice margin by the pressure and forward motion of the ice. Boulton (1996b) hypothesized that fluctuation of the ice margin during glacial advance and retreat cycles would result in migration of depositional and erosional zones, producing erosion surfaces between advance and retreat phase tills at some sites. In the deforming bed model, the processes of sediment advection and the fluctuation of an ice margin during an advance/retreat cycle would produce wedge-shaped till sheets that thin in the up-ice direction (Boulton, 1996a, b). Tills formed in and deposited from a deforming bed would be enriched in silt and clay, because subglacial hydrogeologic conditions associated with fine-grained sediment allow them to fail and be incorporated into the deforming bed more easily. Retreat-phase tills would increase in far-traveled lithologies upward, because of the shifting zone of erosion toward the up-ice direction as the ice retreats.

The following sediment record and till characteristics are predicted to result from subglacial sediment deformation (Alley, 1991; Boulton, 1996a, b; Clark, 1997; Clark and Walder, 1994; Walder and Fowler, 1994): (1) fine-grained, matrix-dominated till beds, (2) uniform till beds that generally lack observable deformation indicators except locally near the base where intraclasts may show evidence of folding and attenuation, (3) sharp basal contacts with subtill sediments that are undeformed or only weakly deformed, (4) advance and retreat phase tills that are separated by clast (cobble/boulder) concentrations and/or erosion surfaces in the accumulation zone and gradational contacts in the ablation zone, (5) clasts with weakly developed striae and striae that do not consistently parallel ice flow direction, (6) pebble fabrics that indicate creep of adjacent diamicton toward channels, (7) large sediment fluxes, (8) thick till sequences preserved in the marginal deposition zone, (9) an increase of far-traveled lithologies upwards in a retreat phase till, and (10) shallow, wide, subglacial channel-fill deposits that may show some evidence of deformation and that have flat, upper contacts with overlying till.

METHODOLOGY

Our study of the landscape and sediment record in central Illinois is based on sedimentology at key exposures. Such exposures, unfortunately, are few and far between in central Illinois; the best exposures are located in quarries and stream cuts (e.g., Wedron, Tuscola, Charleston, and Fairmount Quarries and Clear Creek, Danvers, and Higginsville Sections; Fig. 1). Pebble fabrics are based on measurements of the azimuth and plunge of the long axis of 25 clasts with A/B ratios of at least 2:1. Pebble fabrics are plotted on equal-area lower-hemisphere projections by computer, contoured at 2-σ intervals using the method of Kamb (1959), and evaluated statistically following the eigenvector method of Mark (1973). Generalized cross sections were drawn using all available data, particularly engineering and test borings, water-well records, exposures, and maps of bedrock topography. Time-distance diagrams were constructed using available radiocarbon ages and vertical profiles of lithofacies/glacigenic sequences at key sites to estimate ice-margin fluctuations through time.

CHARACTERISTICS OF THE LANDSCAPE

The Wisconsin Episode landscape of central Illinois consists of a series of subparallel end moraines (Fig. 1), the configuration of which reflects flow out of the Lake Michigan basin. Locally, subglacial topographic highs, controlled by the bedrock geology, have resulted in the development of several sublobes and a complicated moraine pattern (Johnson et al., 1986). Most of the end moraines are true end moraines in the classification of Mickelson et al. (1983); that is, they are composed predominantly of till and formed at an ice margin during the last episode of till deposition. True end moraines in central Illinois generally are either simple or superposed (overridden or partially overridden). Superposed end moraines are common in the broader, multiple-crested morainic systems (e.g., Shelbyville, Bloomington, Illiana, Marseilles; Fig. 1), where the moraines may be made up of tills of multiple readvances or stillstands during which the ice built a new moraine on the proximal slope of an older moraine or in some cases actually overrode the older moraine(s). An example of an overridden moraine is the Valparaiso Moraine in northeastern Illinois (Johnson and Hansel, 1989). It consists of multiple glacigenic sequences made up of tills and proglacial sediments deposited during the Woodstock and Crown Point phases, respectively (Hansel and Johnson, 1992, Fig. 6); much of the relief of the Valparaiso Moraine was inherited from the buried Woodstock Moraine (Hansel, 1983; Hansel and Johnson, 1987).

In map view, some of the moraines of central Illinois are broadly arcuate; others are much more lobate (Fig. 1). Outwash plains are better developed in front of the broadly arcuate moraines; they are rare in front of the more lobate moraines (Hansel and Johnson, 1996, Plate 1). In transverse cross section, the moraines are strongly asymmetric with a steeper distal slope and a long, gentle, ramp-like, proximal slope (Figs. 2, 3). Overall relief (moraine height) varies with different moraines, but ranges generally from 10 to 60 m. Moraine width varies from about 2 km for some of the single moraines to about 20 km for multiple-crested moraines in some of the morainic systems; average width is about 5–7 km. Local relief in the end moraines is not great (<5 m) and is primarily the result of dissection, not hummocks or kettles. Ice-contact slopes are notably lacking in the moraines; as a result, the morainic ridges are subdued and have slopes mostly less than 2%. A blanket of loess averaging a meter thick also may tend to subdue slopes.

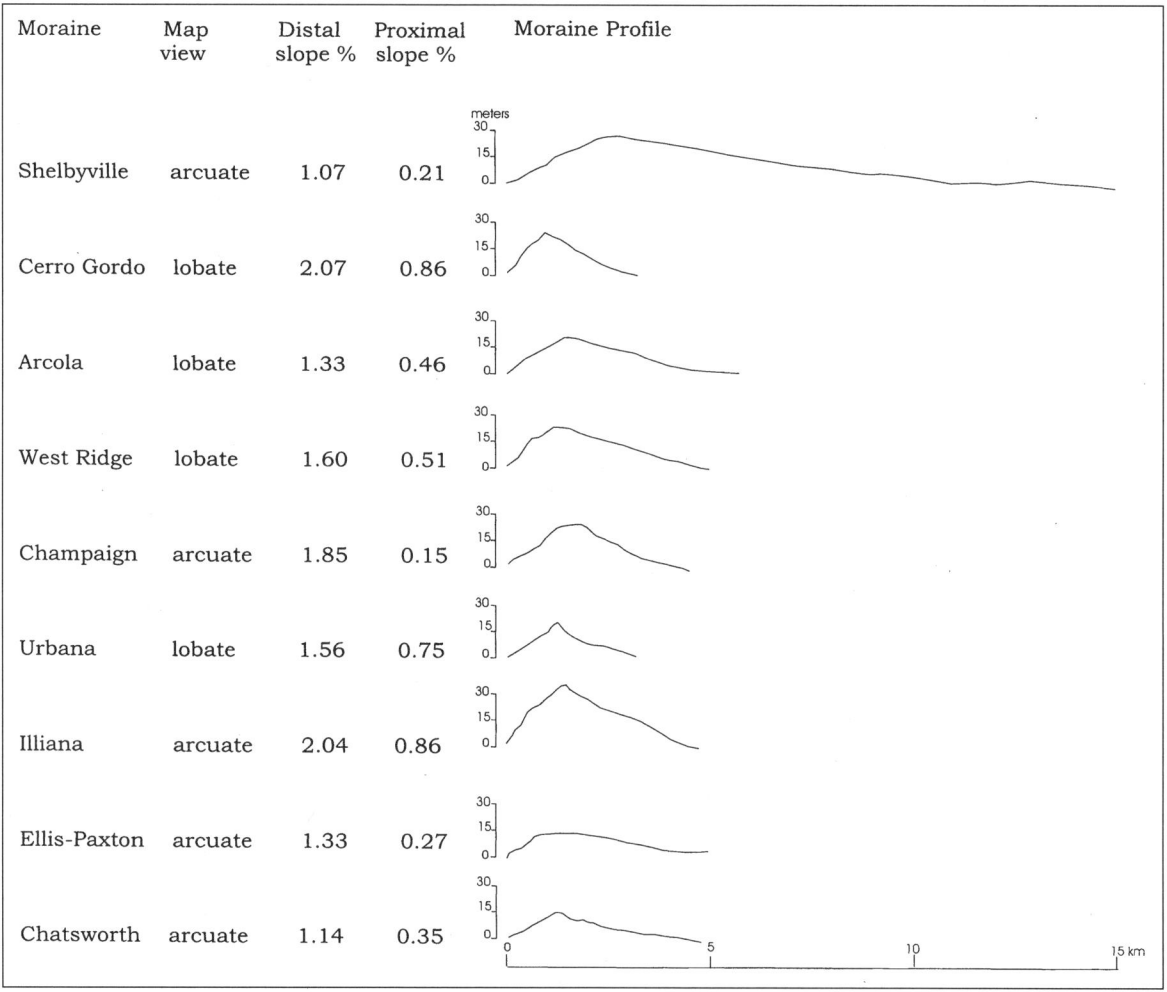

Figure 2. Sample profiles (vertical exaggeration approximately 45×) and distal and proximal slopes (in percent) for arcuate and lobate moraines in central Illinois. Lines of profile shown in Figure 1.

Both end-moraine and till-plain areas in central Illinois are composed predominantly of till that is uniform in texture and composition and attributed to subglacial deposition (Hansel and Johnson, 1996; Johnson, 1976; Mickelson et al., 1983). Examples include till in the Shelbyville Moraine near Shelbyville (Johnson et al., 1971), in the till plain at Charleston Quarry (Gutowski et al., 1991), in the till plain at Wedron Quarry (Johnson and Hansel, 1990), and in end-moraine and till-plain areas in and adjacent to Iroquois County (Moore, 1981) and Vermilion and Coles Counties (Johnson et al., 1972). Supraglacial sediment generally is thin or unidentifiable in either end-moraine or till-plain areas, although it is somewhat more common in the end moraines.

The end moraines are separated by lake plain or low-relief till plain that gradually merges with the gentle, proximal slopes of

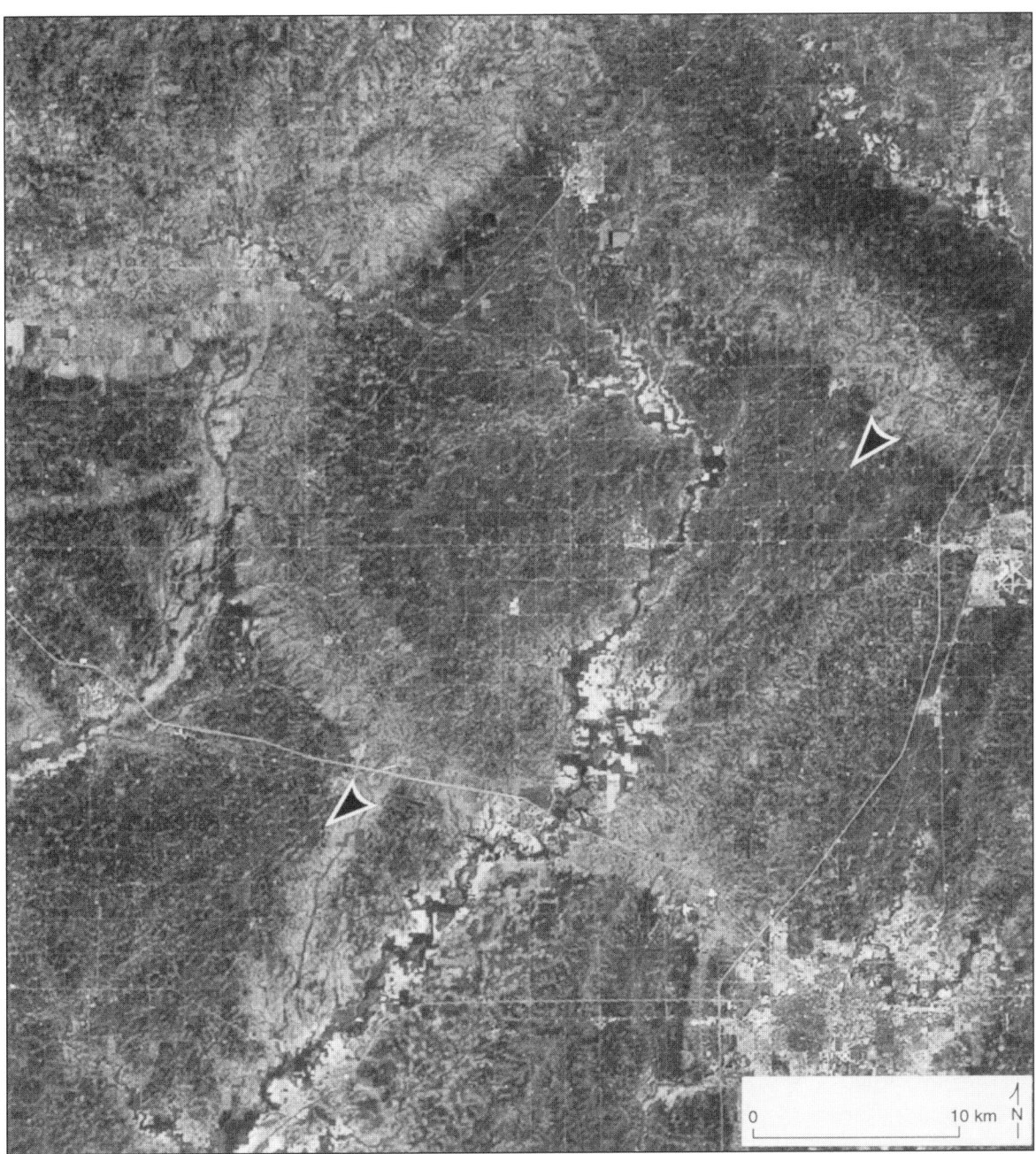

Figure 3. Satellite view of end and ground moraines in east-central Illinois (see rectangle in Fig. 1). Note arcuate end moraines (light tone) that exhibit abrupt distal slopes and long proximal slopes merging with till plain and/or lake plain (dark tone) in the up-ice direction, a reentrant between two sublobes in upper, center part of image, and areas of local small-scale fluting (see arrows). From a Landsat 1 MSS satellite image acquired on June 11, 1978 (Scene ID-LM2024032007816290).

the end moraines (Fig. 3). Till-plain relief varies but most commonly is 3–6 m and related to broad undulations of the land surface. Exceptions include the rare kame that has more relief and steeper slopes. Some till-plain areas appear fluted as evidenced by the linear features parallel to ice flow (i.e., perpendicular to morainic ridges) on the satellite photo in Figure 3. In lake-plain areas, essentially all relief is subdued by the cover of laminated silt and clay (Fig. 4).

SEDIMENT RECORD

The Wisconsin Episode glacial succession in Illinois is comprised of a series of offlapping drift sheets. Except in some parts of the northeastern Illinois, the drift sheets are composed mostly of uniform diamicton. The glacial deposits have been classified into lithostratigraphic units (formations and members), primarily on the basis of lithology (color, texture, composition) and stratigraphic position (e.g., Willman and Frye, 1970; Lineback, 1979; Hansel and Johnson, 1996). The shingled drift sheets pinch out northward beneath successively younger drift sheets (Fig. 4). Locally, tongues of proglacial sorted sediment (outwash sand and gravel and/or lake clay and silt) separate the till units of different drift sheets. Rarely, tongues of proglacial sediment separate till beds of the same till unit.

The generalized cross section in Figure 4 shows that the Wisconsin Episode deposits range from about 10 to 60 m thick along that traverse. The tills form wedge-shaped units that are overthickened in the end moraines. The base of the Wisconsin drift is marked by a relatively smooth erosional contact with paleosols and older glacial drift that had filled bedrock valleys prior to the Wisconsin Episode glaciation. The last interglacial soil (Sangamon Geosol) and in some cases the Wisconsin Episode cold-climate soil (Farmdale Geosol) are preserved beneath proglacial outwash and lake sediment and/or diamicton in much of the outer 50–80 km of the Wisconsin drift (Kempton and Gross, 1971). In the region of the Silurian bedrock high in the northern part of the cross section, most of the older drift and paleosols have been eroded; outwash sand and gravel of the second to the youngest drift sheet (Lemont) overlies bedrock. Generalized cross sections, such as the one shown in Figure 4, and radiocarbon age control (which is sparse between about 20,000 and 15,000 ^{14}C years B.P.) make it possible to construct time-distance diagrams (e.g., Fig. 5).

DIAMICTON CHARACTER AND CONTACT RELATIONSHIPS

The uniform tills of the last glaciation in central Illinois are relatively fine grained. Percent silt and percent clay in the matrix each range from about 30 to 50%, whereas percent sand ranges from about 10 to 40%. The dominant clay mineral in the till matrix is illite, which ranges from about 65 to 80% in unaltered samples. The illite was derived from Paleozoic shales from the Lake Michigan basin and northern Illinois (Willman and Frye, 1970; Glass and Killey, 1987). Magnetic susceptibility is also low, generally ranging from about 20 to 50 × 10^{-5} SI units. Gravel clast lithologies are dominated by Paleozoic sedimentary rocks derived from the Lake Michigan basin and northern Illinois (Willman and Frye, 1970; Mickelson et al., 1983). The composition of the basal part of some till units reflects incorporation of substrate materials (e.g., the Oakland facies of the Tiskilwa Formation; Hansel and Johnson, 1996; which is enriched in organic

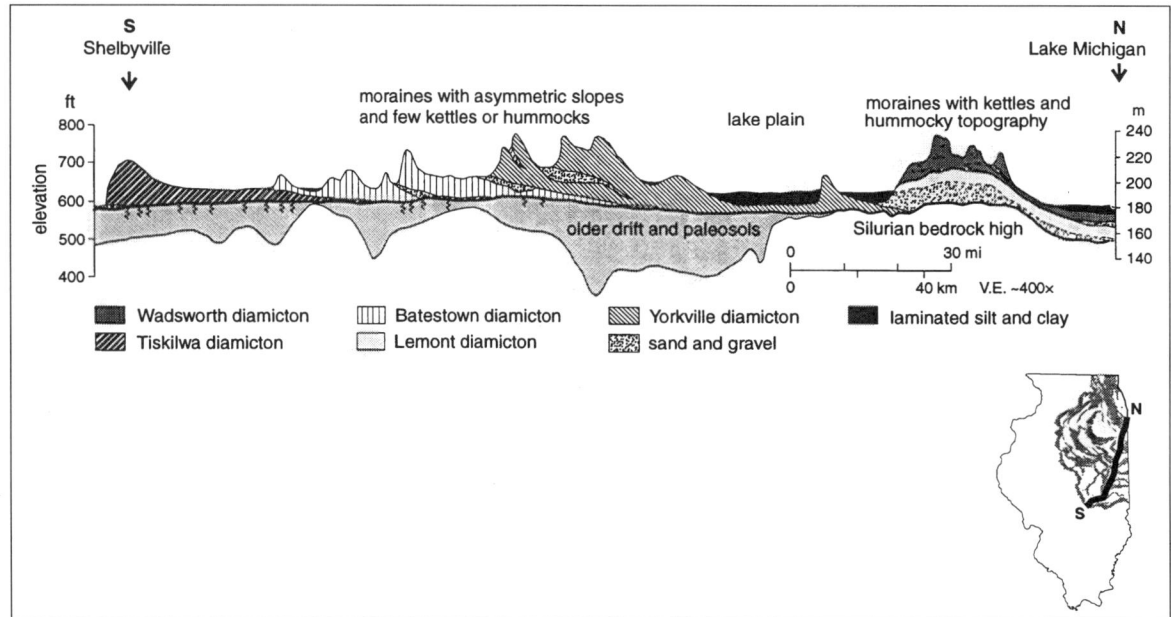

Figure 4. Generalized cross section from Shelbyville to Lake Michigan. Bedrock surface data from Herzog et al. (1994); Quaternary geology from interpretation of geologic records at Illinois State Geological Survey.

silt that is inferred to have been eroded from the landscape as the Wisconsin Episode glacier advanced).

Multiple lithologically distinct till beds, ranging from less than 1 m to several meters thick, commonly occur within a glacigenic sequence (i.e., the deposits of a single ice-margin advance and retreat across a site). These till beds can be distinguished by slight differences in lithology (matrix grain size, gravel clast concentration, color, composition). A good example is shown in the slight difference in matrix grain size between the two homogeneous till beds in sequence II at Wedron Quarry (Fig. 6). Till beds generally consist of diamicton that is homogeneous, except for occasional deformed sand and gravel lenses (Fig. 7a) and subglacial channel-fill deposits (Fig. 7b, c). Planar, erosional contacts are common at the base of till beds, whether at the base of the till sequence (Fig. 8a) or at the contact with a lower till bed (Fig. 8b). Locally, partially enfolded substrate materials (Fig. 8c, d) have been observed along contacts that are elsewhere planar. Substrate materials generally show little evidence of deformation (Fig. 8a). Attenuated lenses of substrate materials are sometimes present near the base of till beds (Fig. 8e). If such lenses are of sorted sediment, they generally lack primary bedding structures. Pebble fabrics are strong (S_1 values of 0.70 or greater), even when attenuated lenses of sorted sediment are present (Fig. 8e). Pebble azimuths generally are oriented parallel to regional ice flow, as determined by end moraine positions, and dip in the up-ice direction.

Locally, stone concentrations are present between and within lithologically distinct till beds (Fig. 8f). Rarely do these stones exhibit the striated, planed off surfaces typical of classic boulder or cobble pavements. Dave Voorhees (written communication,

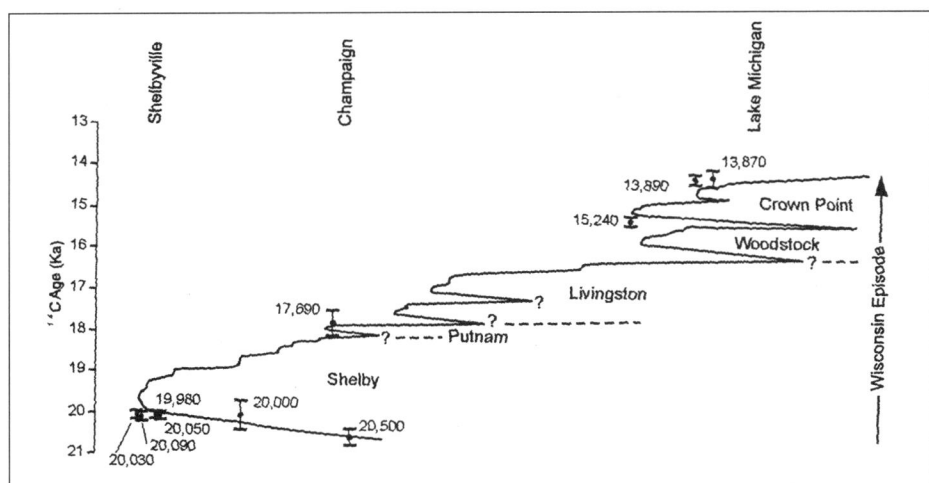

Figure 5. Time-distance diagram showing glacial phases for the Wisconsin Episode Lake Michigan Lobe from Shelbyville to Lake Michigan. Key ^{14}C control is also indicated on diagram.

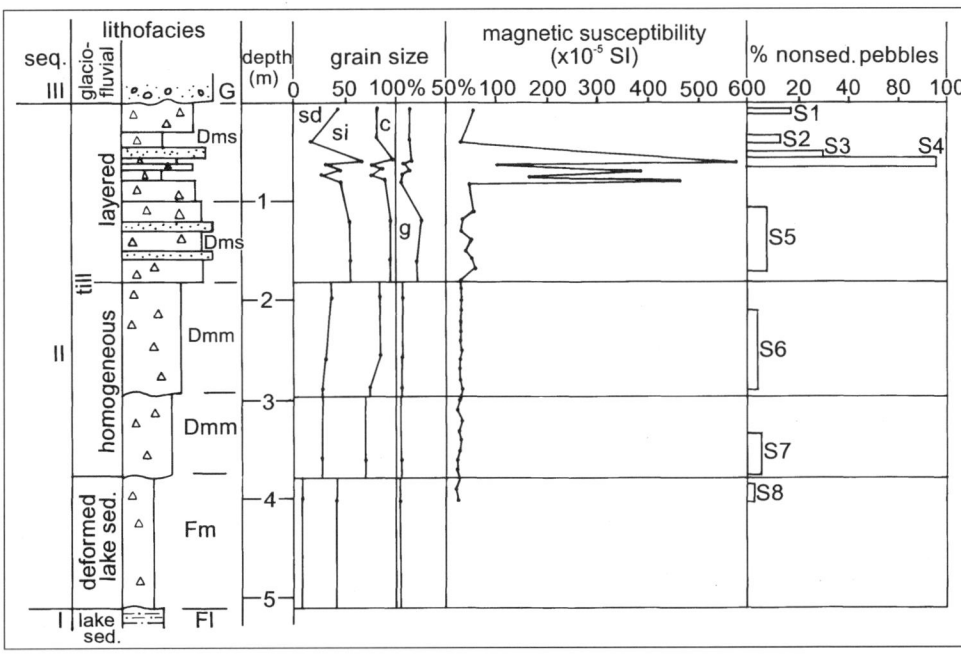

Figure 6. Lithofacies, grain size, magnetic susceptibility, and percentage of nonsedimentary pebbles for samples (S1-S3) from glacigenic sequence II (Batestown Member, Lemont Formation) at Wedron Quarry.

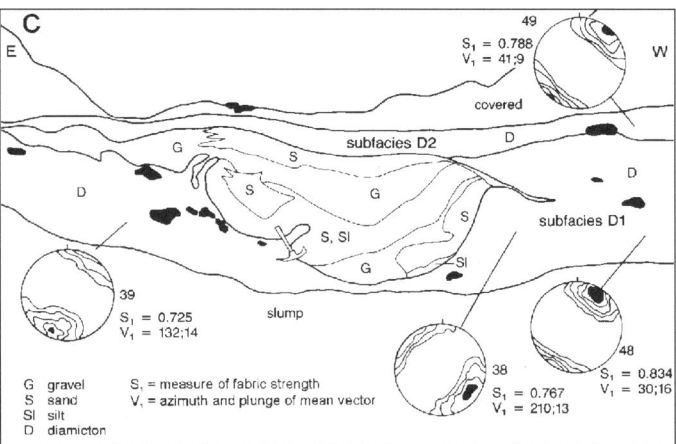

Figure 7 (above and in opposite column). Sand and gravel intraclasts in otherwise uniform diamicton beds: a, deformed sand lens in uniform diamicton of the Batestown till at Wedron Quarry; b, subglacial channel fill with truncated top in between homogeneous till beds of Batestown sequence at Wedron Quarry; c, deformed subglacial channel fill between till beds (subfacies D1 and D2) in the Tiskilwa sequence at Wedron Quarry (see Fig. 8e for explanation of fabric diagrams; from Johnson and Hansel, 1990).

1997) found that the stones are usually rounded and striae are not well developed nor consistent in orientation. Hicock (1991) and Hicock and Dreimanis (1992) reported similar findings for striae on clasts in fine-grained tills in Ontario that they attributed to subglacial deformation.

Subglacial channel-fill deposits, which are locally present between and within the uniform till beds, are small (<0.3 m to several meters wide), are almost always wider than they are thick, exhibit primary bedding and sometimes deformation of beds (including diapirs of diamicton), and have erosional, relatively planar, upper surfaces (Fig. 7b, c). Pebble fabrics measured in till adjacent to a subglacial channel fill at Wedron, Illinois (Fig. 7c), are strong and pebbles dip toward the channel rather than in the up-ice direction as do pebble fabrics farther from the channel, which parallel regional ice flow (Hansel et al., 1987; Johnson and Hansel, 1990).

Near the outer 50 km of the Wisconsin Episode margin, contacts gradational over about a meter have been observed between till beds of the same glacigenic sequences (Hansel et al., 1987). Rarely, layered diamicton facies have been observed at the top of till sequences (e.g., Fig. 6). Matrix composition and gravel lithologies in these layered facies sometimes vary from those in the more uniform till facies. For example, in a layered facies at the top of sequence II (Batestown till) at Wedron Quarry, there are more crystalline rocks representing far-traveled lithologies than in the underlying homogeneous till facies. Drapes of sorted sediment and R-channel deposits were also observed in the layered facies at Wedron Quarry. In central Illinois, a scatter of crystalline boulders on some moraine surfaces and layered sediment observed at the top of the Batestown succession (e.g., at Fairmount Quarry) may correlate to deposits of the layered facies at Wedron Quarry.

DISCUSSION

Significance of landscape and sediment record

The series of end moraines that formed in Illinois during Wisconsin Episode deglaciation requires (1) actively flowing ice to deliver debris to the glacier terminus and (2) conditions whereby the ice margin was stationary for tens to hundreds of years to build up the 30–60 m of sediment in the end moraines. The large number (20) of true end moraines that formed between Shelbyville and Lake Michigan between about 20,000 and 14,000 ^{14}C years ago indicates that the ice lobe stabilized numerous times during its overall retreat. Whereas most of the moraines represent recessional stillstands or slight readvance positions of the ice margin, others, as evidenced by the wedges of proglacial sorted sediment that occur locally between till sheets along their outer margins, represent readvances of the order of tens of kilometers (Hansel and Johnson, 1992, Fig. 6). Figure 5 illustrates that the ice margin readvanced at least six times during the overall retreat from Shelbyville to Lake Michigan.

Figure 8. Contact relationships: a, contact of till with underlying sand and gravel at Charleston Quarry; b, contact (shown by black line) between red (upper) and gray (lower) till beds of the Tiskilwa Formation at the Clear Creek section (note attenuated silt lenses in lower part of beds); c, enfolded contact of diamicton and sand at base of Tiskilwa sequence at Wedron Quarry; d, enfolded contact of diamicton and sand at base of Batestown sequence at Clear Creek; e, attenuated lenses of deformed sand in basal unit of Tiskilwa till at Wedron Quarry; f, stone concentration between two till beds of the Batestown sequence at Wedron Quarry.

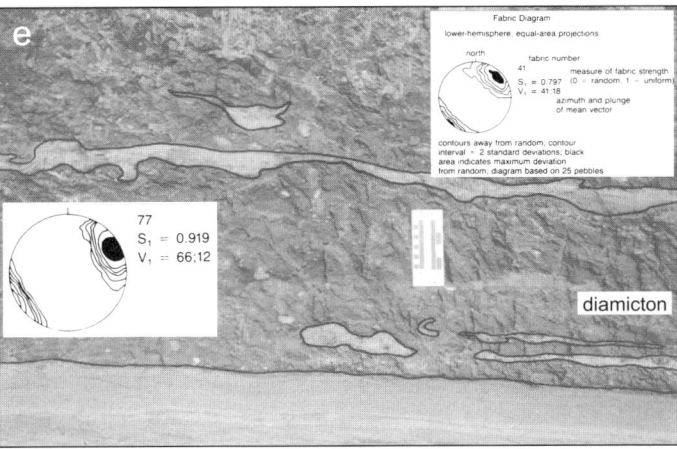

The central Illinois end moraines composed of uniform till contrast strongly with modern end moraines, where uniform till is rare (Johnson and Menzies, 1995). In size, morphology, and composition, the central Illinois end moraines of the Lake Michigan Lobe also are different from moraines in the northern Great Lakes region (Mickelson et al., 1983), such as end moraines of the Green Bay Lobe in southern and central Wisconsin. Clayton and Attig (1997) state that the outermost moraine of the maximum advance of the Green Bay Lobe in Dane County, the Johnstown Moraine, is medium-sized, typically 5–10 m high and 20–100 m wide; whereas the outermost moraines of the maximum advance of the Green Bay Lobe north of Dane County, the Hancock and Almond Moraines, are large, typically 10–20 m high and 0.5–1 km or more wide. Compared to the end moraines in central Illinois (Fig. 2), these Green Bay lobe end moraines are much smaller, have steeper slopes, are less asymmetric (or if asymmetric are more likely to have steeper proximal than distal slopes), and are not composed of uniform till. Instead the Green Bay Lobe end moraines are composed either of supraglacial drift that collapsed and was redeposited as the ice melted (Attig, 1993) or a combination of subglacial meltout till and stratified drift (Lundqvist et al.,1993).

Attig et al. (1989) attribute the landforms in the outer zone of the Green Bay Lobe to presence of permafrost conditions. They maintain that the glacier advanced over permafrost in Wisconsin and that the glacier was frozen to its bed in the marginal zone and thawed in the internal drumlin zone. They call on variation in basal thermal conditions to account for landform differences between the zones in the Green Bay Lobe.

Mickelson (1987) argued that in the southern Great Lakes area, where there is evidence for a deforming bed and very wet conditions during part of the Wisconsin Episode, the frozen-bed zone was absent. However, Johnson (1990) reported the presence of ice-wedge casts and relict patterned ground on the Wisconsin and Illinoian till plains as evidence that permafrost developed in central Illinois beyond the ice margin and on newly exposed glaciated landscape between about 21,000 and 16,000 ^{14}C years B.P. He concluded that permafrost was limited to a narrow, near-continuous zone that fluctuated with the ice margin and lasted no longer than 1,000 years at a site. Thus, it doesn't appear that the presence/absence of permafrost can in itself explain the difference in bed conditions and landforms between the Green Bay Lobe in Wisconsin and Lake Michigan Lobe in Illinois.

The lack of landforms characterized by supraglacial sediment and stratified drift and the small amount of such redeposited sediment on the central Illinois landscape are consistent with relatively clean ice in the southern Great Lakes area during the last deglaciation, as suggested by Mickelson et al. (1983). Even in the marginal zones, there appears to be little evidence for reworking of debris by meltwater during downwasting and backwasting of the glacier. This lack of supraglacial drift is unusual in that the margins of many modern glacier show ample evidence for reworking of supraglacial and ice-marginal debris during ablation. The general lack of supraglacial sediment in central Illinois is hard to explain if a large and continuous supply of debris was transported in basal ice to the glacier margin where it melted out to form the large end moraines. The fluted topography observed on some parts of the till plain in central Illinois (Fig. 3) is consistent with subglacial deformation (Boulton, 1987), but not definitive of any single process of subglacial till deposition (Lundqvist, 1988; Goldthwait, 1988).

In central Illinois the lack of eskers and tunnel channels on the landscape and the absence of R-channel deposits in the uniform till beds are consistent with predictions made by Walder and Fowler (1994) from glaciological theory and discussed by Clark and Walder (1994). They predicted that the subglacial drainage network at the base of gently sloping ice sheets resting on fine-grained deforming sediment should consist of many wide, shallow, probably braided, channels (which they called canals) distributed along the ice-sediment interface; whereas over a rigid bed, a network of relatively few large tunnels in the ice were predicted. The presence between, within, or beneath till beds of small, shallow, channel fills exhibiting planar tops is consistent with predictions for a soft-bedded glacier, as are the sedimentary characteristics of the fills and adjacent tills. The deformation features (such as diapirs of diamicton) observed within some channel-fill deposits (e.g., Fig. 7c) are consistent with a substrate that was soft and deforming when the channel existed, as is the strong orientation of pebbles dipping in the down-ice direction toward the channel-fill deposit, which could reflect creep of diamicton toward the channel (Alley, 1991). A drainage system composed of many shallow, braided canals (as opposed to a few large, arborescent tunnels) could account for the general lack of outwash along the margins of many moraines in central Illinois, because slower velocities would be expected in shallow, wide canals than in large tunnels (Clark and Walder, 1994). The greater cross-sectional area for a canal network would reduce the ability of the subglacial streams to transport large volumes of coarse material. Also, the general lack of subglacial channel-fill deposits in the uniform till beds is consistent with erosion and incorporation of such deposits into a deforming A horizon.

In central Illinois, the presence of overridden outwash distal to and beneath advance/readvance moraines and its general absence distal to recessional moraines is consistent with differences in subglacial hydrology during times of ice sheet instability (glacial advances) and times of ice sheet stability (moraine formation as the ice retreated from the glacial maximum). The variation in the amount of outwash produced cannot be explained by the grain size of the debris in the ice, because all the tills in central Illinois are fairly fine grained.

Sediment flux

The thick Wisconsin drift in Illinois, which averages about 15–20 m in till plain areas and 30–45 m in large end moraines (Soller et al., 1999; Fig. 4), requires a large flux of sediment into the area during the last glaciation. Johnson et al. (1991) calculated an annual transport rate of about 400 m^3 per meter-width of ice flow for a site in the most active part of the Lake Michigan Lobe about 100 km up-ice from the Wisconsin Episode glacial terminus at the Shelbyville Moraine (Fig. 1). This transport rate was based on calculating the volume of till deposited in about 2,500 ^{14}C years for a 10-km-wide flow path and correcting for (1) the volume of surficial loess (averages about 1 m), (2) the volume of material estimated to have been eroded subglacially from within the outer-100-km flow path (varied from 0 to 4 m across the area), and (3) till porosity (assumed 25%). Because the volume of outwash transported from the area was ignored, this estimated rate likely represents a minimum.

A thick and continuous supply of sediment is required to form the thick till sheets shown in Figure 4. If the till sheets were deposited by melting of debris-rich ice, assuming a debris concentration of 10% (or even 15%), a debris-rich ice zone of the order of 100 to several hundred meters would be required. Such thicknesses are not known from modern glaciers (Boulton, 1996a) and in fact might be thicker than the estimated thickness of the Lake Michigan Lobe in central Illinois (Clark, 1992). Alley (1991) concluded that sediment fluxes expected from a deforming-bed model for the southern Laurentide Ice Sheet are consistent with estimated modern fluxes for Antarctic glaciers assumed to be moving over a deforming substrate (Alley et al., 1987), and Jenson et al. (1996) found good agreement between the sediment fluxes predicted by a numerical model for a Lake Michigan Lobe deforming bed and those Johnson et al. (1991) calculated based on field data.

By whatever means sediment was eroded, transported, and deposited in Illinois during the last glaciation, it is clear from the generalized cross section (Fig. 4) that large volumes of sediment were eroded from northern Illinois and redeposited in central Illinois; in the area of the Silurian bedrock high south of Lake Michigan, pre-Wisconsin Episode drift and paleosols as well as proglacial and subglacial deposits of early phases of the Wisconsin Episode are absent. This absence indicates that the glacier may have been flowing over Silurian dolomite in that area during part of some later glacier phases (e.g., late Livingston and early Woodstock phases, Fig. 5). If the Lake Michigan Lobe experienced deforming bed conditions during the Wisconsin glaciation, those conditions were continuous neither temporally nor spatially in Illinois.

Significance of diamicton characteristics and contact relationships

Uniform till beds with planar, erosional basal contacts overlying relatively undeformed substrate, such as are common in the till sequences of central Illinois, are consistent with what Boulton (1996a, b) and Alley (1991) have predicted for till deposited from a deforming bed. The occasional folded contacts (Fig. 8b, c) preserve evidence that the substrate did deform subglacially, as do the lenses of enfolded and attenuated substrate materials within basal till (Fig. 8e). According to Boulton (1987, 1996a, b), only where deposition quickly follows erosion would such evidence for deformation be preserved. The general uniformity of diamicton within the till beds is consistent with sediment that was deformed to relatively high strain before it was deposited, although other origins cannot be ruled out because uniform tills have also been attributed to subglacial meltout.

The presence of subglacial channel-fill deposits with flat, truncated tops beneath younger (often lithologically distinct till beds) is consistent with erosion beneath active ice or at the base of a deforming bed prior to lodgement from ice or from the A horizon of a deforming bed (Johnson and Hansel, 1990; Alley, 1991; Boulton, 1996a); however, multiple till beds with subglacial channel-fill deposits and erosion surfaces between them are incompatible with a simple meltout origin. Also consistent with a deforming bed origin for the uniform till adjacent to a channel-fill deposit in the lower sequence at Wedron Quarry in central Illinois (Fig. 7c) are (1) the strong orientation of pebbles dipping down-ice and toward the channel immediately adjacent to the channel fill, and (2) the presence of diapirs of diamicton and soft deformation features within the channel fill. Soft, deformable till could readily creep toward channels and undergo diapiric deformation at the channel bases because of high pore water pressures (Alley, 1991; Clark and Walder, 1994).

The clast concentrations that locally occur between till beds in central Illinois are consistent with those predicted by Boulton (1996b) to develop in association with erosion surfaces between advance and retreat tills. He explained the concentration of large clasts at the A/B interface during erosion to the failure of the relatively low-density deforming horizon to lift the relatively denser, large clasts above the A/B plane. Evidence that these clast concentrations were not formed as classic boulder/cobble pavements includes that the clast concentrations (1) rarely have striae that are strong and consistently oriented, (2) straddle the contact between till beds or occur at the base of the upper till bed, and (3) generally lack faceted upper surfaces with striae parallel to ice flow. Where clast concentrations do show upper faceted surfaces and striae parallel to ice flow, it is possible that the deforming bed thinned and the base of the ice slid over the concentrated clasts.

The strong pebble fabrics measured in the uniform till beds are not definitive of any single type of till deposition. Strong pebble fabrics have been measured in tills attributed to meltout, lodgement, ploughing, and deformation origins (e.g., Lawson, 1979; Johnson and Hansel, 1990; Clark and Hansel, 1989; Hart, 1994; Hooyer and Iverson, 1997); however, strong fabrics in deformation tills have generally been attributed to a thin deforming bed (Hart, 1994; Hooyer and Iverson, 1997). We have measured strong pebble fabrics parallel to regional ice flow both in uniform till facies and in the lower part of till beds that contain attenuated lenses of sorted sediment that we attribute to enfolding at the base of a deforming bed (Johnson and Hansel, 1990; Fig. 8e).

The lack of deformation or weak deformation in subtill sediments has been used to argue both against (Kemmis, 1986, 1981; Clayton et al., 1989, 1987; Ehlers, 1996) and for (Boulton, 1987,

1996a, b; Alley, 1991) the deforming bed model. Boulton (1987, 1996a, b) argued that a soft deforming layer would act as a buffer between the stiffer ice and the underlying sediment, minimizing the shear stress on the underlying sediment of the B horizon. If the ice flowed directly over the underlying sediment, a larger shear stress would develop at the interface and would be more likely to generate deformation in the sediment. Alley (1991) reached similar conclusions; he argued that higher water pressures would be required for basal sliding than for deformation.

The contrast of the homogeneous till beds in sequence II at Wedron Quarry with the layered till bed in the upper part of the sequence (Fig. 6) is consistent with sedimentary facies that may reflect different processes of glacial erosion, transport, and deposition. The presence of R-channel fills with primary bedding, draped layers over large clasts, and distinct layers with high magnetic susceptibility and far-traveled lithologies, particularly mafic clasts, is consistent with a meltout origin for the layered facies. Hansel and Johnson (1992) interpreted the upper part of the layered facies to reflect meltout of far-traveled englacial debris and the lower part to reflect meltout of more-local debris (Lake Michigan basin/northern Illinois Paleozoic rock). The homogeneous till beds beneath the layered facies are more compact, more massive, and much more uniform in grain size, magnetic susceptibility, and gravel clast lithology than is diamicton in the layered facies. The lithology of these till beds reflects a Lake Michigan basin/northern Illinois source. The presence of small, wide, shallow channel-fill deposits with flat truncated tops located both in the deformed clay beneath the lower till bed and in the lower till bed beneath the upper homogeneous till bed and the presence of stone concentrations in the base of the upper homogeneous till bed (but not at the base of the lower till bed overlying clay) are both consistent with deposition from a deforming bed. The two homogeneous till beds are similar, varying only in that the lower bed is consistently clayier (Fig. 6). The erosion surface between the two till beds, which is locally marked by a cobble concentration, is consistent with the erosion surface predicted by Boulton (1996b) to occur in the accumulation zone between advance and retreat tills deposited from a deforming bed during a single advance and retreat cycle. The lack of strong striae with a preferred orientation on the cobbles is consistent with their rotation and concentration at a descending A/B interface when the site was in the zone of erosion. Using a deforming-bed model, the upper till bed would represent subsequent deposition during the retreat, which would leave the cobble concentration to mark the maximum extent the A/B interface descended during erosion. The characteristics of the channel-fill deposits at this contact and below the lower till bed are consistent with those predicted (Clark and Walder, 1994; Walder and Fowler, 1994) to form in association with a deforming bed.

Model of till deposition and moraine formation in central Illinois

Reconnaissance observations by one of us (A.K.H.) on an INQUA Till Commission field trip in 1994 to Fláajökull in Iceland offer a possible modern analogue for ice-marginal conditions in Illinois during the last glaciation. An end moraine composed of till currently is being constructed at the margin of Fláajökull, but along portions of its margin the glacier itself appears to be relatively free of basal debris (Fig. 9a, b). We infer that the moraine is being built as debris beneath the glacier is transported up and deposited on the proximal slope. The thickness of the subglacial debris layer varies around the margin, but ranges to as much as about 20 cm. Where exposed, flutes were observed on the proximal slope till surface. Evidence for seasonal reworking of sediment on the proximal slope by ice push has been observed. The distal slope of the moraine is steeper and approximates the angle of repose of the material. Mass-wasting processes are active along the distal slope. All these characteristics of the end moraine forming at the margin of Fláajökull appear consistent with those of end moraines in central Illinois. The till that makes up the end moraine in Iceland is much sandier than till in the end moraines in central Illinois, however.

We herein propose a model of end-moraine formation for central Illinois (Fig. 10). End moraines, composed predominantly

Figure 9. Photos of margin of Fláajökull in Iceland: a, end moraine forming beneath relatively clean ice; b, moraine formation at the glacier margin (people are on moraine crest and van is at base of distal slope).

of uniform till, are hypothesized to have formed when the glacier was in mass balance with a relatively stationary ice margin. In our model, debris is inferred to have been transported by deformation in the bed. This debris gained shear strength and was deposited incrementally on a long, gentle ramp on the up-glacier side of the moraine as ice velocity slowed toward the ice margin. As the glacier margin retreated, repeated accretion of debris from the base of the deforming bed would result in the formation of the gentle proximal slope that merges with till plain in the up-glacier direction. Buildup of the moraine proper would have resulted in the steeper distal slope that likely was the site of mass wasting and localized fluvial activity. In the deforming-bed model presented here, the low-relief till plain, characteristic of the inter-end-moraine area in central Illinois is inferred to have formed by basal accretion from a deforming bed during times of ice margin retreat.

The cause of the variation in plan view of the broadly arcuate moraines and the more lobate ones is consistent with earlier suggestions by Johnson et al. (1986) that the arcuate moraines formed after regional ice-margin advances that affected the entire Lake Michigan Lobe, whereas the lobate moraines are primarily recessional moraines within a sublobe. Other than variation in the amount of proglacial fluvial sediment, we are unaware of significant variations in till lithology or sedimentology between broadly arcuate and more lobate moraines (Moore, 1981). Although more abundant glaciofluvial sedimentation would be consistent with a glacial surge (Menzies, 1995), the landscape and sediment record are not compatible with glacial surge followed by regional stagnation. Unlike in the Des Moines Lobe area in north-central Iowa (Kemmis et al., 1981; Kemmis, 1991), ice stagnation features and reworked supraglacial sediment are uncommon in the Lake Michigan Lobe area of central Illinois. Thus, we argue that the arcuate moraines formed after ice-margin advances with a positive mass balance. Basal sliding may have contributed more to glacial movement during times of instability and ice-margin advance than during more stable times of moraine formation, however. For example, on the basis of data collection beneath Storglaciaren, a mostly soft-bedded, temperate glacier in Sweden, Iverson et al. (1995) showed that glacier flow accelerated when basal water pressure increased enough to raise the glacier from its bed.

The advance/readvance end moraines commonly are large, broad, and multiple crested; the ice margin must have stabilized at those positions for hundreds of years during which sediment was continuously delivered to the ice margin, perhaps by subglacial deformation. The smaller, more lobate moraines are inferred to have formed after times of negative mass balance when the ice-margin withdrew to a more stable position and developed a lower ice-surface profile prior to the formation of a recessional moraine. The lack of outwash distal to these moraines is consistent with the development of a canal system that was unable to transport large clasts, except where water collected in few large proglacial meltwater channels.

The reasons for the dynamics of the Wisconsin Episode

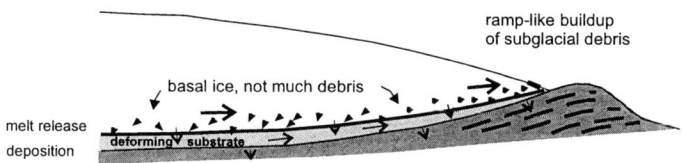

Figure 10. Model of moraine formation for central Illinois. Melting at the base of the glacier releases water and debris to the substrate, which consists of soft sediment (older drift and soil material). This water soaks into the substrate, which becomes saturated, loses shear strength, and deforms by pervasive shear as ice is transferred toward the glacier margin. In the marginal zone, the ice is thin and relatively free of debris because of basal melting farther up-ice. Debris delivered to the ice margin is deposited as till as the substrate dewaters. Where the ice margin remains stable for tens to hundreds of years, enough debris accumulates to form a moraine that consists of a broad, low ridge. The proximal slope of the moraine is low, gentle, and ramplike, whereas the distal slope may initially approximate the angle of repose and be the site of debris flows.

Lake Michigan Lobe in central Illinois are not understood. The lobe seems to have fluctuated between times of instability when the ice-margin advanced or retreated, probably fairly rapidly, and times of stability when the ice-margin was stationary for tens of hundreds of years and flow delivered sediment to build broad, low-relief end moraines. The impact of climate variation on ice-margin fluctuation and subglacial sediment characteristics and hydrology of the lobe is not known. Although the impact of changes in accumulation and ablation rates on ice sheet dynamics is uncertain, numerical modeling of a soft-bedded Lake Michigan Lobe by Jenson et al. (1996) indicates that relatively small changes in sediment viscosity, such as might result from changing from frozen to unfrozen conditions (or vice versa), can account for more rapid changes in ice-margin fluctuation than can larger changes in accumulation or ablation.

CONCLUSIONS

Glacial landscape characteristics and sediment relationships in central Illinois are entirely consistent with theoretical predictions for a subglacial deformation origin. Asymmetric end moraines comprised of uniform till are compatible with a deforming bed advecting sediment toward the ice margin where it is deposited incrementally as ice velocity decreases. Although till that has the characteristics of meltout till locally is present, such till has distinct sedimentary characteristics that are very different from those of the uniform till beds that comprise most of the landforms in central Illinois. The sedimentary characteristics and contact relationships of uniform till beds of a single advance/retreat cycle are consistent with those predicted by Boulton (1996a, b) to develop at the southern margin of midlatitude continental glaciers. The general lack of supraglacial sediment on the landscape is also more consistent with deposition from a deforming bed beneath relatively clean ice than from meltout from debris-rich ice.

Nevertheless, other origins for the till and the landforms of central Illinois cannot be ruled out because the deforming-bed model of glacial erosion, transport, and deposition remains largely theoretical and the formation of uniform till is not limited to a unique process (e.g., Hicock, 1990). Neither is it clear from modern glacier studies how thick and extensive subglacial deforming beds are.

ACKNOWLEDGMENTS

We thank Richard B. Alley, Peter U. Clark, Thomas V. Lowell, David M. Mickelson, and Barbara J. Stiff for helpful discussions and Howard C. Hobbs, E. Donald McKay, and Slawek M. Tulaczyk for thoughtful and constructive reviews. Barbara J. Stiff also assisted with computer graphics. This research was supported by the National Science Foundation under Grant EAR-9204838.

REFERENCES CITED

Alley, R. B., 1991, Deforming bed origin for southern Laurentide till sheets?: Journal of Glaciology, v. 37, no. 125, p. 67–76.

Alley, R. B., Blankenship, D. D., Bentley, C. R., and Rooney, S. T., 1987, Till beneath Ice Stream B, 3. Till deformation: evidence and implications: Journal of Geophysical Research, v. 92, no. B9 p. 8921–8929.

Attig, J. W., 1993, Pleistocene geology of Taylor County, Wisconsin: Wisconsin Geological and Natural History Survey Bulletin 90, 25 p.

Attig, J. W., Mickelson, D. M., and Clayton, L., 1989, Late Wisconsin landform distribution and glacier-bed conditions in Wisconsin: Sedimentary Geology, v. 62, p. 399–405.

Begét, J. E., 1986, Modeling the influence of till rheology on the flow and profile of the Lake Michigan lobe, southern Laurentide ice sheet, U.S.A.: Journal of Glaciology, v. 32, no. 111, p. 234–241.

Boulton, G. S., 1987, A theory of drumlin formation by subglacial deformation, in Menzies, J., and Rose, J., eds., Drumlin Symposium: Rotterdam, Balkema, p. 25–80.

Boulton, G. S., 1996a, The origin of till sequences by subglacial sediment deformation beneath mid-latitude ice sheets: Annals of Glaciology, v. 22, p. 75–84.

Boulton, G. S., 1996b, Theory of glacial erosion, transport, and deposition as a consequence of subglacial sediment deposition: Journal of Glaciology, v. 42, no. 140, p. 43–62.

Boulton, G. S., and Dobbie, K. E., 1993, Consolidation of sediments by glaciers: Relations between sediment geotectonics, soft-bed glacier dynamics and subglacial ground-water flow: Journal of Glaciology, v. 39, no. 131, p. 26–44.

Boulton, G. S., and Hindmarsh, R. C. A., 1987, Sediment deformation beneath glaciers: Rheology and geological consequences: Journal of Geophysical Research, v. 92, no. B9, p. 9059–9082.

Boulton, G. S., and Jones, A. S., 1979, Stability of temperate ice caps and ice sheets resting on beds of deformable sediment: Journal of Glaciology, v. 24, no. 90, p. 29–43.

Bowen, D. Q., 1978, Quaternary geology: Oxford, Pergamon Press, 221 p.

Clark, P. U., 1992, Surface form of the southern Laurentide Ice Sheet and its implications to ice sheet dynamics: Geological Society of America Bulletin 104, p. 595–605.

Clark, P. U., 1994, Unstable behavior of the Laurentide Ice Sheet over deforming sediment and its implications for climate change: Quaternary Research, v. 41, p. 19–25.

Clark, P. U., 1997, Chapter 6, Sediment deformation beneath the Laurentide Ice Sheet, in Martini, I. P., ed., Late glacial and postglacial environmental changes: Quaternary, Carboniferous-Permian, and Proterozoic: New York, Oxford University Press, p. 81–97.

Clark, P. U., and Hansel, A. K., 1989, Clast ploughing, lodgement and glacier sliding over a soft glacier bed: Boreas, v. 18, p. 201–207.

Clark, P. U., and Walder, J. S., 1994, Subglacial drainage, eskers, and deforming beds beneath the Laurentide and Euransian ice sheets: Geological Society of America Bulletin, v. 106, p. 304–314.

Clayton, L., and Attig, J. W., 1997, Pleistocene geology of Dane County, Wisconsin: Wisconsin Geological and Natural History Survey Bulletin 95, 64 p.

Clayton, L., Teller, J. T., and Attig, J. W., 1985, Surging of the southwestern part of the Laurentide Ice Sheet: Boreas, v. 14, no. 3, p. 235–241.

Clayton, L., Attig, J. W., and Mickelson, D. M., 1987, Evidence against deformable bed material as the cause of southern Laurentide ice streams and surges: INQUA 1987 Programme with Abstracts, p. 145.

Clayton, L., Mickelson, D. M., and Attig, J. W., 1989, Evidence against pervasively deformed bed material beneath rapidly moving lobes of the southern Laurentide ice sheet: Sedimentary Geology, v. 62, p. 203–208.

Ehlers, J., 1996, Quaternary geology: New York, John Wiley and Sons, 578 p.

Glass, H. D., and Killey, M. M., 1987, Principles and applications of clay mineral composition in Quaternary stratigraphy: Examples from Illinois: USA, in van der Meer, J. J. M., ed., Tills and glaciotectonics: Rotterdam, A. A. Balkema, p. 117–126.

Goldthwait, R. P., 1988, Classification of geomorphic features, in Goldthwait, R. P., and Matsch, C. L., eds., Genetic Classification of Glacigenic Deposits: Rotterdam, Balkema, p. 267–277.

Gutowski, V., Borries, S., Boyer, R., and Hoffman, K., 1991, A Pleistocene section at Charleston Stone Quarry, Coles County, Illinois, in Jorstad, R. B., ed., The General, Environmental and Economic Geology and Stratigraphy of East-Central Illinois, Guidebook for the 55th Annual Tri-State Geological Field Conference: Charleston, Illinois, Eastern Illinois University, p. 42–47.

Hansel, A. K., 1983, The Wadsworth Till Member of Illinois and the equivalent Oak Creek Formation of Wisconsin, in Mickelson, D. M., and Clayton, L., eds., Late Pleistocene history of southeastern Wisconsin: Geoscience Wisconsin, v. 7, p. 1–16.

Hansel, A. K., and Johnson, W. H., 1987, Ice-marginal sedimentation in a late Wisconsinan end moraine complex, northeastern Illinois, USA, in van der Meer, J. M. J., ed., Tills and glaciotectonics: Rotterdam, A. A. Balkema, p. 97–104.

Hansel, A. K., and Johnson, W. H., 1992, Origin and significance of a layered diamicton facies at Wedron, Illinois: Geological Society of America Programs with Abstracts, v. 24, no. 4, p. 19.

Hansel, A. K., and Johnson, W. H., 1996, Wedron and Mason Groups: Lithostratigraphic Reclassification of the Deposits of the Wisconsin Episode, Lake Michigan Lobe Area: Illinois State Geological Survey Bulletin 104, 116 p.

Hansel, A. K., and Johnson, W. H., 1997, Ice sheet dynamics of the Lake Michigan Lobe during Wisconsin Episode retreat in Illinois: Geological Society of America Abstracts and Programs, v. 29, no. 4, p. 19.

Hansel, A. K., Johnson, W. H., and Socha, B. J., 1987, Sedimentological characteristics and genesis of basal tills at Wedron, Illinois, in Kujansuu, R., and Saarnisto, M., eds., INQUA Till Symposium, Finland 1985: Geological Survey of Finland Special Paper 3, p. 11–21.

Hansel, A. K., Johnson, W. H., and Voorhees, D. H., 1993, Subglacial till of deformation origin from the last glacial episode in central Illinois: Geological Society of America Abstracts with Programs, v. 25, no. 6, p. A–393.

Hart, J. K., 1994, Till fabric associated with deformable beds: Earth Surface Processes and Landforms, v. 19, p. 15–32.

Herzog, B. L., Stiff, B. J., Chenoweth, C. A., Warner, K. L., Sieverling, J. B., and Avery, C., 1994, Buried bedrock surface of Illinois: Illinois State Geological Survey, Illinois Map 5, scale 1:500,000.

Hicock, S. R., 1990, Genetic till prism: Geology, v. 18, p. 517–519.

Hicock, S. R., 1991, On subglacial stone pavements in till: Journal of Geology,

v. 99, p. 607–619.

Hicock, S. R., and Dreimanis, A., 1992, Deformation till in the Great Lakes region: Implication for rapid flow along the south-central margin of the Laurentide Ice Sheet: Canadian Journal of Earth Sciences, v. 29, p. 1565–1579.

Hooyer, T. S., and Iverson, N. R., 1997, Laboratory studies of fabric development in shearing till: Geological Society of America Abstracts with Programs, v. 29, no. 4, p. 21.

Iverson, N. R., Hooke, R. B., Hanson, B., and Jansson, P., 1995, Flow mechanism of glaciers on soft beds: Science, v. 267, p. 80–81.

Jenson, J. S., Clark, P. U., MacAyeal, D. R., Ho, C. L., and Vela, J. C., 1996, Numerical modeling of subglacial sediment deformation: Implications for the behavior of the Lake Michigan Lobe, Laurentide Ice Sheet: Journal of Geophysical Research, v. 101, no. B4, p. 8717–8728.

Johnson, W. H., 1976, Quaternary stratigraphy in Illinois: Status and current problems, *in* Mahaney, W. C., ed., Quaternary stratigraphy of North America: Stroudsburg, Pennsylvania, Dowden, Hutchinson, & Ross, Inc., p. 161–196.

Johnson, W. H., 1990, Ice-wedge casts and relict patterned ground in central Illinois and their environmental significance: Quaternary Research, v. 33, no. 1, p. 51–72.

Johnson, W. H., and Hansel, A. K., 1989, Age, stratigraphic position, and significance of the Lemont drift, northeastern Illinois: Journal of Geology, v. 97, p. 301–318.

Johnson, W. H., and Hansel, A. K., 1990, Multiple Wisconsinan glacigenic sequences at Wedron, Illinois: Journal of Sedimentary Petrology, v. 60, p. 26–41.

Johnson, W. H., and Hansel, A. K., 1995, The glacial landscape of central Illinois: The product of a deforming bed: Geological Society of America Abstracts with Programs, v. 27, no. 6, p. A-170.

Johnson, W. H., Glass, H. D., Gross, D. L., and Moran, S. R., 1971, Glacial Drift of the Shelbyville Moraine at Shelbyville, Illinois: Illinois State Geological Survey Circular 459, 23 p.

Johnson, W. H., Follmer, L. R., Gross, D. L., and Jacobs, A. M., 1972, Pleistocene stratigraphy of east-central Illinois: Illinois State Geological Survey Guidebook 9, 97 p.

Johnson, W. H., Moore, D. W., and McKay, E. D., III, 1986, Provenance of late Wisconsinan (Woodfordian) till and origin of the Decatur Sublobe, east-central Illinois: Geological Society of America Bulletin, v. 97, no. 9, p. 1098–1105.

Johnson, W. H., Hansel, A. K., and Stiff, B. J., 1991, Glacial transport rates, late Wisconsinan Lake Michigan Lobe in central Illinois: Implications for transport mechanisms and ice dynamics: Geological Society of America Abstracts with Programs, v. 23, no. 5, p. A-61.

Johnson, W. H., and Menzies, J., 1995, Chapter 3, Pleistocene supraglacial and ice-marginal deposits and landforms, *in* Menzies, J., ed., Past glacial environments: Sediments, forms and techniques, Glacial environments: Volume 2: Oxford, Butterworth-Heinemann, p. 137–160.

Kamb, W. B., 1959, Ice petrographic observations from Blue Glacier, Washington, in relation to theory and experiment: Journal of Geophysical Research, v. 64, p. 1891–1901.

Kemmis, T. J., 1981, Importance of the regelation process to certain properties of basal tills deposited by the Laurentide ice sheet in Iowa and Illinois, U.S.A.: Annals of Glaciology, v. 2, p. 147–152.

Kemmis, T. J., 1986, Properties of sediment beds beneath Quaternary tills in the midcontinent U.S.A.: Eos (Transactions, American Geophysical Union), v. 67, no. 44, p. 947.

Kemmis, T. J., 1991, Glacial landforms, sedimentology, and depositional environments of the Des Moines Lobe, northern Iowa [Ph.D. thesis]: University of Iowa, 384 p.

Kemmis, T. J., Hallberg, G. R., and Lutenegger, A. J., 1981, Depositional environments of glacial sediments and landforms on the Des Moines Lobe, Iowa: Iowa Geological Survey Guidebook Series, no. 6, 132 p.

Kempton, J. P., and Gross, D. L., 1971, Rate of advance of the Woodfordian (Late Wisconsinan) glacial margin in Illinois—Stratigraphic and radiocarbon evidence: Geological Society of America Bulletin, v. 82, no. 11, p. 3245–3250.

Lawson, D. E., 1979, A comparison of the pebble orientations in ice and deposits of the Matanuska Glacier, Alaska: Journal of Geology, v. 87, p. 620–645.

Lineback, J. A., 1979, Quaternary deposits of Illinois (Map): Illinois State Geological Survey, scale 1:500,000.

Lundqvist, J., 1988, Glacigenic processes, deposits, and landforms, *in* Goldthwait, R. P., and Matsch, C. L., eds., Genetic classification of glacigenic deposits: Rotterdam, Balkema, p. 3–16.

Lundqvist, J., Clayton, L., and Mickelson, D. M., 1993, Deposition of the late Wisconsin Johnstown moraine, south-central Wisconsin: Quaternary International, v. 18, p. 53–59.

Mark, D. M., 1973, Analysis of axial orientation data, including till fabrics: Geological Society of America Bulletin, v. 84, p. 1369–1374.

Menzies, J., 1995, The dynamics of ice flow, *in* Menzies, J., ed., Modern glacial environments: Processes, dynamics and sediments: Oxford, Butterworth-Heinemann, p. 139–196.

Mickelson, D. M., 1987, Central Lowlands, *in* Graf, W. L., ed., Geomorphic systems of North America: Geological Society of America Centennial Special Volume 2, p. 111–118.

Mickelson, D. M., Acomb, L. J., and Bentley, C. R., 1981, Possible mechanism for rapid advance and retreat of the Lake Michigan lobe between 13,000 and 11,000 years B.P.: Annals of Glaciology, v. 2, p. 185–186.

Mickelson, D. M., Clayton, L., Fullerton, D. S., and Borns, H. W., Jr., 1983, The Late Wisconsin glacial record of the Laurentide ice sheet, *in* Wright, H. E., Jr., ed., Late-Quaternary environments of the United States, Volume 1, The late Pleistocene: Minneapolis, The University of Minnesota Press, p. 3–37.

Moore, D. W., 1981, Stratigraphy of till and lake beds of late Wisconsinan age in Iroquois and neighboring counties, Illinois [Ph.D. thesis]: Urbana, Illinois, University of Illinois, 200 p.

Soller, D. R., Price, S. D., Kempton, J. P., and Berg, R. C., 1999, Three-dimensional geologic maps of Quaternary sediments in east-central Illinois: USGS Miscellaneous Investigations Series I-2669, 3 atlas sheets, scale 1:500,000 (in press).

Sugden, D. E., and John, B. S., 1976, Glaciers and landscape: New York, John Wiley and Sons, 376 p.

Walder, J. S., and Fowler, A., 1994, Channelized subglacial drainage over a deformable bed: Journal of Glaciology, v. 40, p. 3–15.

Willman, H. B., and Frye, J. C., 1970, Pleistocene stratigraphy of Illinois: Illinois State Geological Survey Bulletin 104, 204 p.

MANUSCRIPT ACCEPTED BY THE SOCIETY OCTOBER 8, 1998

Reconstruction of the Green Bay Lobe, Wisconsin, United States, from 26,000 to 13,000 radiocarbon years B.P.

Patrick M. Colgan
Department of Geology, Northeastern University, 14 Holmes Hall, Boston, Massachusetts 02115; pcolgan@lynx.neu.edu

ABSTRACT

The distribution of glacial landforms, sediments, and advance-retreat chronology allow the reconstruction of the Green Bay Lobe from 26,000 to 13,000 B.P. (radiocarbon years before present). During the last glacial maximum about 18,000 B.P., the lobe was as much as 1,000 m thick over Green Bay. The terminus was steep and average basal shear stress was relatively high (as much as 25 kPa). During retreat a series of ice-margin positions are recorded in end moraines and ice-contact fans. Retreat may have begun as early as 16,000 B.P. (about 19,000 calendar years) or as late as 13,000 B.P. (about 15,500 calendar years). During some retreat phases, ice near the terminus was stagnant and surface slope and basal shear stresses were very low (<5 kPa). Small drumlins (<1 km long) and remolded drumlins suggest periods of stability or minor advance during overall retreat.

Permafrost was present in Wisconsin from 26,000 to 13,000 B.P. During the glacial maximum and deglaciation the glacier bed was thawed and only a narrow frozen-toe was present (<5 km wide). Retreat rates were about 50–100 m/yr before 13,000 B.P. and probably increased after this time. Equilibrium-line altitude was lower than 1,200 m during the glacial maximum and higher than this during deglaciation. Ice-surface-lowering rates were high during deglaciation (1.7–7 m/yr).

After 13,000 B.P. the lobe retreated into the lowland and was fronted by a large proglacial lake. Fine-grained lake sediment may have acted as a bed lubricant as ice-thrust features, till draped over eskers and drumlins, and discontinuous end moraines suggest that the lobe surged repeatedly after 13,000 B.P. After 9,900 B.P. the lobe retreated into southern Ontario.

INTRODUCTION

Reconstructions of ice-surface profiles, bed conditions, and glacier dynamics provide quantitative information about Pleistocene glaciers. Most reconstructions are based on landforms and sediments, and our present knowledge of modern glacier environments. In this paper, I describe how glacial landforms and sediments are used to reconstruct ice-surface profiles, bed conditions, and the dynamics of the Green Bay Lobe (Fig. 1) during the period from 26,000 to 13,000 B.P. (radiocarbon years before present).

METHODS

Landforms were mapped from aerial photographs (1:20,000 and 1:40,000 scale), topographic maps (1:24,000 and 1:100,000 scale), and existing glacial geologic maps (1:100,000 scale) for the southern end of the Green Bay lowland between latitude 42.5° N and 44° N, and longitude 88° W and 90° W (Fig. 2). Flow directions were reconstructed from striae, streamlined landforms, eskers, and end moraines. In all cases the correlation of particular flow indicators to a phase is interpretive. Profiles are reconstructed using the methods of Mathews (1974), Ridky and

Colgan, P. M., 1999, A reconstruction of the Green Bay Lobe, Wisconsin, United States, from 26,000 to 13,000 radiocarbon years B.P., *in* Mickelson, D. M., and Attig, J. W., eds., Glacial Processes Past and Present: Boulder, Colorado, Geological Society of America Special Paper 337.

Figure 1. Green Bay Lobe at its maximum position.

Bindschadler (1990), and Clark (1992). Mass-balance calculations estimate equilibrium line altitude (ELA) and balance velocities (see Colgan, 1996).

GEOLOGIC SETTING

Ice lobes of the Laurentide Ice Sheet were controlled by broad lowlands such as the Green Bay lowland, which begins in northern Michigan and trends to the southwest through Green Bay and Lake Winnebago (Fig. 3). Its eastern margin is bounded by an escarpment of Silurian dolomite. This escarpment rises 50–200 m out of Green Bay and Lake Winnebago. Softer Ordovician shale has been excavated from the axis of the lowland. The western margin consists of an upland of Precambrian igneous and metamorphic rock, overlain by Cambrian sandstone and Ordovician dolomite. Total lowland relief varies from 100 to 350 m with the highest relief in the north and lowest relief in the south (Figs. 3 and 4). Relief in the adjacent Lake Michigan basin is approximately 200–400 m. Before deglaciation, relief was greater, because sediments younger than 13,000 B.P. are present in both basins.

STRATIGRAPHY AND CHRONOLOGY

The oldest glacial unit is the Horicon Formation, a yellowish brown, sandy till (70% sand, 20% silt, 10% clay) with interbedded glaciofluvial and lacustrine deposits, and thin supraglacial sediment overlying basal till. The Horicon is present throughout the glaciated area covered by the Green Bay Lobe (Fig. 2). The Kewaunee Formation overlies the Horicon Formation and is found only in the northern part of the study area near Lake Winnebago (Fig 2). It is a reddish brown till (40% sand, 40% silt, 20% clay) that contains interbedded glaciofluvial and lacustrine deposits.

Material underlying till and outwash that yielded radiocarbon dates older than 26,000 B.P. suggest that initial advance into Wisconsin occurred after this time (Mickelson et al., 1984; Attig et al., 1985). In southeastern Wisconsin, dates from basal sediment in bogs show that ice may have been at its maximum as late as 13,150 ± 150 B.P. A date of 12,950 ± 100 B.P. obtained from lake sediment on top of the Horicon Formation near Valders, Wisconsin, suggests that retreat from the maximum may have been rapid (Maher and Mickelson, 1996). Alternatively, these dates may simply record the change from tundra to boreal forest at about 13,000 B.P. (Clayton et al., 1992). Dates on wood from the Two Creeks interval (11,800 ± 200 B.P.) indicate retreat and readvance at least three times between 12,950 and 11,800 B.P. After 9,900 B.P. ice retreated out of Wisconsin (Clayton, 1984).

ICE LOBE RECONSTRUCTIONS

Flow history

At its maximum the Green Bay Lobe was 125 km wide and more than 200 km long. Reconstructions of the Laurentide Ice Sheet suggest that flow to the lobe originated near a divide west of James Bay (Denton and Hughes, 1981; Boulton et al., 1985; Dyke and Prest, 1987). Ice flowed across the eastern end of Lake Superior and was channeled through the lowland to where it terminated near Madison, Wisconsin, about 1,400 km from the divide.

Figure 5 summarizes flow directions for seven pre-13,000 B.P. phases and one post-13,000 B.P. phase. During all but the earliest phase, flow was concentrated along the lowland axis as a separate flow from the adjacent Lake Michigan Lobe. The pattern of streamlined landforms indicates diverging flow out of the lowland. The existence of the Kettle Moraine, an interlobate complex composed of ice-contact and glaciofluvial sediments, indicates that the boundary between the Lake Michigan and Green Bay Lobes was east of the Silurian escarpment.

Most of the ice originating south of the Hudson ice divide (Dyke and Prest, 1987) was probably channeled through Lake Michigan. The present basin is eroded to a depth of 100 m below sea level, suggesting concentrated flow there. Green Bay is eroded to a depth of about 25 m above present sea level. The relative elevations of Green Bay and Lake Michigan lowlands suggest that the two lobes flowed separately through much of the last glaciation and that ice in Lake Michigan had greater velocity and erosional power. The depth of erosion also reflects lithologic and structural controls. Once a lowland was established, channeling of ice may have created a positive feedback that enhanced erosion in the larger basin.

Ice likely flowed directly out of Lake Michigan and into the Green Bay lowland sometime before the Johnstown-Hancock phase (Fig. 5B) or during the earlier Arnott phase (Fig. 5A). Drumlins located in eastern Fond du Lac County (Fig. 6) reflect streamlining by ice that flowed over the Silurian escarpment to

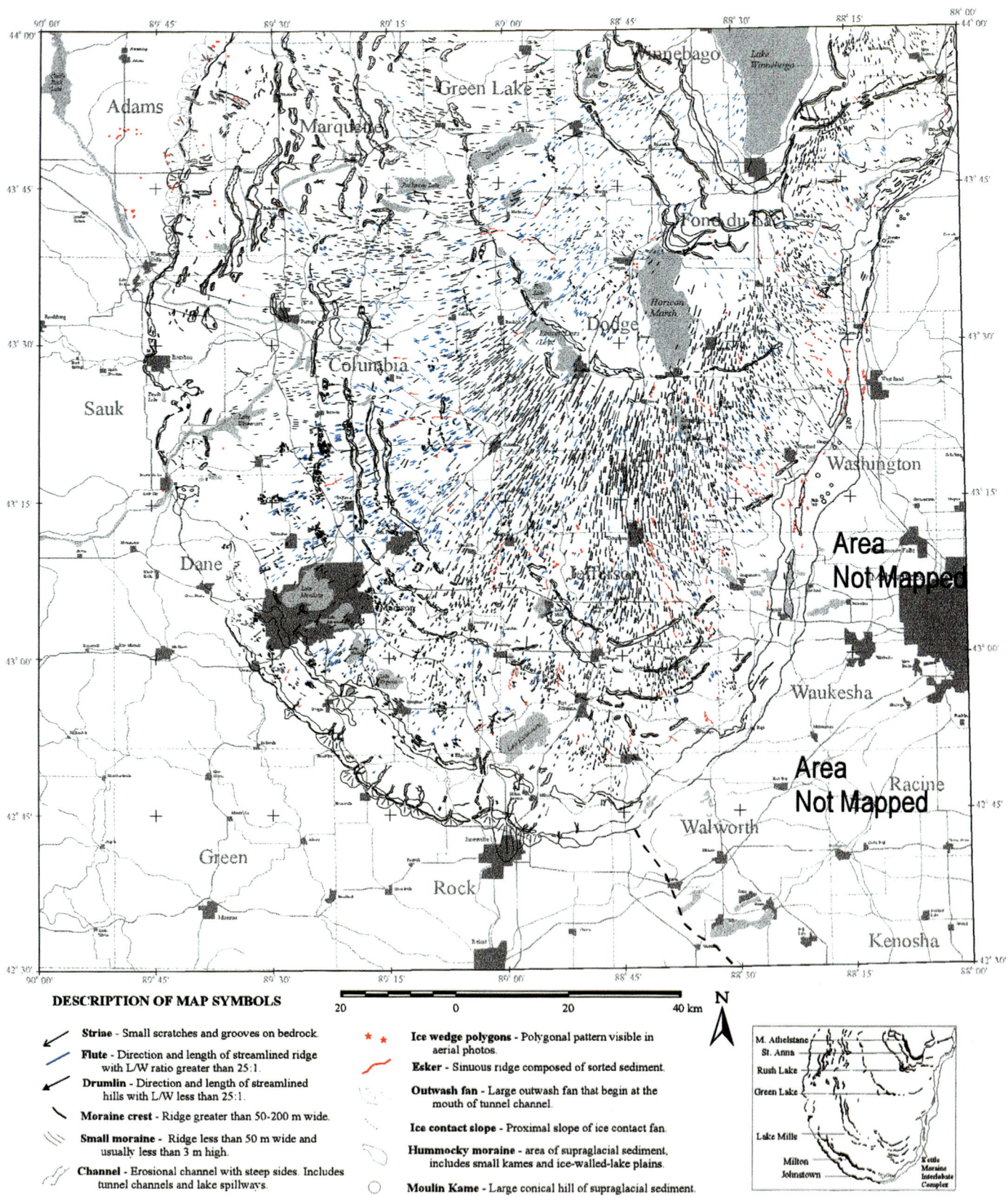

Figure 2. Glacial landforms of the southern Green Bay Lobe region (from Colgan, 1996).

Figure 3. Shaded relief image of Green Bay and Lake Michigan basins. Major features include: H, Hancock moraine; J, Johnstown moraine; W, Lake Winnebago; N, Niagaran escarpment; K, Kettle interlobate complex. Image was created from USGS 30 arc-sec digital elevation data. Spacing of elevations is approximately 1 km. Illumination direction is approximately 315° and sun angle is 25°.

the west (Fig. 5A; Alden, 1932; Thwaites and Bertrand, 1957) Subsequent flows eroded and remolded these hills into forms with a northwest to southeast trend (Fig. 5B–G).

Drumlins diverging away from the lowland's axis indicate that the shape of the terminus was controlled by lowland topography during subsequent phases. Small drumlins (<1 km long) formed during retreat, were superimposed on large drumlins (1–6 km long) that formed when ice was at its maximum position (Colgan and Mickelson, 1997). During the St. Anna phase (Fig. 5G) the position of the terminus was constrained by the Silurian escarpment. The bedrock relief of this escarpment (about 200 m) suggests that thin ice was all that remained within the lowland by this time.

Post-13,000 B.P. advances that deposited till of the Kewaunee Formation were also deflected by the Silurian escarpment and therefore must have been as thin lobes (see Socha et al, this volume, chapter 14). These advances resulted in little erosion, but rather deposited sediment in the basin (Thwaites, 1943; Thwaites and Bertrand, 1957; Evenson et al., 1977; McCartney and Mickelson, 1982; Need, 1985). Advances and retreats of Lake Michigan Lobe at this time also accom-

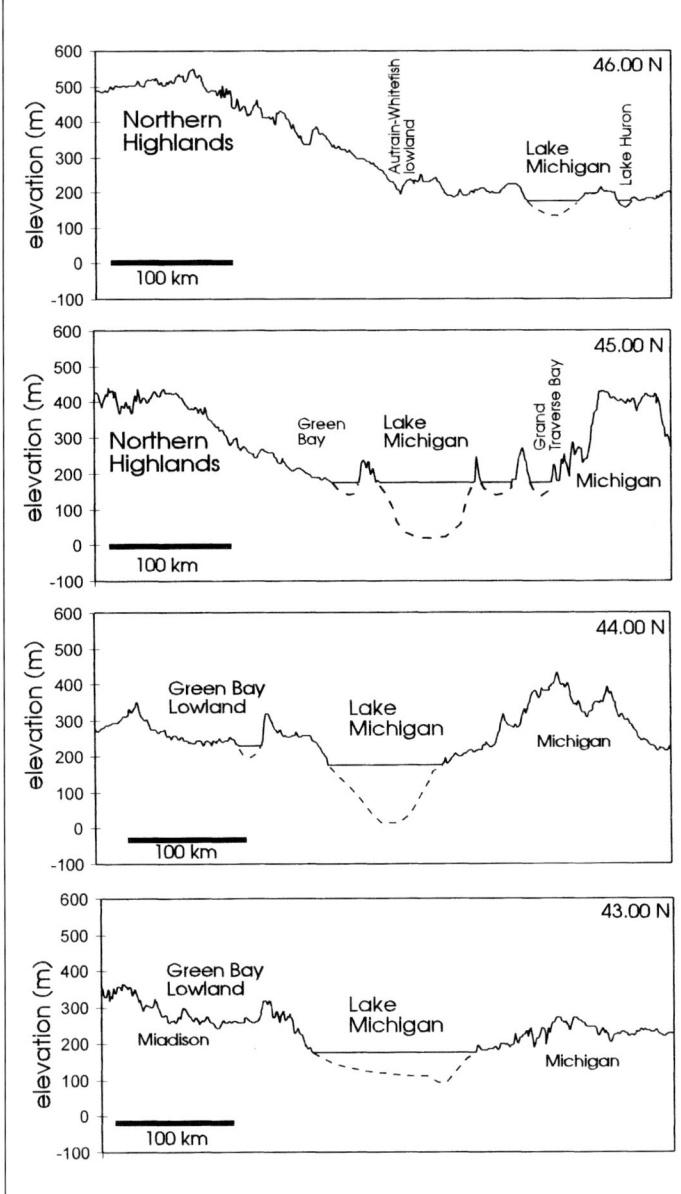

Figure 4. Four east-west topographic profiles across the Green Bay and Lake Michigan basins. Profiles begin at 90° W in central Wisconsin and end at 84° W in Michigan. Profiles run along lines of latitude and are oriented perpendicular to flow lines of Green Bay and Lake Michigan Lobes. Maximum depth of Lake Michigan occurs between profiles at 44° N and 45° N, where the lake bottom is approximately 75 m below sea level. Maximum depth of Green Bay occurs between profiles of 45° N and 46° N and is approximately 100 m above sea level. Land surface elevations are in meters and were extracted from a digital elevation model with elevations spaced approximately 1 km apart. Lake bottom profiles are approximate and drawn from bathymetric charts.

plished little erosion of older units (Acomb et al., 1982). Six till units in the Oak Creek Formation (time equivalent to upper Horicon Formation) suggest that the Lake Michigan Lobe advanced repeatedly between 14,500 and 13,500 B.P. (Simp-

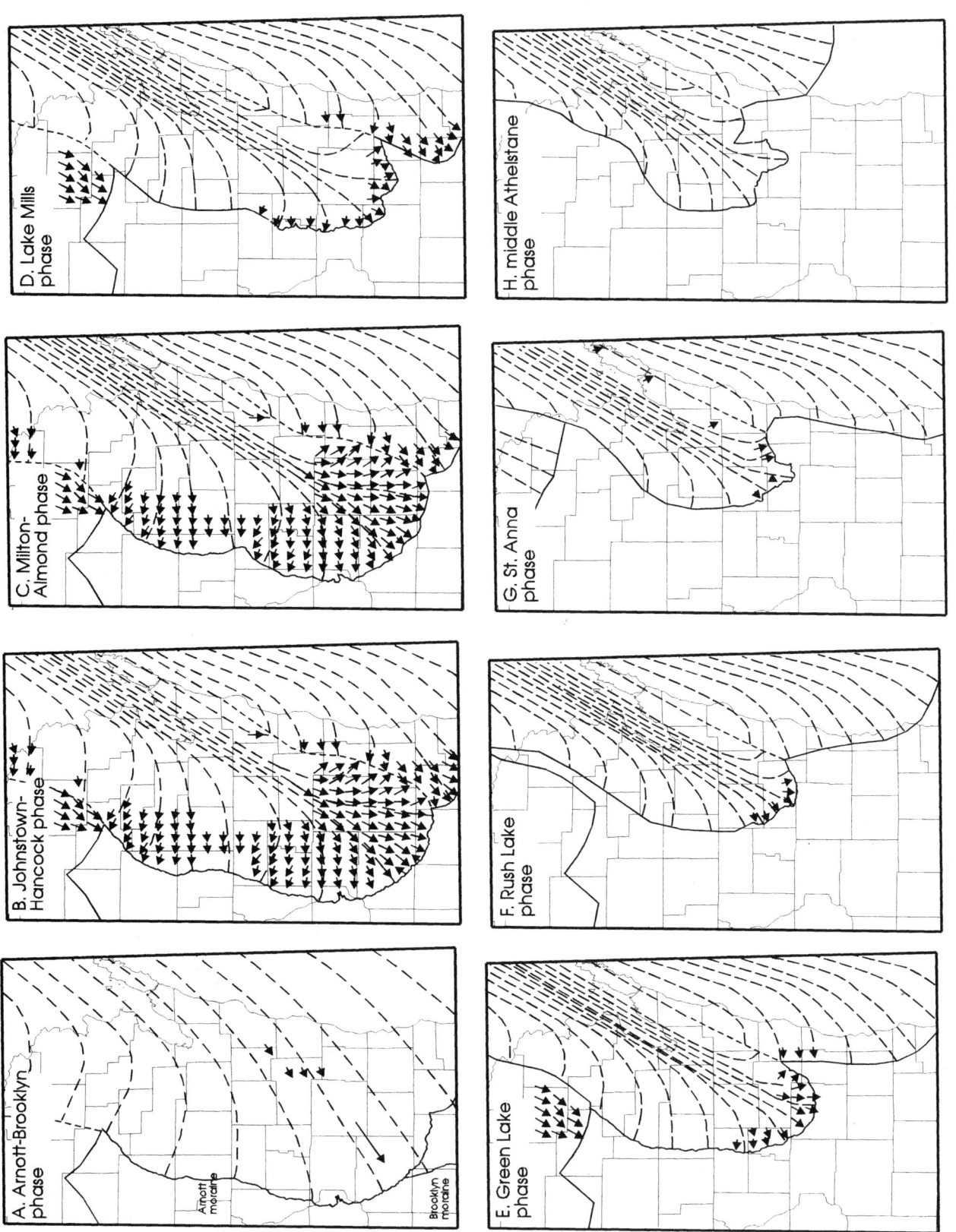

Figure 5. Reconstruction of flow directions of the Green Bay Lobe, 26,000 to 13,000 B.P. Flow directions reconstructed from drumlins and striae are shown as arrows, inferred flow lines are shown as dotted lines. A. Arnott-Brooklyn phase; B. Johnstown-Hancock phase; C. Milton-Almond phase; D. Lake Mills phase; E. Green Lake phase; F. Rush Lake phase; G. St. Anna phase; H. middle Athelstane phase.

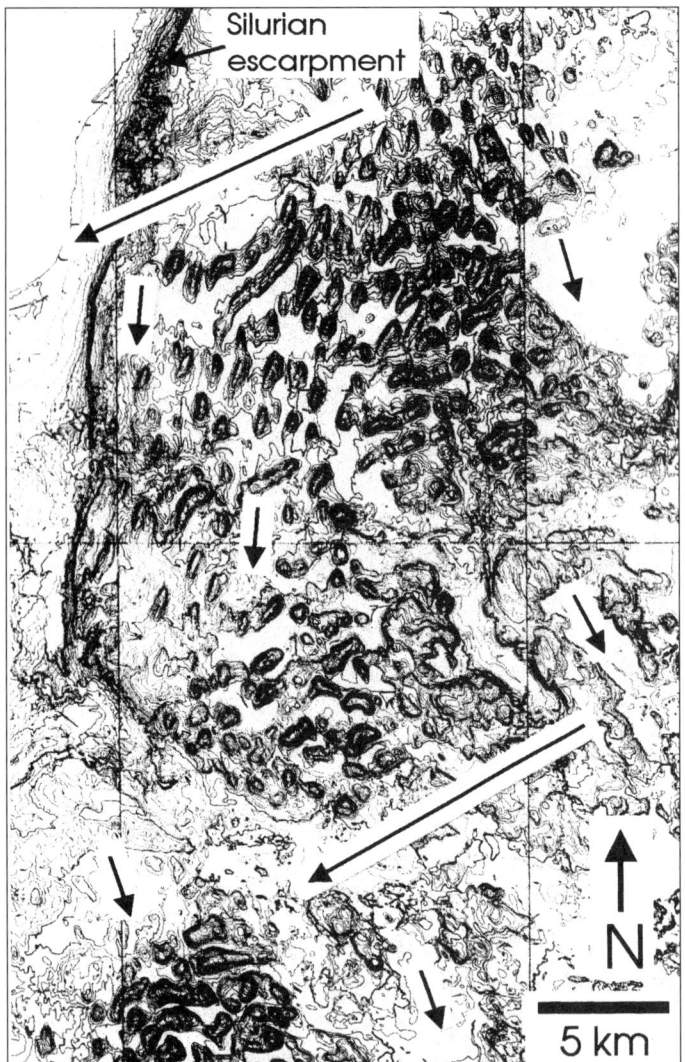

Figure 6. Elongate hills located in Fond du Lac County (10-ft contours from Eden and St. Peter 1:24,000 Quadrangles). Early flow was from northeast to southwest (large arrows). These hills were later streamlined by flow that was out of the Green Bay lowland from north to south-southeast (small arrows).

kins, 1989). This may be because ice was thin and wet-based (Mayewski et al., 1981; Simpkins, 1989) or that it may have developed a floating, partially calving margin (Mickelson et al., 1981; Coleman et al., 1989; Simpkins, 1989).

The flow direction and radial pattern of the Green Bay Lobe did not change between the Johnstown-Hancock (Fig. 5B) and St. Anna phases (Fig. 5G). Local changes in flow occurred near the terminus due to thinning and subsequent topographic control. This suggests that retreat before 13,000 B.P. was progressive, with only minor readvances and that there were no major reorganizations of ice within this sector of the Laurentide Ice Sheet before 13,000 B.P. A major retreat (>300 km) and a series of readvances (as much as 120 km) after 13,000 B.P. suggest reorganizations within the ice-drainage basin. Some have suggested surges for these readvances (Prest, 1969; Wright, 1971; Mickelson et al., 1981). Others suggest a response to rapid climate change (Evenson et al., 1977).

Ice-surface profiles and basal shear stress

The surface slopes of glaciers and ice sheets reflect the basal resistance of their beds. Glaciers that slide over smooth, well-lubricated beds have gently sloping ice-surface profiles, whereas glaciers with rough or frozen beds have steeper ice-surface profiles. Estimates of average basal shear stress (τ_b) in modern valley glaciers vary between 50 and 200 kPa (Paterson, 1994). Gentle surface slopes occur near ice domes, over subglacial lake basins, in outlet glaciers, ice streams, and ice plains (Cooper et al., 1982; Bentley, 1987; Bindschadler et al., 1987). While most glaciers with gently sloping ice surface profiles move slowly (<10 m/yr), some have high surface velocities (>500 m/yr). My reconstructions suggest that the Green Bay Lobe had a gentle ice surface profile during deglaciation, and a steeper profile may have been present sometime before the Johnstown-Hancock phase.

A reconstruction of the Green Bay Lobe based on moraine elevations and flow lines indicates that ice was approximately 250 m thick, 25 km from the terminus (Clark, 1992). This is similar to estimates made by Alden (1911) based on ice-marginal features in the Baraboo Hills. My reconstruction (Fig. 7) of the margin in the Baraboo Hills area, is slightly steeper than both of these. All of these methods assume that moraine crest elevations and flow lines can be used to reconstruct ice-surface form lines (Mathews, 1974). Reconstructions of valley glaciers are better constrained, because trim lines and lateral moraines provide unequivocal evidence of the former ice-surface slope. Because southern Laurentide glacier lobes were also confined to broad lowlands and glacial troughs they may also be reconstructed this way (Thorson, 1980; Clark, 1992), although reconstructions are sensitive to interpretations of flow direction near the margin.

In many lowland lobes, the highest elevations of moraine crests decrease systematically down glacier, parallel to the midline of the lobe. Mathews (1974) concluded that this slope approximated the surface slope of the glacier midline. Moraine elevations along the western margin of the Green Bay lowland indicate low midline slopes. These slopes do not reflect the slope near the terminus, but the slope of the midline of the lobe approximately 50 to 150 km from the terminus. Moraines deposited before 13,000 B.P. have slopes of 1.0 ± 0.3 to 1.8 ± 0.3 m/km over distances of 50–150 km. Elevations of the Johnstown-Hancock, Milton-Almond, and Green Lake–Bowler moraines have the steepest slopes. The innermost Lake Mills moraine (Fig. 5D) has a slope of 1.0 ± 0.3 m/km. The Middle Athelstane moraine (Fig. 5H), has a slope of 0.7 ± 0.3 m/km.

The slope of the western margin of the Green Bay Lobe is similar to the slopes of moraines deposited by other Laurentide glacier lobes (e.g., Superior Lobe, 1.2 m/km, Wright et al., 1973; western Canada, less than 2.0 m/km, Mathews, 1974; Wisconsin Valley Lobe, 4.0 m/km, Ham, 1994). Ice-surface

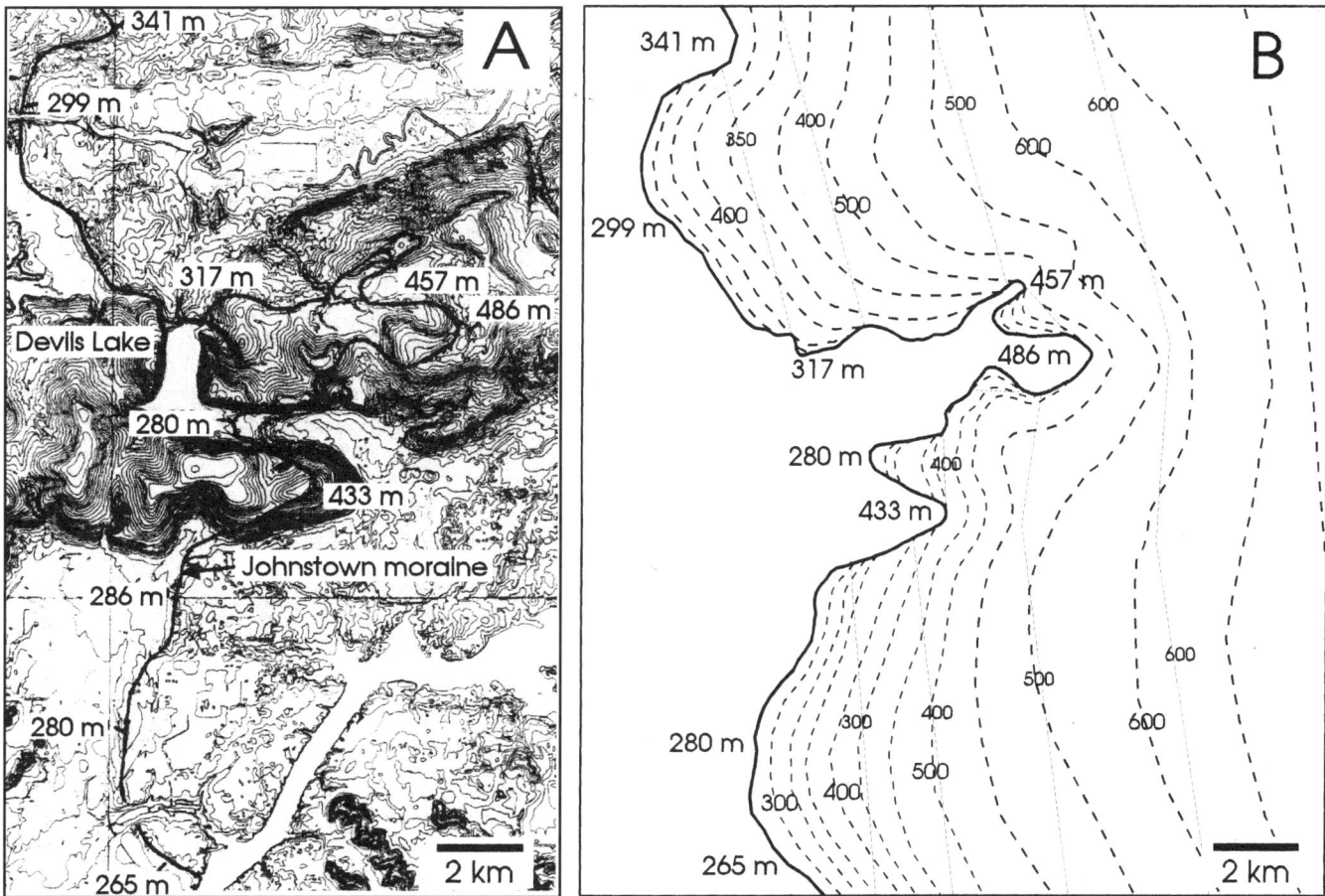

Figure 7. A, Outline of Johnstown moraine in the Baraboo Hills (10-ft contours taken from North Freedom, Baraboo, Sauk City, and Sauk Prairie 1:24,000 Quadrangles). Highest moraine elevation is approximately that of highest elevation in Baraboo Hills (Sauk Point, 486 m). Lowest elevation of moraine crest is located just west of Wisconsin River (elevation ~265 m). B, Reconstruction of ice margin in the Baraboo Hills area. Bold dotted lines are ice-surface contours drawn from moraine elevations that assume convergence of flow toward the Baraboo Hills and diverging flow in the small sublobes to the north and south. Thin dashed line and gray elevations are for lowest possible reconstruction assuming no convergence or divergence of flow lines. Since there are few ice-flow indicators in the area it is difficult to determine which model is correct. Flow was probably converging because meltwater deposits are common in the reentrant in the margin. This suggests that subglacial and supraglacial meltwater was focused toward the reentrant. This would require an ice-surface topography that focused water towards the Baraboo Hills.

profiles produced using the force-balance method (Ridky and Bindschadler, 1990) also demonstrate that the midline slope of the Green Bay Lobe could not have been much steeper than slopes of the western margin (Colgan, 1996).

The profiles produced here provide reasonable estimates of ice thickness and basal shear stresses within 100 km of the terminus (Fig. 8). The difference between profiles, produced by each method, within 50 km of the terminus is less than 150 m. Based on these profiles, average basal shear stresses (τ_b) were between 7 and 25 kPa, except during the Lake Mills phase when τ_b was about 5 kPa (within 100 km of the terminus). Ice-surface slopes and τ_b may have increased upglacier as occurs in ice streams or they may have decreased if flow lines were diverging from a dome or divide (Cooper et al., 1982; Bentley, 1987; Paterson, 1994). The steep profiles near the terminus and the low slopes of the western moraines suggest the later case.

Reconstructed equilibrium line altitude and balance velocity

A profile (Fig. 9) was constructed using the force balance method of Ridky and Bindschadler (1990) in order to estimate the equilibrium line altitude (ELA) and balance velocity of the Green Bay Lobe. I used the 18,000 B.P. flow line reconstruction of Dyke and Prest (1987) and a τ_b of 25 kPa for the first 500 km of the profile. This is the steepest profile that is compatible with the geomorphic evidence during the Johnstown-Hancock phase. It is possible that ice near the margin was thicker before this time. Longitudinal spreading of the margin is assumed for the last 100 km of the flow line based on the divergence of drumlins. For the rest of the flow line, divergence is assumed to be linear and dependent on distance from the ice divide (Ridky and Binschadler, 1990, Fig. 4). For the remaining 900 km of the 1,400-km flow line, a τ_b of 50 kPa

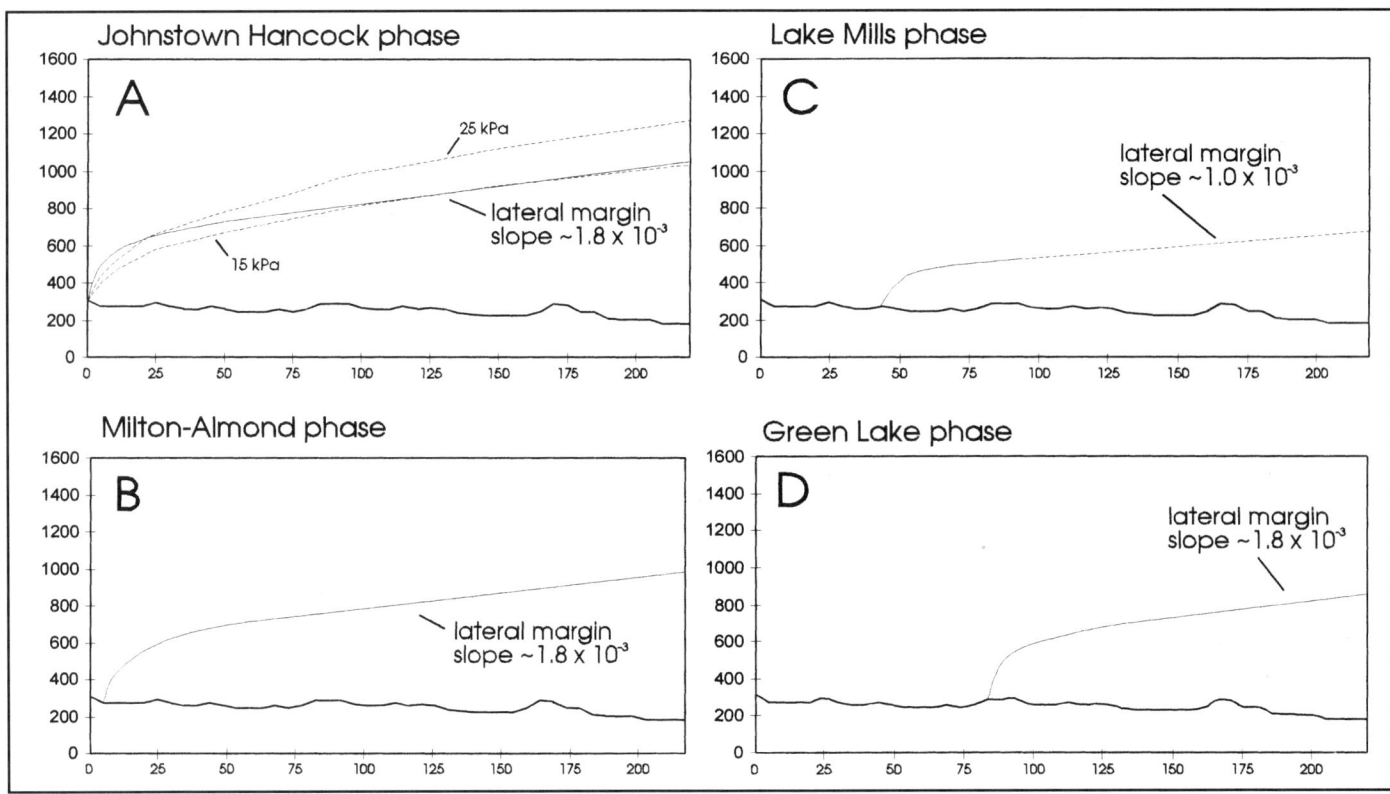

Figure 8. Reconstructed profiles of the Green Bay Lobe. Solid lines are profiles derived from estimates of slope near terminus (terminus to 50 km) and slopes of moraine crests along the western margin of the lobe (50–200 km from terminus). A, Johnstown-Hancock phase. Dashed lines show two profiles reconstructed using force balance method (Ridky and Bindschadler, 1990) assuming a constant basal shear stress of 15 and 25 kPa; B, Milton-Almond phase; C, reconstruction of innermost Lake Mills position; D, Green Lake phase.

was assumed. Bed elevations were derived from 30-arc-sec digital elevation data at intervals of approximately 20 km. Isostatic depression of the ice sheet was based on reconstructions by Clayton and Attig (1989) for the marginal 300 km and those of Clark et al. (1994) for the rest of the profile. These reconstructions suggest that maximum isostatic depression was 700 m at James Bay and that the ice sheet was not fully compensated for at 18,000 B.P.

The shape of the profile is similar to the "bowler hat" model of Boulton et al. (1985), because it specifies low τ_b over sedimentary rocks in lowlands, and high τ_b over crystalline rocks of the Canadian Shield, but it differs in that it produces thicker ice margins. This is because their model assumed a τ_b of less than 5 kPa. The reconstructed maximum ice sheet elevation of about 3,600 m is higher than the Denton and Hughes (1981, Fig. 8) minimum model along the same profile.

Three ablation-accumulation patterns were used to provide a range of possibilities for the southern margin of the Laurentide Ice Sheet during the last glacial maximum: (1) a "cold continental" pattern represents glaciers in the Canadian Arctic, Svalbard, and northern Greenland (Fig. 9A); (2) an "intermediate" pattern represents polythermal glaciers typical of climates in southern Greenland and northern Iceland (Fig. 9B), and (3) a "maritime" pattern (Fig. 9C) that represents a high ablation gradient, similar to glaciers in southeast Alaska, southern Iceland, and Norway (Boulton et al., 1985). In each model, a maximum accumulation rate of 0.25 m/yr was assumed. This is an intermediate value between average values for Greenland (0.30 m /yr) and Antarctica (0.143 m /yr; Paterson, 1994). Rates may have been higher in a zone just above the equilibrium line, but other areas near the ice divide with lower rates because the "elevation desert effect" would compensate (Oerlemans and van der Veen, 1984). Once the profile and ablation-accumulation pattern were established, balance calculations were made for several trial ELA values until accumulation balanced ablation (to within 1% of total mass). Balance velocity was then calculated at each grid node assuming continuity and a steady profile.

The cold-continental pattern predicts an ELA of about 1,200 m and a maximum ablation rate of 0.30 m/yr at the terminus (Fig. 9). The maximum balance velocity is 150 m/yr and occurs approximately 140 km from the terminus. The ratio of accumulation area to total area (A/Ao, where A is total accumulation area and Ao is total area of flow line, which is shaped like a pie-shaped wedge) is 0.54 and net ablation occurs over the last 260 km of the flow line. The intermediate mass-balance pattern (Fig. 9B) predicts a lower ELA of about 750 m and maximum ablation rate of 1.4 m/yr (Fig. 9). The maximum balance velocity is about 300 m/yr. The A/Ao ratio is 0.71. Net ablation occurs

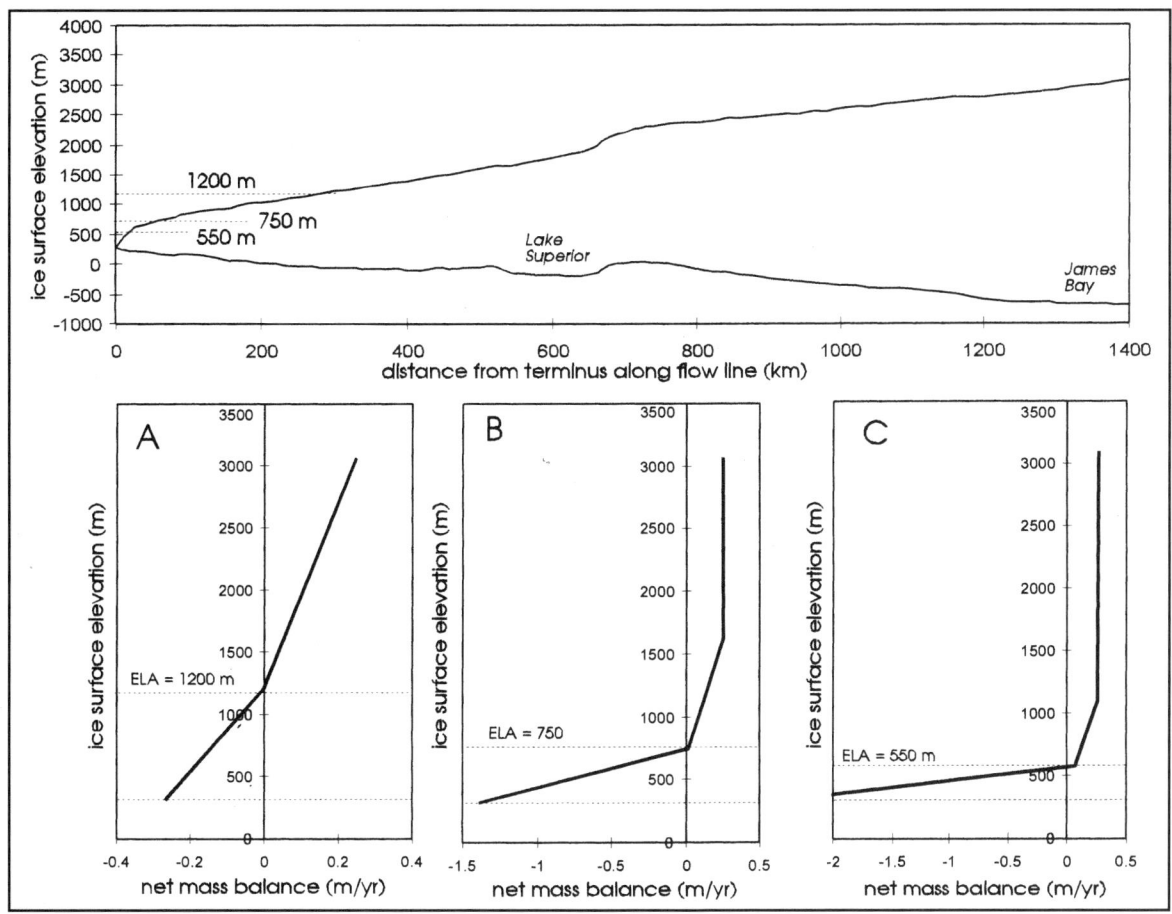

Figure 9. Ice-surface profile (at top) and ablation-accumulation gradients (bottom graphs) used in mass-balance calculations. A, cold polar mass-balance pattern; B, intermediate pattern of a polythermal glaciers; C, maritime pattern.

along the last 50 km of the flow line. This is the result of the divergence in the flow band within 140 km of the terminus, which greatly increases the area of ablation due to lateral spreading. A third reconstruction assumes a high accumulation-ablation gradient similar to that used by Booth (1986) to reconstruct the balance velocity of the 18,000-B.P. Puget Lobe. The mass-balance pattern was reconstructed from late Wisconsin valley glaciers in the Pacific Northwest. This model predicts an ELA of about 550 m (Fig. 9C), a maximum velocity of about 350 m/yr, and a maximum ablation rate of 3.0 m/yr. Ablation occurred over less than 30 km of the flow line and the A/Ao was 0.88.

The predicted ELAs are very sensitive to the divergence of flow lines near the ice margin and to the size of the accumulation area. This is because changing the divergence increases or decreases the size of the ablation area in the model. Models that assume no divergence predict ELAs about 400 m higher. The position of maximum velocity is also very sensitive to where divergence begins. In all the models though, maximum velocity occurs just up glacier from where the flow band begins to diverge.

There is no geomorphic evidence to constrain the ELA of the Green Bay Lobe. The maximum elevations of lateral moraines in valley glaciers are sometimes cited as a minimum ELA because moraines are only deposited below the equilibrium line. The maximum elevation of moraines associated with the Green Bay Lobe are less than 600 m elevation. They would have been 50–100 m lower because of crustal depression, which is lower than the reconstructed ELAs of the continental and intermediate mass-balance distributions. This suggests that the maritime pattern predicts an unreasonably low ELA for this profile.

If the accumulation rates used in the continental and intermediate models are reasonable, then an increase in the ELA above about 1,200 m would have produced a glacier in negative balance. Paterson (1994) notes that glaciers with a significant portion of their surface elevation near that of the ELA should be very sensitive to small changes in ELA. This must have been the case with thin lobes along the southern margin of the Laurentide Ice Sheet, and may explain their rapid, progressive retreat. Once downwasting began, more and more of the lobe would have been below the ELA and therefore contributed to a negative balance. Several climate models have suggested that by 18,000 B.P. much of the Laurentide Ice Sheet would have been in a negative mass balance (Gates, 1976; Manabe and Broccoli, 1985; Kutzbach 1987). The low ELAs produced here support this interpretation.

Estimate of ice surface lowering rates during retreat

Numerous small moraines were formed by the retreating Green Bay Lobe. These moraines are less than 1 m high and between 5 and 10 m wide. They are visible in black-and-white aerial photos and can be traced for as much as 2 km (see Colgan, 1996). It is possible that many of these moraines formed annually or nearly so. If so they may be used to estimate the amount of ice-surface lowering necessary to produce the observed spacing (Boulton, 1986). Ham (1994) used this method to determine the ice-surface-lowering rate of the Wisconsin Valley Lobe in northern Wisconsin. He estimated that ice-surface-lowering rates were between 1.9 and 5.7 m/yr at 16,000 B.P. This assumes an ice-surface slope of 3°–4° within 1 km of the terminus. The slope near the terminus of the Green Bay Lobe during the Lake Mills phase, when small moraines were forming, was similar to this or may have been slightly lower. Slopes of 2°–4° and a retreat rate of 25 m/yr, result in an ice-surface-lowering rate of between 0.87 and 1.7 m/yr. Slopes of 2°–4° and a retreat rate of 100 m/yr results in an ice-surface-lowering rate of between 3.5 and 7 m/yr. The spacing between ridges is between 50 and 100 m, suggesting average ice-surface-lowering rates of 1.7–7 m/yr. Since ice was thin and probably moving slowly near the terminus, it is also possible that these rates are equal to ablation rates.

These estimates are similar to ablation rates measured along the northern margin of Vatnajökull in Iceland and along the southern margin of the Barnes Ice Cap. Both of these ice caps are retreating at rates between 25 and 50 m/yr (Hooke, 1973; Sigurdsson, 1991). Retreat rate and ice-surface-lowering rates are nearly twice the rates predicted for steady-state conditions. This is consistent if the glacier was in negative mass balance and retreating when the small moraines were deposited. Larger compound moraines (such as those deposited during the Lake Mills phase) suggest that ablation rates were lower for short periods (perhaps because of cooler summers for a decade or so).

If the estimates of ice-surface lowering are correct, then the ELA rose above 1,200 m during deglaciation. During retreat, a mass balance pattern with a steep ablation-accumulation gradient is required, because such a pattern produces high ablation rates at the margin as well as sufficient ice-surface velocities near the terminus to produce small winter readvances.

Glacier bed and ice-margin temperatures

Evidence for permafrost along the terminus. Relict ice-wedge polygons formed in Illinois (Johnson, 1990) between 21,000 and 16,000 B.P., and in Wisconsin between 28,000 and 13,000 B.P. (Attig et al., 1989). Ice-wedge polygons on the deglaciated surface (see Fig. 2) suggest that permafrost was present at least locally along the retreating terminus of the Green Bay Lobe. The distribution of continuous permafrost today is restricted to areas where the mean annual temperature is at least $-5 - -8$ °C (Brown, 1970; Pewe, 1983). The southern limit of discontinuous permafrost extends to the present 0 °C isotherm. Because ice-wedge polygons formed on the deglaciated surface, mean annual temperatures must have been below 0 °C near the terminus as the lobe retreated. Because ice wedge-casts and relict polygons are not found on surfaces younger than 13,000 B.P. in Wisconsin, mean annual temperatures probably warmed to above 0 °C by this time.

Fossil insects and pollen demonstrate that tundra was present in the Upper Midwest from 28,000 to 13,000 B.P. (Schwert and Ashworth, 1988, 1992; Garry et al., 1990; Schwert, 1992; Torpen and Schwert, 1996). Pollen and plant macrofossils indicate that tundra was present near Wolf Creek, southeastern Minnesota, from 20,500 to 14,700 B.P. (Birks, 1976); in southeastern Iowa from 18,000 to 16,500 B.P. (Baker et al., 1986); and in Wisconsin as late as 13,000 B.P. (Maher and Mickelson, 1996). Fossil beetles suggest that near Wedron, Illinois, mean January temperatures at 21,500 B.P were 21–24 °C colder than present (~ 6 °C), and summer temperatures were 11 °C cooler than present (~24 °C; Elias, 1996).

Evidence for a thawed glacier bed and a narrow frozen zone at the terminus. Drumlins, flutes, striae, and eskers suggest that the bed of the Green Bay Lobe was unfrozen during deglaciation, even though permafrost was present in front of the glacier. Broad zones of hummocky sediment (2–5 km wide) do suggest that a narrow zone (<5 km wide) of freezing-on developed near the terminus or that nonsteady oscillations of the ice margin occurred. The zone of freezing-on disappeared or narrowed (<<1 km wide) during retreat and the margin may have been very thin or stagnant. This would have allowed the winter cold wave to penetrate at the terminus and may have allowed a very narrow (<100 m wide) frozen-bed zone to form in winter.

Attig et al. (1989) hypothesize that during the Johnstown-Hancock phase, a 5- 20-km-wide, frozen-bed zone was present at the terminus, based on the presence of tunnel channels (spaced at approximately 10-km intervals), broad hummocky moraines, and an absence of drumlins. During retreat they argue that the frozen-bed zone disappeared and narrow moraines, drumlins, striae, and eskers formed under warmer ice margin conditions. While this hypothesis explains the distribution of supraglacial sediments it is unlikely that a wide frozen-bed zone formed because of the extreme glaciological conditions this requires (Hooke, 1977; Mooers, 1990; Paterson, 1994). These conditions include very low mean annual surface temperatures (<–36 °C), dead ice, very thin ice (<100 m), or very low ablation rates (<0.1 m/yr).

These conditions would be unlikely for the Green Bay Lobe even at its maximum. Summer temperatures above 0 °C (~9 °C), large amounts of summer radiation, and warm, moist Gulf air, would have resulted in ablation rates significantly higher than 0.1 m/yr along much of the southern Laurentide margin at the last glacial maximum. Climate models indicate significant summer melting along the southern margin of the Laurentide Ice Sheet at 18,000 B.P. (Gates, 1976; Manabe and Broccoli, 1985; Kutzbach, 1987). Ice-surface profiles suggest that ice was at least 150 m thick within 3 km of the margin during the Johnstown-Hancock phase and >250 m thick within 10 km of the terminus. Mass-balance reconstructions suggest that at its late glacial maximum, ablation rates were between 1.0 and 1.6 m/yr, if the glacier were in balance. During retreat, ablation rates increased to at least 1.7 to 3.5 m/yr. High rates of ablation would eliminate the possibility of a wide frozen-bed zone in a steady-state or slowly retreating glacier. Another factor that would reduce the width of a frozen-bed zone is the transport of heat in basal meltwater (Clarke et al., 1984; Mooers, 1990)

Observations of modern glaciers confirm that a wide frozen-bed zone is not present even in cold polar glaciers. Colbeck et al. (1974) noted that the glacier bed is temperate at a depth of 200 m along the southwest margin of the Greenland Ice Sheet. In this area the mean annual temperature is below −6 °C. Temperate basal ice is also present near the terminus of White Glacier under an ice thickness of 100–150 m. Mean annual temperature is between −15 and −11 °C in this part of Axel Heiburg Island, Northwest Territories, Canada (Blatter, 1987). Temperate ice is also present near the margin of Trapridge Glacier under similar ice thickness and a mean annual temperature between −6 and −8 °C (Clarke et al., 1984). Since temperate ice is present even under very thin ice thickness (<150 m) and very cold conditions (mean annual temperatures of −15 °C), it is unlikely that a frozen bed was present under thicker ice near the margin of the Green Bay Lobe. It is possible that a frozen-bed zone could have developed when the ice margin advanced rapidly over an area of permafrost in the glacier forefield (Mooers, 1990).

Evidence for basal freeze-on of debris at the terminus. The hummocky moraines built during the Johnstown-Hancock and Milton-Almond phases do reflect enhanced incorporation of material in englacial and supraglacial positions near the terminus (Attig et al., 1989). This can be accomplished with a narrow zone (<1 km) of cold basal ice or an oscillating ice margin. Freeze-on of basal water at the terminus under ice thickness less than 100 m has been observed in polar glaciers (Boulton, 1970a, b, 1971). Boulton described this process occurring near the thin margins of glaciers in Svalbard, where mean annual temperatures are about −4 °C. Incorporation of large amounts of debris has also been documented along the terminus of Whiskey Glacier, a small outlet glacier on James Ross Island, Antarctica (Chinn and Dillion, 1987). Basal freeze-on that incorporates basal debris near the terminus has also been documented along the Barnes Ice Cap (Hooke, 1973), the Greenland Ice Sheet (Hooke, 1970; Sugden et al., 1987a), in outlet glaciers in Antarctica and South Georgia Island (Sugden et al., 1987b; Souchez et al., 1988), and even in a temperate glacier in southeast Alaska (Strasser et al., 1993).

Flow regimes

The dominance of streamlined landforms and moraines suggest that steady compressive flow was the dominant flow regime of the Green Bay Lobe during retreat. While it cannot be entirely ruled out, there is no evidence for an extensive surge advance of the lobe to its maximum position. Radiocarbon dates provide no control on the rate of advance. Therefore, hypotheses advocating an initial surge advance of the Green Bay Lobe can only be considered speculative. The radial pattern of drumlins and flutes does suggest that surging did not affect large parts of the lobe. In surging glaciers it is common for there to be discontinuities in flow velocity over short distances (forming shear margins and moraines). This has been documented in both modern and ancient glaciers (Kamb et al., 1985; Thorinson, 1969; Raymond et al., 1987; Sharp et al., 1988, 1994; Ham, 1994). Strike-slip boundaries are present along the margins of minor sublobes that may have surged into small proglacial lake basins during the Lake Mills, Rush Lake, and St. Anna phases (Fig. 5D–F), but they probably do not require surge of the entire lobe. Similar surges have affected the margins of large ice caps in Iceland (Thorinson, 1969) and the Canadian Arctic (Holdsworth, 1973). In these examples the terminus advanced less than 5 km and only a portion of the margin was affected.

Rate of deglaciation

Radiocarbon dates on material overlying the Horicon Formation provide contradictory data about the rate of retreat. Dates on organic sediment obtained from Devils Lake in Sauk County, Wisconsin, have been interpreted as indicating that the Green Bay Lobe was near its maximum as late as 13,000 to 14,000 B.P. (Maher, 1982). Dates on organic material in lacustrine sediments associated with the early Athelstane advance suggest that the Green Bay Lobe advanced to a position near Valders, Wisconsin, following a major retreat out of the basin by 13,000 B.P. (Maher and Mickelson, 1996). Assuming that both interpretations are correct requires a total retreat and readvance of at least 450 km in about 500–1,500 calendar years (the absolute error in radiocarbon dates). This would require average retreat rates of 300 to 900 m/yr to satisfy this chronology. On the other hand it is also possible that the Devils Lake dates do not record the beginning of retreat, but are too young by 1,000 to 3,000 years. They may simply reflect the degradation of permafrost and arrival of spruce by about 13,000 to 14,000 B.P. (Clayton et al., 1992). Permafrost features on the deglaciated surface also suggest that deglaciation occurred before 13,000 B.P. and that enough time was available to form ice-wedge polygons.

Geomorphic evidence suggests that deglaciation proceeded rapidly, but progressively. Small ridges between the Johnstown and Lake Mills moraines suggest retreat rates of about 50 m/yr during early deglaciation. At this rate the Green Bay Lobe could have retreated completely out of the map area in about 3,600 years. The time required to retreat completely out of the basin would be about 8,000 years, assuming a constant retreat rate of 50 m/yr. Since this is also incompatible with the relative chronology, it is likely that retreat rates increased as the ice margin retreated into glacial lake Oshkosh (Thwaites, 1943; Need, 1985).

DISCUSSION

On the basis of landform and sediment distribution, paleoclimate data, and my reconstructions, it is clear that there are no complete modern analogs for the Green Bay Lobe, although it bears similarity to parts of glaciers in Iceland and southern Alaska, and the terminal parts of ice streams. Polar and subpolar glaciers are plausible analogs for the Green Bay Lobe's terminus at its maximum, because they commonly produce englacial debris near the terminus (Boulton, 1970b; Hooke, 1970; Chinn and Dillion, 1987). Polar and subpolar glaciers are poor models for deglaciation because much of the southern margin of the Laurentide Ice Sheet after 18,000 B.P. experienced warm summers (with high insolation), which produced high ablation rates and large volumes of meltwater.

Temperate valley glaciers and ice caps may provide a good analog for the ablation area, but their small size limits further comparison. Some have compared the southeastern margin of the Laurentide Ice Sheet to Malaspina Glacier in southeastern Alaska (Gustavson and Boothroyd, 1987). This glacier has low ice-surface slopes (<14 m/km), high ablation rates (0.4–4 m/yr for debris covered and clean ice), and evidence of both steady and surging flow. Geomorphic features currently being formed by Malaspina Glacier include flutes, eskers, and hummocky moraines, although the high mean annual air temperature (6–12 °C) and high precipitation rate (~3.2 m/yr) suggest a climate and glacier regime much less severe than that of the Green Bay Lobe from 26,000 to 16,000 B.P.

Glaciers in Iceland have produced a distribution of landforms similar to those of the Green Bay Lobe. These include drumlins, flutes, eskers, small push moraines, and end moraines (Krüger and Thomsen, 1984; Krüger, 1993, 1994; Boulton, 1986). The mean annual temperature near the northern margin of Mýrdalsjökull, where small annual moraines and drumlins are actively forming, is −1–0 °C (Krüger, 1993). Permafrost is not found in this area and the glacier has been retreating at a rate of 8–25 m/yr since the 1920s. As on other glaciers in Iceland, high winter accumulation is offset by high summer ablation (Boulton, 1986; Krüger, 1993). Although winter temperatures were colder along the margin of the Green Bay Lobe, summer temperatures and ablation rates may have been similar to those in southern Iceland today. The landforms and sediments created during deglaciation are also similar to those produced by outlet glaciers in Glacier Bay, Alaska, that were cut off from their accumulation areas and subsequently stagnated (Goldthwait, 1974; Syverson et al., 1994).

The size of Laurentide ice lobes suggests a comparison to West Antarctic ice streams, which have similar ice-surface slopes, ice thicknesses, and lengths (Patterson, 1997). Ice streams by definition are streams of rapidly moving ice (50–1,000 m/yr) bounded by slowly moving ice (<10 m/yr; Bentley, 1987). Similar boundaries between fast- and slow-moving ice may have existed upstream of the Green Bay Lobe and other lobes. Flow paths reconstructed from drumlins suggest that much of the ice was channeled down the Green Bay lowland.

Ice-surface slopes of the Green Bay Lobe are similar to the surface slopes of distal ice streams in West Antarctica. For example, the slope of Ice Stream B near camp UpB is about 2.7 m/km. Slopes near the grounding line become even lower (~0.35 m/km) because basal shear stress decreases towards the grounding line (Shabtaie et al., 1987). In the case of the Green Bay Lobe basal shear stress was probably increasing towards the margin as in outlet glaciers.

However, several major differences suggest that lowland lobes were fundamentally different from ice streams. Ice streams are bounded by slowly moving ice, ice lobes are bounded by topography. Flow was diverging near the margins of lowland lobes in the Central Lowlands similar to those of modern outlet glaciers. Ice lobes and outlet glaciers increase in slope near the margin (Cooper et al., 1982; Clarke, 1987). West Antarctic ice streams end in ice shelves and therefore decrease in slope down glacier. The climate regime of present ice streams is also very different from the marginal areas of the Laurentide Ice Sheet. Present ice streams experience little surface melting and no surface water is able to penetrate to the glacier bed. Most ablation occurs by calving at the edge of ice shelves. Clark (1992) compared lowland glacier lobes to the low sloping "ice plains" at the distal ends of West Antarctic ice streams. Considering the low ice-surface profiles reconstructed here this analogy seems appropriate, yet this cannot be confirmed until we know more about the dynamics of ice flow upstream from glacier lobes. If an ice stream was feeding the Green Bay Lobe, we should be able to find some evidence for this north of the study area.

CONCLUSIONS

The Green Bay Lobe was a broad, thin outlet glacier of the Laurentide Ice Sheet. The total distance from the margin to the divide was approximately 1,400 km. Ice-surface profile reconstructions suggest that basal shear stress was less than 25 kPa (within 150 km of the terminus) for all phases before 14,000 B.P. Very low shear stress (<5–10 kPa) was present during retreat phases, and during phases after 13,000 B.P. when the lobe may have surged repeatedly into the lowland.

Permafrost was present along the retreating terminus of the Green Bay Lobe, yet most of the glacier bed was temperate, except perhaps for a narrow zone (<5 km wide) near the terminus during the glacial maximum. During retreat this zone narrowed or completely disappeared. During the Johnstown and Milton phases, meltwater discharged through a series of tunnel channels. During retreat this system switched to an esker system. Low ice-surface profiles and well-preserved esker systems suggest that ice may have became stagnant near the end of the Lake Mills phase (~14,000 B.P.).

There are no modern analogs for the Green Bay Lobe. Possible analogs for the terminal environment include modern glaciers in Iceland, Alaska, and Norway. Mass-balance calculations suggests that maximum velocity (150–350 m/yr) was focused along the axis of the lowland. A low equilibrium-line altitude (ELA; 750–1,200 m), and ablation rates of 0.5–1.5 m/yr are required for a steady-state flow regime during the Johnstown-Hancock phase. During retreat, ablation rates probably doubled to at least 1.7–3.5 m/yr. The gentle ice-surface profile during deglaciation made the lobe extremely sensitive to changes in ELA and explains the rapid progressive retreat from the 18,000 B.P. position.

ACKNOWLEDGMENTS

I thank David M. Mickelson for his contributions, support, and criticism of the ideas in this paper. Lee Clayton, Lou J. Maher, Charles R. Bentley, and John E. Kutzbach also provided valuable feedback and editing. Reviewers Mark D. Johnson, Carrie J. Paterson, and John W. Attig helped to greatly improve the paper. Support for this project was provided by the Lewis G. Weeks Fund, Department of Geology and Geophysics at the University of Wisconsin–Madison, and NSF grant EAR-9627798. One summer of fieldwork was supported by an ARCO summer scholarship and Sigma Xi. Additional support came from the Graduate School of the University of Wisconsin–Madison.

REFERENCES CITED

Acomb, L. J., Mickelson, D. M., and Evenson, E. B., 1982, Till stratigraphy and late glacial events in the Lake Michigan Lobe of eastern Wisconsin: Geological Society of America Bulletin, v. 93, p. 289–296.

Alden, W. C., 1911, Radiation of glacial flow as a factor in drumlin formation: Geological Society of America Bulletin, v. 22, p. 33–34.

Alden, W. C., 1932, Glacial geology of the central United States: 16th International Geological Congress, Guidebook, v. 26, 54 p.

Attig, J. W., Clayton, L., and Mickelson, D. M., 1985, Correlation of late Wisconsin glacial phases in the western Great Lakes area: Geological Society of America Bulletin, v. 96, p. 1585–1593.

Attig, J. W., Mickelson, D. M., and Clayton, L., 1989, Late Wisconsin landform distribution and glacier-bed conditions in Wisconsin: Sedimentary Geology, v. 62, p. 399–405.

Baker, R. G., Rhodes, R. S., III, Schwert, D. P., Ashworth, A. C., Frest, T. J., Hallberg, G. R., and Janssens, J. A., 1986, A full-glacial biota from southeastern Iowa, USA: Journal of Quaternary Science, v. 1, p. 91–107.

Bentley, C. R., 1987, Antarctic ice streams: a review: Journal of Geophysical Research, v. 92, p. 8843–8858.

Bindschadler, R. A., Stephenson, S. N., MacAyeal, D. R., and Shabtaie, S., 1987, Ice dynamics at the mouth of Ice Stream B: Journal of Geophysical Research, v. 92, p. 8885–8894.

Birks, H. J. B., 1976, Late-Wisconsinan vegetational history at Wolf Creek, central Minnesota: Ecological Monographs, v. 46, p. 395–429.

Blatter, H., 1987, On the thermal regime of an Arctic valley glacier: A study of White Glacier, Axel Heiberg Island, N.W.T., Canada: Journal of Glaciology, v. 33, p. 200–211.

Booth, D. B., 1986, Mass balance and sliding velocity of the Puget Lobe of the Cordilleran ice sheet during the last glaciation: Quaternary Research, v. 25, p. 269–280.

Boulton, G. S., 1970a, The deposition of subglacial and melt-out till at the margins of certain Svalbard glaciers: Journal of Glaciology, v. 9, p. 231–245.

Boulton, G. S., 1970b, On the origin and transport of englacial debris in Svalbard Glaciers: Journal of Glaciology, v. 9, p. 213–228.

Boulton, G. S., 1971, Till genesis and fabric in Svalbard, Spitzbergen, in Goldthwait, R. P., ed., Till, A Symposium: Columbus, Ohio, Ohio State University Press, p. 41–72.

Boulton, G. S., 1986, Push-moraines and glacier-contact fans in marine and terrestrial environments: Sedimentology, v. 33, p. 677–698.

Boulton, G. S., Smith, G. S., Jones, A. S., and Newsome, J., 1985, Glacial geology and glaciology of the last mid-latitude ice sheets: Journal of Geological Society of London, v. 142, p. 447–474.

Brown, R. J. E., 1970, Permafrost in Canada: Toronto, University of Toronto Press, 234 p.

Chinn, T. J. H., and Dillion, A., 1987, Observations on a debris-covered polar glacier "Whiskey Glacier," James Ross Island, Antarctic Peninsula, Antarctica: Journal of Glaciology, v. 33, p. 300–310.

Clarke, G. K. C., 1987, Fast glacier flow: Ice streams, surging, and tidewater glaciers: Journal of Geophysical Research, v. 92, p. 8835–8841.

Clarke, G. K. C., Collins, S. G., and Thompson, D. E., 1984, Flow, thermal structure, and subglacial conditions of a surge-type glacier: Canadian Journal of Earth Science, v. 21, p. 232–240.

Clark, J. A., Hendricks, M., Timmermans, T. T., Struck, C., and Hilverda, K. J., 1994, Glacial isostatic deformation of the Great Lakes region: Geological Society of America Bulletin, v. 106, p. 19–31.

Clark, P. U., 1992, Surface form of the southern Laurentide Ice Sheet and its implications to ice-sheet dynamics: Geological Society of America Bulletin, v. 104, p. 595–605.

Clayton, L., 1984, Pleistocene geology of the Superior region, Wisconsin: Wisconsin Geological and Natural History Survey, Information Circular, scale 1:250,000, 40 p.

Clayton, L., and Attig, J. W., 1989, Glacial Lake Wisconsin: Geological Society of America Memoir 173, 80 p.

Clayton, L., Attig, J. W., Mickelson, D. M., and Johnson, M. D., 1992, Glaciation of Wisconsin: Wisconsin Geological and Natural History Survey, Educational Series 36, 4 p.

Colbeck, S. C., and Gow, A. J., 1974, The margin of the Greenland Ice Sheet at Isua: Journal of Glaciology, v. 13, p. 155–165.

Colgan, P. M., 1996, The Green Bay and Des Moines Lobes of the Laurentide Ice Sheet: Evidence for stable and unstable glacier dynamics 18,000 to 12,000 B.P. [Ph.D. thesis]: University of Wisconsin, Madison, 274 p.

Colgan, P. M., and Mickelson, D. M., 1997, Genesis of streamlined landforms and flow history of the Green Bay Lobe, Wisconsin, USA: Sedimentary Geology, v. 111, p. 7–25.

Coleman, S. M., Clark, J. A., Clayton, L., Hansel, A. K., and Larsen, C. E., 1989, Deglaciation, lake levels, and meltwater discharge in the Lake Michigan basin: Quaternary Science Reviews, v. 13, p. 879–890.

Cooper, A. P. R., McIntyre, N. F., and Robin, G. deQ., 1982, Driving stresses in the Antarctic Ice Sheet: Annals of Glaciology, v. 3, p. 59–62.

Denton, G. H., and Hughes, T. J., 1981, The last great ice sheets: New York, John Wiley, 484 p.

Dyke, A. S, and Prest, V. K., 1987, Late Wisconsinan and Holocene history of the Laurentide Ice Sheet: Geographie Physique et Quaternaire, v. 41, p. 237–263.

Elias, S., 1996, Bugs in the (climate) system: Insect evidence of the nature and timing of the climate warming in North America: American Quaternary Association, 14th Biennial Meeting, Abstracts, Flagstaff, p. 12.

Evenson, E. B., Mickelson, D. M., and Farrand, W. R., 1977, Stratigraphy and correlation of the late Wisconsinan glacial events in the Lake Michigan basin: Geographie Physique et Quaternaire, v. 31, p. 53–59.

Garry, C. E., Schwert, D. P., Baker, R. G., Kemmis, T. J., Horton, D. G., and Sullivan, A. E., 1990, Plant and insect remains from the Wisconsinan interstadial/stadial transition at Wedron, northcentral Illinois: Quaternary Research, v. 33, p. 387–399.

Gates, W. L., 1976, The numerical simulation of ice-age climate with a global general circulation model: Journal of Atmospheric Sciences, v. 33, p. 1844–1873.

Goldthwait, R. P., 1974, Rates of formation of glacial features in Glacier Bay, Alaska, in Coates, D. R., ed., Glacial geomorphology: Binghamptom, New York, State University of New York, p. 163–185.

Gustavson, T. C., and Boothroyd, T. C., 1987, A depositional model for outwash, sediment sources, and hydrologic characteristics, Malaspina Glacier, Alaska: A modern analog of the southeastern margin of the Laurentide Ice Sheet: Geological Society of America Bulletin, v. 99, p. 187–200.

Ham, N. R., 1994, Glacial geomorphology and dynamics of the Wisconsin Valley Lobe of the Laurentide Ice Sheet, Lincoln County, Wisconsin [Ph.D. thesis]: Madison, University of Wisconsin, 235 p.

Holdsworth, G., 1973, Evidence of a surge on Barnes Ice Cap, Baffin Island: Canadian Journal of Earth Sciences, v. 10, p. 1565–1574.

Hooke, R. L., 1970, Morphology of the ice-sheet margin near Thule, Greenland: Journal of Glaciology, v. 9, p. 303–324.

Hooke, R. L., 1973, Flow near the margin of the Barnes Ice Cap, and the development of ice-cored moraines: Geological Society of America Bulletin, v. 84, p. 3929–3948.

Hooke, R. L., 1977, Basal temperature in polar ice sheets: A qualitative review: Quaternary Research, v. 7, p. 1–13.

Johnson, W. H., 1990, Ice-wedge casts and relict patterned ground in central Illinois and their environmental significance: Quaternary Research, v. 33, p. 51–72.

Kamb, B., Raymond, C. F., Harrison, W. D., Englehardt, H. F., Echelmeyer, K. A., Humphrey, N. F., Brugman, M. M., and Pfeffer, T., 1985, Glacier surge mechanism: 1982–1983 surge of Variegated Glacier, Alaska: Science, v. 227, p. 469–479.

Krüger, J., 1993, Moraine-ridge formation along a stationary ice front in Iceland: Boreas, v. 22, p. 101–109.

Krüger, J. 1994, Glacial processes, sediments, landforms, and stratigraphy in the terminus region of Mýrdalsjökull, Iceland: Folia Geographica Danica, v. 21, 233 p.

Krüger, J. and Thomsen, H. H., 1984, Morphology, stratigraphy, and genesis of small drumlins in front of the glacier Mýrdalsjökull, south Iceland: Journal of Glaciology, v. 30, p. 94–105.

Kutzbach, J. E., 1987, Model simulations of the climatic patterns during the deglaciation of North America, in Ruddiman, W. F., and Wright, H. E., Jr., eds., North America and adjacent oceans during the last deglaciation: Boulder, Colorado, Geological Society of America, The Geology of North America, v. K-3, p. 425–446.

Maher, L. J., Jr., 1982, The palynology of Devils Lake, Sauk County, Wisconsin, in Knox, J. C., Clayton, L., and Mickelson, D. M., eds., Quaternary history of the driftless area: Wisconsin Geological and Natural History Survey Field Trip Guide Book Number 5: University of Wisconsin-Extension, Geological and Natural History Survey, p. 119–135.

Maher, L. J., Jr., and Mickelson, D. M., 1996, Palynology and radiocarbon evidence for deglaciation events in the Green Bay Lobe, Wisconsin: Quaternary Research, v. 46, p. 251–259.

Manabe, S., and Broccoli, A. J., 1985, The influence of continental ice sheets on the climate of the ice age: Journal of Geophysical Research, v. 90, p. 2167–2190.

Mathews, W. H., 1974, Surface profile of the Laurentide Ice Sheet in its marginal areas: Journal of Glaciology, v. 13, p. 37–43.

Mayewski, P. A., Denton, G. H., and Hughes, T. J., 1981, Late Wisconsin ice sheets of North America, in Denton, G. H., and Hughes, T. J., eds., The last great ice sheets: Toronto, John-Wiley & Sons, p. 67–131.

McCartney, M. C., and Mickelson, D. M., 1982, Late Woodfordian and Greatlakean history of the Green Bay Lobe, Wisconsin: Geological Society of America Bulletin, v. 93, p. 297–302.

Mickelson, D. M., Acomb, L. J., and Bentley, C. R., 1981, Possible mechanism for the rapid advance and retreat of the Lake Michigan Lobe between 13,000 and 11,000 B.P.: Annals of Glaciology, v. 2, p. 185–186.

Mickelson, D. M., Clayton, L., Baker, R. W., Mode, W. N., and Schneider, A. F., 1984, Pleistocene stratigraphic units of Wisconsin: Wisconsin Geological and Natural History Survey, Miscellaneous Paper 84-1, 15 p.

Mooers, H. D., 1990, Ice-marginal thrusting of drift and bedrock: thermal regime, subglacial aquifers, and glacial surges: Canadian Journal of Earth Sciences, v. 27, p. 849–862.

Need, E., 1985, Pleistocene geology of Brown County, Wisconsin: Wisconsin Natural History and Geological Survey, Information Circular #48, 19 p.

Oerlemans, J., and van der Veen, C. J., 1984, Ice sheets and climate: Dordrecht, Netherlands, D. Reidel, 217 p.

Paterson, W. S. B., 1994, The physics of glaciers: New York, Pergamon, 480 p.

Patterson, C. J., 1997, Southern Laurentide ice lobes were created by ice streams: Des Moines Lobe in Minnesota, USA: Sedimentary Geology, v. 111, p. 249–261.

Pewe, T. L., 1983 The periglacial environment in North America during Wisconsin time, in Porter, S. C., ed., Late-Quaternary environments of the United States: Minneapolis, University of Minnesota Press, v. 1, p. 157–189.

Prest, V. K., 1969, Retreat of Wisconsin and Recent ice in North America: Geological Survey of Canada, Map 1257A, scale 1:5,000,000.

Raymond, C., Johannesson, T., Pfeffer, T., and Sharp, M., 1987, Propagation of a glacier surge into stagnant ice: Journal of Geophysical Research, v. 92, p. 9037–9049.

Ridky, R. W., and Bindschadler, R. A., 1990, Reconstruction and dynamics of the late Wisconsin "Ontario" ice dome in the Finger Lakes region, New York: Geological Society of America, v. 102, p. 1055–1064.

Schwert, D. P., 1992, Faunal transitions in response to an ice age: the late Wisconsinan record of Coleoptera in the north-central United States: The Coleopterists Bulletin, v. 46, p. 68–94.

Schwert, D. P., and Ashworth, A. C., 1988, Late Quaternary history of northern beetle fauna of North America: a synthesis of fossil and distributional evidence: Memoirs of the Entomological Society of Canada, v. 144, p. 93–107.

Schwert, D. P., and Ashworth, A. C., 1992, Fossil insect assemblages of the last glacial maximum (21,000–16,500 B.P.) in the Upper Midwest: Geological Society of America, Abstracts with Programs, v. 24, p. 63.

Shabtaie, S., Whillans, I. M., and Bentley, C. R., 1987, The morphology of ice streams A, B, and C, West Antarctica, and their environs: Journal of Geophysical Research, v. 92, p. 8865–8883.

Sharp, M., Lawson, W., and Anderson, R. S., 1988, Tectonic processes in a surge-type glacier: Journal of Structural Geology, v. 10, p. 499–515.

Sharp, M., Jouzel, J., Hubbard, B., and Lawson, W., 1994, The character, structure and origin of the basal ice layer of a surge-type glacier: Journal of Glaciology, v. 40, p. 327–340.

Sigurdsson, O., 1991, Glacier fluctuations in Iceland: Jökull, v. 40, p. 169–174.

Simpkins, W. W., 1989, Genesis and spatial distribution of variability in the lithostratigraphic, geotechnical, hydrogeological, and geochemical properties of the Oak Creek Formation in southeastern Wisconsin [Ph.D. thesis]: Madison, University of Wisconsin, 394 p.

Souchez, R., Lorrain, R., and Tison, J. L., 1988, Co-isotopic signature of two mechanisms of basal-ice formation in Arctic outlet glaciers: Annals of Glaciology, v. 10, p. 163.

Strasser, J. C., Lawson, D. E., Evenson, E. B., Gosse, J. C., and Alley, R. B., 1993, Frazil ice growth at the terminus of the Matanuska Glacier, Alaska, and its implications for sediment entrainment in glaciers and ice sheets: Geological Society of America, Abstracts and Programs, Northeastern Section, v. 24, p. 78.

Sugden, D. E., Knight, P. G., Livesey, N., Lorrain, R. D., Souchez, R. A., Tison, J. L., and Jouzel, J., 1987a, Evidence for two zones of debris entrainment beneath the Greenland Ice Sheet: Nature, v. 328, p. 238–241.

Sugden, D. E., Clapperton, C. M., Gemmel, J. C., and Knight, P. G., 1987b, Stable isotopes and debris in basal glacier ice, south Georgia, Southern Ocean: Journal of Glaciology, v. 33, p. 324–329.

Syverson, K. M., Gaffield, S. J., and Mickelson, D. M., 1994, Comparison of esker morphology and sedimentology with former ice-surface topography, Burroughs Glaciers, Alaska: Geological Society of America Bulletin, v. 106, p. 1130–1142.

Thorson, R. M., 1980, Ice-sheet glaciation of the Puget Lowland, Washington, during the Vashon Stade (late Pleistocene): Quaternary Research, v. 13, p. 303–321.

Thorinson, S., 1969, Glacier surges in Iceland with special reference to Bruarjökull: Canadian Journal of Earth Sciences, v. 6, p. 875–882.

Thwaites, F. T., 1943, Pleistocene of part of northeastern Wisconsin: Geological Society of America Bulletin, v. 54, p. 87–144.

Thwaites, F. T., and Bertand, K., 1957, Pleistocene geology of the Door Peninsula, Wisconsin: Geological Society of America, v. 68, p. 831–880.

Torpen, H. J., and Schwert, D. P., 1996, Environmental significance of insect remains from sub-till organics of the Des Moines Lobe, Iowa: Geological Society of America, Abstracts with Programs, North-Central Section, v. 28, p. 67–68.

Wright, H. E., 1971, Retreat of the Laurentide Ice Sheet from 14,000 to 9,000 years ago: Quaternary Research, v. 1, p. 316–330.

Wright, H. E., Matsch, C. L., and Cushing, E. J., 1973, Superior and Des Moines Lobes: Geological Society of America, Memoir 136, p. 153–185.

MANUSCRIPT ACCEPTED BY THE SOCIETY OCTOBER 8, 1998

Ice-surface profiles and bed conditions of the Green Bay Lobe from 13,000 to 11,000 ^{14}C-years B.P.

Betty J. Socha
Department of Geology & Geophysics, University of Wisconsin, 1215 Dayton Street, Madison, Wisconsin 53706;
bjsocha@geology.wisc.edu

Patrick M. Colgan
Geology Department, Northeastern University, 14 Holmes Hall, Boston, Massachusetts 02115; pcolgan@lynx.dac.new.edu

David M. Mickelson
Department of Geology & Geophysics, University of Wisconsin, 1215 W. Dayton Street, Madison, Wisconsin 53706;
mickelson@geology.wisc.edu.

ABSTRACT

Between 13,000 and 11,000 ^{14}C-years B.P. the Chilton and Glenmore advances of the Green Bay Lobe deposited reddish brown clay till in eastern Wisconsin. Reconstruction of ice-surface slopes for these two advances using end-moraine elevations indicates that ice was thin and gently sloping, and that driving stresses were low. During the pre-Twocreekan (i.e., >12,500 ^{14}C-years B.P.) Chilton advance, ice in the main part of the lobe, in the Lake Winnebago lowland, had an average surface slope of about 2m/km, and was <200 m thick within 35 km of the terminus. The driving stress was about 2 kPa. Ice that protruded southeastward about 20 km into the Brillion basin as a small sublobe had an average surface slope of about 3m/km, and was <150 m thick. The driving stress was about 3 kPa. Ice that advanced over the Silurian escarpment had an average surface slope of about 11m/km, was <150 m thick, and had a driving stress of about 12 kPa within a few kilometers of the terminus. The post-Twocreekan Glenmore advance did not extend as far into Lake Winnebago and the Brillion basin, nor as far onto the Silurian escarpment as the earlier Chilton advance. The Glenmore ice had average slopes of 2–3m/km, a driving stress of about 2 kPa, and a thickness of <150 m within 15 km of the terminus. The driving stress for both was somewhat lower than earlier advances of the Green Bay Lobe, which had driving stresses of 5 to 25 kPa. The rebound history of the area is complex and no adjustments have been made for isostatic deformation. The ice configurations during the Chilton and Glenmore advances reflect basal sliding and/or subglacial sediment deformation, probably due to the high subglacial water pressure that developed in the lake sediment and underlying shale and dolomite in the Lake Winnebago lowland and the Brillion basin. It is possible that the margin configurations do not reflect the terminus of a steady state glacier, but rather the terminus of a surging glacier. We conclude that the low sloping ice at the margin of the Green Bay Lobe indicates bed conditions similar to those found beneath an ice plain of a West Antarctic ice stream.

Socha, B. J., Colgan, P. M., and Mickelson, D. M., 1999, Ice-surface profiles and bed conditions of the Green Bay Lobe from 13,000 to 11,000 14C-years B.P., *in* Mickelson, D. M., and Attig, J. W., eds., Glacial Processes Past and Present: Boulder, Colorado, Geological Society of America Special Paper 337.

INTRODUCTION

The surface morphology of parts of the southern Laurentide Ice Sheet have been reconstructed from the slope of moraine crests (Wright, 1972; Mathews, 1974), from both the slope of moraine crests and indicators of ice-flow direction (Clark, 1992; Colgan, 1996), or by specifying the driving stress (Boulton et al., 1985; Fisher et al., 1985; Hooke and Mooers, 1986; Mooers, 1989). Reconstructions based on geomorphic evidence indicate thin, low sloping ice. Studies that specified low driving stresses (Fisher et al., 1985; Clark et al., 1996) produced a thinner ice sheet that is more similar to that indicated by geomorphic evidence than those studies that specified higher driving stresses (Hooke and Mooers, 1986; Mooers, 1989).

Reconstructions of the ice-surface slope, based on geomorphic evidence, as opposed to assigning a value for shear stress, provide information about past glacier-bed conditions and flow regimes assuming that surface slopes are in equilibrium with the basal resistance of the bed. Gently sloping ice indicates a smooth, well-lubricated bed, whereas steeply sloping ice indicates more resistance to basal motion, perhaps due to a rough, well-drained, or frozen bed. Basal shear stress in valley glaciers usually ranges from 50 to 150 kPa, and in ice sheets ranges from 0 to 100 kPa with a mean of about 50 kPa (Paterson, 1994). Low slopes are found near ice domes, in outlet glaciers, and on ice streams (Bentley, 1987). The surface velocity of gently sloping ice is generally low (<10 m/yr), but some glaciers such as Ice Stream B in West Antarctica move much more rapidly (>500 m/yr). The rapid movement of Ice Stream B has been associated with the presence of a deforming bed (Blankenship et al., 1986) or with basal sliding in addition to bed deformation (Engelhardt and Kamb, 1998; Engelhardt et al., 1990; Tulaczyk, this volume, Chapter 15).

The purpose of this study is to reconstruct the ice-surface morphology of the Green Bay Lobe from end-moraine elevations and ice-flow direction indicators for the Chiton and Glenmore advances that took place between 13,000 and 11,000 ^{14}C-years B.P. These advances took place during a period of rapid deglaciation when the margin was fronted by a large proglacial lake. We then relate the reconstructed ice-surface profiles of the Green Bay Lobe and shear stresses derived from them to inferred bed conditions and dynamics of the lobe. This study demonstrates that the configuration of the margin of the Green Bay Lobe was complex and dynamic. The geomorphic evidence presented here supports models of ice-lobe behavior based on surging glaciers and distal portions of ice streams rather than traditional models based on valley glaciers and polar ice caps.

METHODS

A description of the graphical reconstruction method that we used and a discussion of assumptions and potential sources of error is provided in Clark (1992), and reviewed by Colgan (1996) and Colgan and Mickelson (1997). A major assumption

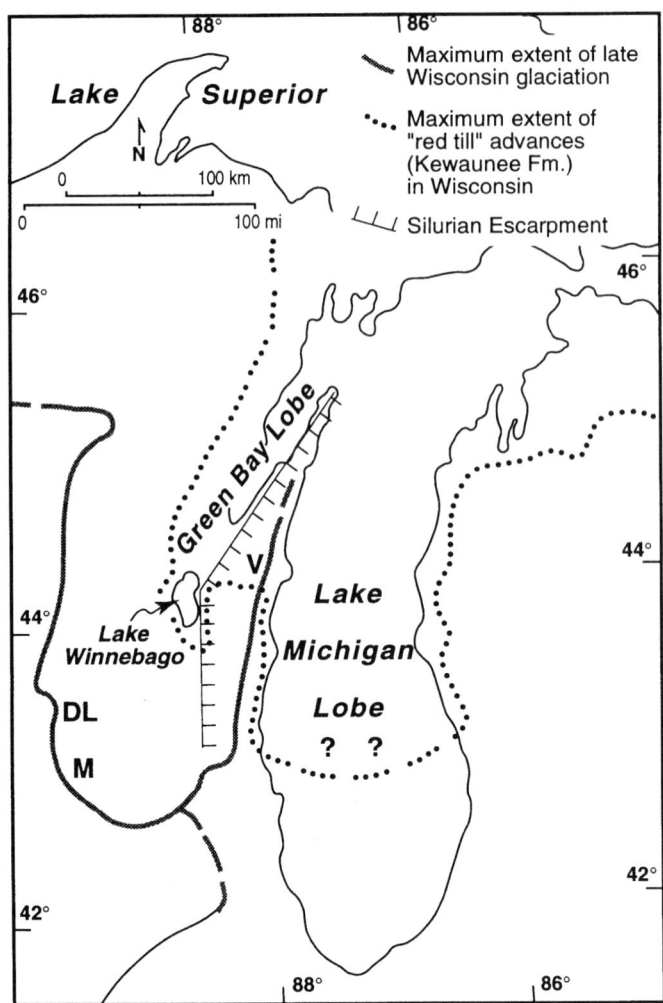

Figure 1. Terminal moraines and area of reddish brown till in the Green Bay and Lake Michigan Lobes. DL, Devils Lake; M, Madison; V, Valders.

is that the highest elevation on a moraine corresponds to the elevation of the ice surface at that point, at the time the moraine formed. It is also important that the indicators of ice-flow direction used to draw flowlines all formed at one time, and that lines of equal elevation on the ice surface (formlines) are perpendicular to the flowlines (Clark, 1992). In modern glaciers with low slopes and low driving stresses (<10 kPa), it is possible for flowlines to be oblique to formlines (Bindschadler et al., 1987). In this case, when reconstructions are made by drawing the formlines at right angles to flowlines, *maximum* values for ice thickness and calculated basal shear stress result. The stress calculation assumes that the ice is perfectly plastic, and the glacier is in a steady state.

The steps taken in the reconstruction method are first to compile the geologic data from which to interpret the extent and flow direction of the ice, and then to map the elevations of moraine crests from topographic maps. This is done by picking

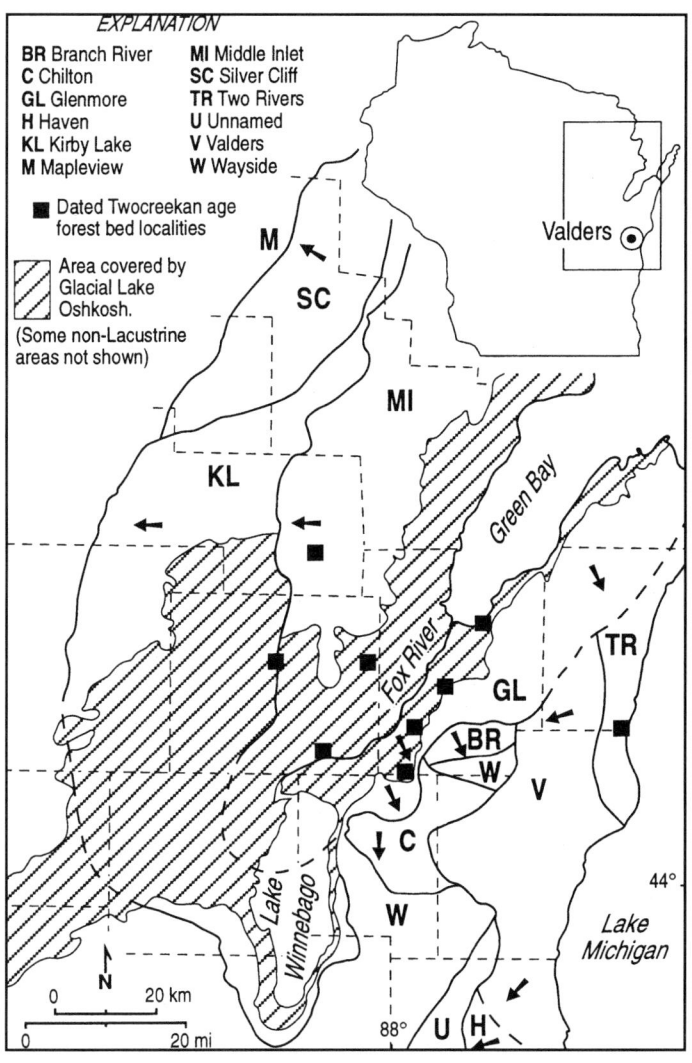

Figure 2. Lithostratigraphic units in eastern Wisconsin. Names of lithostratigraphic units are listed in the Explanation. Arrows indicate approximate direction of ice flow.

the highest elevation, generally every few kilometers along the moraine. Flow indicators are mapped from air photos and maps, with care taken to identify with which moraine the flow indicators are associated. Flowlines are then drawn parallel to the flow-direction indicators, and formlines are drawn at right angles to the flowlines. Finally, contour lines of elevation on the ice surface are drawn by projecting the moraine elevations parallel to the formlines (Clark, 1992).

Contours on the reconstructed ice surface are used to estimate the surface slope and the thickness of the ice. These are then both used to calculate the driving stress from the equation

$$\tau_d = \rho g h (\sin \alpha) \quad (1)$$

where τ_d is the driving stress, ρ is the density of ice, g is the acceleration due to gravity, h is the thickness of the ice, and α is the surface slope of the ice (Paterson, 1994). Driving stress (τ_d) is balanced by basal shear stress (τ_b) with corrections made for longitudinal-stress gradient (G) and stress from side drag (S) so that

$$\tau_d = \tau_b + 2G + S \quad (2)$$

(Paterson, 1981). $\tau_d = \tau_b$ when G and S are negligible (Paterson, 1981). G is negligible when driving stresses are based on ice-surface reconstructions over horizontal distances that are greater than 10 to 20 times the ice thickness, and driving stress is greater than 10 kPa (Clark, 1992). If G is ignored for low driving stresses, the results are minimum values with errors of up to several tens of percent (Clark, 1992). Side drag (S) is insignificant except for valley glaciers (Raymond, 1980) and ice streams (Alley et al., 1987), and can probably be ignored in reconstructions of ice-surface profiles for lobes of an ice sheet (Clark, 1992).

TABLE 1. STRATIGRAPHIC UNITS IN THE GREEN BAY AND LAKE MICHIGAN LOBES

Radiocarbon Years B.P.	Time Stratigraphy	Western Green Bay Lobe	Eastern Green Bay Lobe	Western Lake Michigan Lobe
		Kewaunee Formation	Kewaunee Formation	Kewaunee Formation
	Greatlakean	Middle Inlet Member	Glenmore Member	Two Rivers Member
12,500 to 11,500	Twocreekan	Two Creeks Forest Bed	Two Creeks Forest Bed	Two Creeks Forest Bed
	Late Woodfordian	Kirby Lake Member	Chilton Member	Valders Member
				Haven Member
		Silver Cliff Member	Branch River Member	Ozaukee Member
~ 13,000	Port Huron			
25,000 to 13,000		Horicon Formation	Horicon Formation	Holy Hill Formation
		Wayside Member		

GEOLOGIC SETTING

At its maximum extent the Green Bay Lobe was about 125 km wide and 200 km long (Fig. 1). Flowlines for ice in the Green Bay Lobe begin at the probable ice divide near James Bay, Canada, about 1,400 km northeast of the terminus of the lobe (Dyke and Prest, 1987). Ice flowed southwestward across the east end of Lake Superior and then was channeled southwestward through the Green Bay lowland. The Green Bay Lobe advanced into the Green Bay lowland in northeastern Wisconsin several times during the late Wisconsin.

The shape of the lobe was asymmetrical because rocks underlying the Green Bay basin dip gently eastward and differential erosion has produced a ridge—the Silurian escarpment. This ridge, which is about 100 m high, bounds the east side of Lake Winnebago, the Fox River lowland and Green Bay (Fig. 1). The west side of the lobe advanced up the regional slope over generally low terrain with little local relief—much of which is due to the bedrock topography.

The lobe reached its maximum extent just southwest of Madison, Wisconsin, sometime between 25,000 and 16,000 B.P. (Fig. 1) and deposited sandy, brown till of the Horicon Formation (Mickelson et al., 1984; Colgan and Mickelson, 1997). The Green Bay Lobe may have begun retreat as early as 18,000 B.P. (Clayton et al., 1992) or as late as 13,000–12,500 B.P. (Maher and Mickelson, 1996). If retreat was late, there could have been only a very short time interval between readvances that deposited reddish brown clay tills of the Kewaunee Formation (Maher and Mickelson, 1996). Between retreat from the terminal moraine marking the maximum extent and subsequent deposition of reddish brown clay till, the ice margin apparently retreated more than 350 km to allow the level of Lake Michigan to drop and the transport of red sediment from Lake Superior into the Green Bay basin. Ice then readvanced over 250 km and deposited the Silver Cliff and Kirby Lake tills on the west side of the Green Bay Lobe and the Branch River and Chilton tills on the east side of the Green Bay Lobe (Table 1; Fig. 2). The Green Bay Lobe margin then retreated at least 150 km and an organic horizon and a spruce forest, the Two Creeks Forest Bed, developed (McCartney and Mickelson, 1982; Maher and Mickelson, 1996). Spruce trees from the Two Creeks Forest Bed, most frequently radiocarbon dated at 12,050 to 11,750 B.P., indicate that a forest grew and was drowned by meltwater and rising lake levels in front of the readvancing ice margin (Kaiser, 1994). Tills of the Middle Inlet and Glenmore Members, and, in the Lake Michigan basin, till of the Two Rivers Member of the Kewaunee Formation, overlie the forest bed (Table 1; Fig. 2).

The average radiocarbon age of the Two Creeks Forest Bed (about 11,800 B.P.) has a calibrated age of about 13,800 calendar years ago (Maher and Mickelson, 1996). The radiocarbon age of leaves and twigs in pond sediments beneath the Chilton equivalent (Valders) till (12,965 ± 200 B.P.) has a calibrated age of about 15,000 calendar years B.P. These age constraints suggest rapid oscillations of the Green Bay Lobe margin with advance and retreat rates on the order of about 0.5 km/yr for the period 15,000 to 13,800 calendar years B.P. This period of margin oscillation coincides with a period of rapid climate change (abrupt warming at about 15,000 calendar years B.P., followed by cooling until about 13,000 calendar years B.P.) indicated in the Greenland ice cores (Cuffey et al., 1995). The Green Bay Lobe's dynamic behavior during the period of general deglaciation may have been a stable response due to mass balance changes associated with climate, or the dynamics may have been nonsteady behavior (such as surging) due to internal instability that produced margin advances unrelated to climatic change. Glacier bed conditions (rapid sliding and bed deformation) may have contributed to rapid margin oscillations (Alley, 1991; Clark, 1994), including surge advances (Colgan, 1996). We reconstructed the ice-surface morphology for two advances—the post-Twocreekan Glenmore and pre-Twocreekan Chilton advances—during this period of rapid margin oscillations (and possibly rapid climate change) to determine basal driving stresses and evaluate the role of driving stress in the dynamic behavior of the ice margin.

The Chilton ice margin was contemporaneous with the Valders margin of the Lake Michigan Lobe (Table 1; Fig. 3); (McCartney and Mickelson, 1982). McCartney and Mickelson (1982) and Need (1983, 1985) recognized a sandier Branch River till on the highland east of the Fox River basin that was thought to be older than the Chilton advance. Both of these were overridden by the post-Twocreekan Glenmore advance (Table 1; Fig. 2). It appears now, however, that the end moraine composed of Branch River till is continuous with the end moraine composed of Chilton till, and that the tills are the same age and simply represent facies change and minor overriding where the margin rose from the basin over the escarpment. Elevations along the moraine crest rise progressively up-ice supporting this interpretation.

On the west side of the lobe, the textural differences between the Silver Cliff and Kirby Lake tills (time equivalents to the Branch River and Chilton tills; Table 1) may also represent differences in source and proximity to the basin axis and not significantly different advances. The sandier Silver Cliff till was deposited slightly earlier than the more clayey Kirby Lake till locally as seen in several exposures, but the major grain-size difference is likely caused by proximity to the fine sediments in the basin axis. Our present interpretation of the Branch River—Chilton and Silver Cliff—Kirby Lake units is that differences in grain size are due to local substrate difference with little or no difference in age and with little retreat between the two advances. In the center of the lobe the ice was overriding and entraining fine-grained lake sediment, while at the margins the ice encountered sandy till and bedrock. The upland areas were probably better drained and the basal resistance to ice flow was likely greater than in the lowland. This difference in bed conditions may have led to minor variations in margin activity such as local advances and therefore the local overriding supported by the stratigraphy.

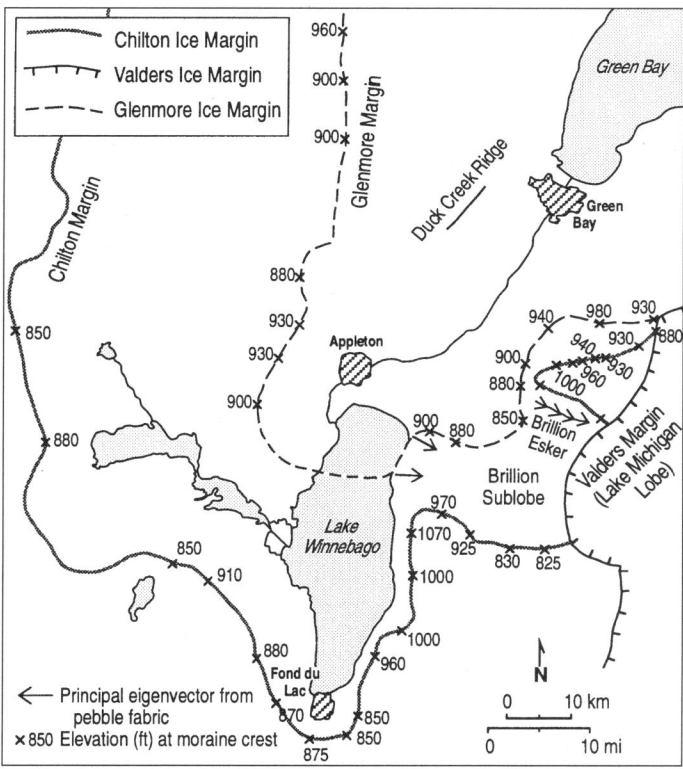

Figure 3. Ice extent and flow-direction indicators.

Elevations of the crest of the end moraine deposited during the Chilton advance range from about 280 m in the lowland to 360 m on top of the Silurian escarpment (Fig. 3). Ice flowed into the Lake Winnebago lowland, which is underlain by shale and lake sediment, and into the Brillion basin, which was formed by a breach in the Silurian escarpment (Fig. 3). The uppermost bedrock in the Brillion basin is predominantly shale.

The correlative Valders advance of the Lake Michigan Lobe also had a different margin configuration than preceding or successing advances. A lobe of ice extended toward the southwest from Lake Michigan depositing thin clayey Valders till more than 10 km beyond the previous western limit of Lake Michigan Lobe ice, yet its extent in the main part of the Lake Michigan basin was less than earlier advances. The Valders till is not recognized as a distinct unit very far north, and it has been suggested that this tongue of ice represents a minor surge out of the Lake Michigan basin (Acomb et al., 1982).

The terminal moraine of the post-Twocreekan Glenmore advance has been mapped in the field on the east side of the lobe (Fig. 3), but has not been mapped in the field in the Lake Winnebago lowland or on the west side of the lobe. We mapped the moraine in these areas using 1:24,000-scale topographic maps, and note that the moraine is discontinuous and difficult to trace because it is buried by younger lake sediment or is difficult to distinguish from bedrock relief. The Glenmore margin did not go as far south into the Lake Winnebago lowland or the Brillion basin, nor as far south on the upland as the Chilton ice.

DATA SOURCES

Figure 3 shows the locations of the ice margins, moraine elevations, and flow direction indicators used in the reconstruction. The ice extent used in the reconstruction was taken from Thwaites (1943), Thwaites and Bertrand (1957), McCartney and Mickelson (1982, unpublished field data), and Need (1983, 1985). This information was supplemented with inspection of 1:24,000-scale, U.S. Geological Survey topographic maps, and field investigations in the central and southeastern parts of the lobe.

Flow direction indicators used to constrain flowlines include till fabric data, the orientation of flutes, small moraines, shear zone deposits, drumlins, and eskers. Ice-flow direction is assumed to be parallel to the orientation of the principal eigenvector determined from the orientation of the long axis of 25 pebbles in till. Small moraines (i.e., moraines that are less than 10 m wide and 2 m high) present near the Chilton margin, are evident on topographic maps, air photos, and in the field. Flutes on the reddish brown till surface south of Lake Winnebago are visible on air photos. The only esker used in the reconstruction is the Brillion esker, which is located in the Brillion Sublobe and associated with the Chilton advance (McCartney and Mickelson, 1982). In general, flow-direction indicators are sparse in the reddish brown till areas of the Green Bay Lobe. Where flow direction indicators are lacking the flowlines were drawn at right angles to the margin.

The Duck Creek Ridge, located southwest of Green Bay (Fig. 3) has been interpreted as debris that accumulated in a shear zone along the western margin of the Green Bay Lobe during the Glenmore advance (Need, 1985). The Duck Creek Ridge is about 10 km long, 200 to 300 m wide, and as much as 35 m high. It is composed of a complex of sand, silt, and clay lake sediment of the Middle Inlet and Kirby Lake Members (Need, 1983). The bedding is commonly disturbed and the ridge parallels adjacent drumlins that are interpreted to have formed during the Glenmore advance (Need, 1985). The orientations of the ridge and the adjacent drumlins were used as indicators of ice-flow direction in the reconstruction of the Glenmore ice surface.

Striations on the bedrock upland, and drumlins in the southeastern part of the lobe covered by the Chilton advance, were not used in the Chilton reconstruction. These striations are overlain by Horicon till (Table 1) and the drumlins are composed primarily of Horicon till (with possibly a bedrock core). Therefore, these both appear to be associated with earlier advances of the Green Bay Lobe.

Regional depth-to-bedrock information from Trotta and Cotter (1973) was incorporated into the estimates of ice thickness, but no adjustment was made for isostatic rebound. The rebound history in the region is not well known and is probably complex (Clark et al., 1994). However, it is likely that the uplift was slightly less at the ice margin than toward the interior of the lobe, so that the reconstructed profiles based on end moraines, if anything, are slightly steeper than the ice actually was.

ICE-SURFACE RECONSTRUCTIONS

Chilton advance

The reconstructed ice surface for the Chilton advance rises from about 260 m (850 ft) at the south end of Lake Winnebago to about 330 m (1080 ft) where it overtopped the Silurian escarpment (Fig. 4). The lobe is asymmetrical because of flow against the escarpment. At the southeast end of Lake Winnebago, the slope of the moraine against the escarpment is about 8m/km within 3.4 km of the terminus (Colgan, 1996). Near the north end of the lake, the Brillion Sublobe protruded through a breach in the Silurian escarpment. North of the Brillion basin, the ice came up onto the escarpment, but did not completely override it. A window of older Horicon till, the Wayside Member (Table 1), is present on the upland area (mostly in drumlins). Although the relief is low, with a maximum of about 100 m, the ice margin was deflected by the topography both here and along Lake Winnebago, suggesting very thin ice.

The reconstruction indicates that within 35 km of the terminus in the Lake Winnebago lowland, the average slope of the ice was 2 m/km, the thickness was <200 m, and the driving stress was about 2 kPa. Ice protruding southeast about 20 km into the Brillion basin was slightly steeper sloped. The surface slope in the Brillion basin was typically about 3 m/km, the thickness was <150 m, and the driving stress was about 3 kPa. Ice that came up on the Silurian escarpment had an average surface slope of about 11 m/km, was <150 m thick, and had a driving stress of about 12 kPa within a few kilometers of the terminus.

Glenmore advance

Based on the segment of Glenmore Moraine that has been mapped, it appears that the ice during the Glenmore advance (Fig. 5) had similar characteristics to the ice during the Chilton advance (Fig. 4). In the main lobe in the Lake Winnebago lowland and in the Brillion Sublobe, the estimated average surface slope within 15 km of the terminus was 2 to 3 m/km, the thickness was <150 m, and the driving stress was about 2 kPa.

DISCUSSION

The driving stresses calculated for both the Chilton and Glenmore advances are extremely low. Sources of error in the calculations include not accounting for isostatic rebound, but this would likely reduce the ice-surface gradient. The thicknesses may be underestimated since there was likely an initial

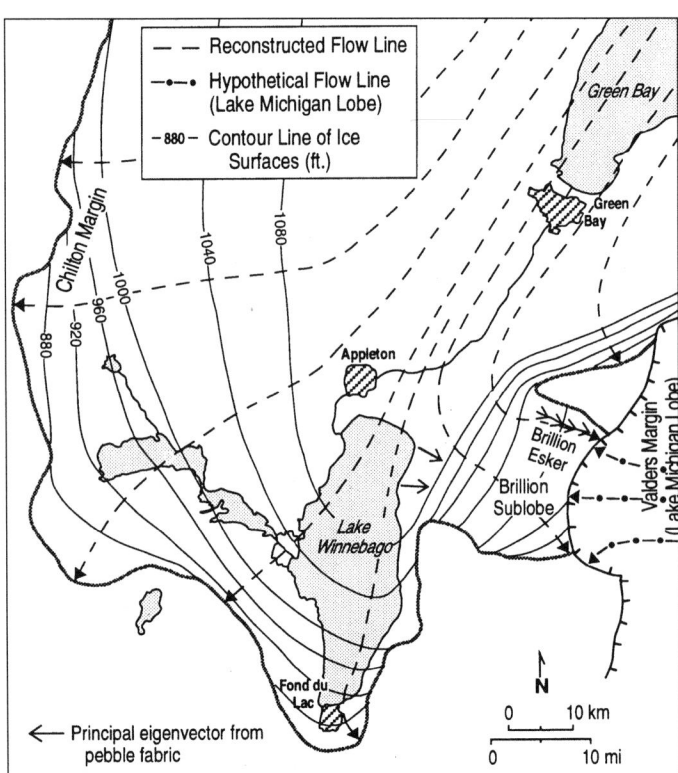

Figure 4. Reconstruction for the Chilton advance.

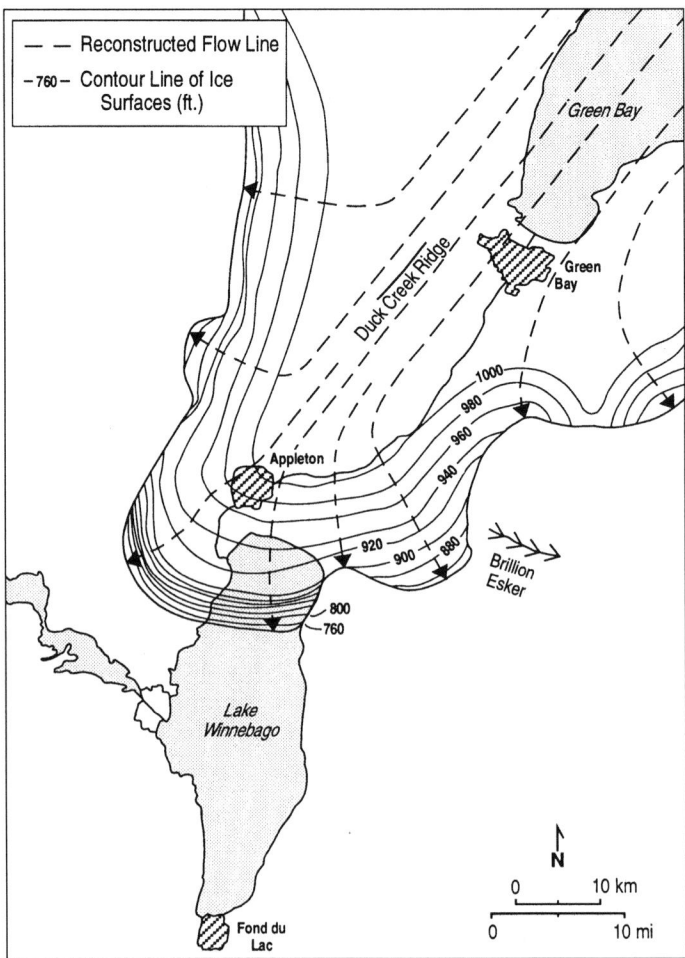

Figure 5. Reconstruction for the Glenmore advance.

thickening at the terminus that is not accounted for. However, if the ice were 50–100 m thicker at the terminus, the basal shear stress would not change much because the stress calculation is more sensitive to surface slope than ice thickness. Side drag (S) was also ignored and may be significant along the Silurian escarpment. For these low driving stresses, ignoring the longitudinal-stress gradient (G) may have produced minimum values with errors of several tens of percent (Clark, 1992). But even if the calculated driving stress is underestimated by a factor of two, the resultant driving stresses remain very low. Qualitatively, the fact that the ice was deflected by the low bedrock escarpment, supports the thin-ice reconstruction. We infer from the reconstructions that the bed was smooth and well lubricated, consistent with a margin advancing over lake sediment in the basins and then on to the uplands depositing clay till or over-riding older clay till. Where the upland area consisted of sandy till and bedrock, the bed was likely better drained, and ice-surface slopes were steeper and the driving stress was slightly higher.

Reconstructed ice-surface profiles for the reddish brown till advances are lower than those for earlier advances of the Green Bay Lobe. The margins were relatively steep during the Johnstown phase (18,000 to 16,000 B.P.), and Milton, Green Lake, and early Rush Lake phases (16,000 to 14,000 B.P.), and driving stresses were 15 to 25 kPa (Colgan and Mickelson, 1997). During the younger Lake Mills and Rush Lake retreat phases, the driving stresses were <10 kPa (Colgan and Mickelson, 1997). Drumlins formed in association with the higher driving stresses, whereas formation of flutes and remolding of larger drumlins took place beneath gently sloping ice with lower driving stresses (Colgan and Mickelson, 1997).

Clark (1992) suggested that the surface slopes, driving stresses, and rapid flow rates of the Des Moines, James, and Lake Michigan Lobes were similar to the distal ends of ice streams draining the West Antarctic ice sheet. We infer from our reconstruction of the ice-surface profiles, estimated ice thicknesses, and calculated basal shear stresses that the subglacial conditions at the margin of the Green Bay Lobe during the Chilton and Glenmore advances were probably similar to those near the grounding line of Ice Stream B, West Antarctica. The low driving stresses calculated indicate basal conditions similar to Ice Stream B's ice plain, which is the transition zone between grounded ice in the ice stream and floating ice in the ice shelf. Ice Stream B's ice plain, a region about 130 km from the grounding line, has a surface slope of about 0.4 m/km, and a driving stress of about 2 to 7 kPa (Bindschadler et al., 1987). This driving stress is comparable to the calculated driving stress for the Green Bay Lobe during deglaciation.

The calculated driving stresses, and the estimated flow rates inferred from the stratigraphic record for the Green Bay Lobe, are similar to those of a West Antarctic ice stream. However, the Green Bay Lobe was thinner that these ice streams and the climate of North America from 13,000 to 11,000 ^{14}C-years B.P. was very different than the present climate of Antarctica. We therefore recognize that ice streams are not a close analog for the large lowland lobes of the Laurentide Ice Sheet, but we know of no better analog. We limit our analogies to basal conditions and suggest that the basal conditions of the Green Bay Lobe and the distal portion of modern ice streams that terminate in the Ross Sea were likely similar. The margin of the Green Bay Lobe in the basin, terminating in a proglacial lake, behaved very much like the distal part of a modern ice stream terminating in the sea. The Green Bay Lobe terminated in shallow water that was probably never deep enough to float the ice, so slopes were low, but not zero. The ice was grounded but the effective pressure was probably very low. High subglacial water pressure would likely have developed as the ice advanced into the proglacial lake and overrode lake sediment and shale. Ice deformation would have been an insignificant component of flow. The motion of the glacier was likely due to some combination of basal sliding, and sediment deformation. Supporting field evidence of these conditions may be present, and detailed work is underway to evaluate the geologic record, especially tills and the bedrock surface for indications of basal sliding or bed deformation.

The margin configurations for the Chilton and Glenmore advances may not represent a steady-state glacier, but rather a post-surge terminus, such as suggested by Wright (1971) and Mickelson et al. (1981) for the post-14,000-B.P. readvances of the Lake Michigan and the Green Bay Lobes. The change from a relatively stable ice margin to a surge event may be associated with the textural difference between the Branch River and Chilton tills. The ice on the upland (depositing sandy Branch River till) may have been in a relatively stable position, while ice in the center of the lobe (depositing clayey Chilton till) surged out into the lowland. The mechanisms of surging are poorly understood for soft bed glaciers (Paterson, 1994), but it is possible that changing bed conditions associated with the saturated clay till in the lowland could have facilitated a surge. It may be that the unusual configuration of the Valders advance out of the Lake Michigan basin also reflects a surge, perhaps triggered by the same change in basal conditions as the Chilton advance. Without a better understanding of the mechanisms of surging in soft bed glaciers, and a chronology much more detailed than radiocarbon methods allow, the relationship between changing bed conditions and surging of ice-sheet lobes cannot be tested.

CONCLUSIONS

Ice-surface reconstructions for the Chilton and Glenmore advances of the Green Bay Lobe between 13,000 and 11,000 ^{14}C-years B.P. indicate that the ice was very thin, with low surface slopes, and very low driving stresses. High subglacial water pressure likely resulted in bed deformation, basal sliding, or both. The bed conditions may have been similar to those of Ice Stream B's ice plain, where the ice is grounded just above hydrostatic equilibrium. Possibly the bed conditions during this period of rapid margin oscillations played a role in the dynamic behavior of the ice lobe that is similar to the role that the bed conditions of Ice Stream B play in determining the rapid flow rates of the ice stream. Rapid climate change between 13,000 and 11,000 ^{14}C-years B.P. may have coincided with bed conditions conducive to margin oscillations.

This study supports previous worker's (Wright, 1972; Mathews, 1974; Fisher et al., 1985; Clark, 1992) findings that the southern marginal of the Laurentide Ice Sheet was low sloping and thin and yet, as shown by this study in the Green Bay Lobe, the marginal area was a complex and dynamic environment. The Green Bay Lobe's dynamic behavior during the period of general deglaciation may have been due to mass balance changes associated with climate, or the dynamics may have been nonsteady behavior, due to internal instability that produced margin advances unrelated to climatic change. At least locally, the advances of the Green Bay Lobe seem to have been facilitated by the substrate.

Models of ice lobes assuming parabolic profiles similar to modern outlet glaciers, such as those by Hooke and Mooers (1986) and Mooers (1989), are not supported by geomorphic evidence, whereas models of ice lobe incorporating surging and analogies to ice plains are compatible with geomorphic evidence. Lobes of the Laurentide Ice Sheet may have acted much like large outlet glaciers during the glacial maximum, but it is apparent that other styles of advance and retreat were operating during the rapid deglaciation that began about 15,000 calendar years B.P.

ACKNOWLEDGMENTS

This research was partially funded by National Science Foundation grant number EAR-9627798 and by the U.S. Geological Survey EDMAP program. We thank Peter Clark, John Attig, and Staci Ensminger for their review comments.

REFERENCES CITED

Acomb, L. J., Mickelson, D. M., and Evenson, E. B. 1982, Till stratigraphy and late glacial events in the Lake Michigan Lobe of eastern Wisconsin: Geological Society of America Bulletin, v. 93, p. 289–296.

Alley, R. B., 1991, Deforming-bed origin for the southern Laurentide till sheets?: Journal of Glaciology, v. 37, p. 67–76.

Alley, R. B., Blankenship, D. D., Bentley, C. R., and Rooney, S. T., 1987, Till beneath Ice Stream B: 3. Till deformation: Evidence and implications: Journal of Geophysical Research, v. 92, p. 8921–8929.

Bentley, C. R., 1987, Antarctic ice streams: A review: Journal of Geophysical Research, v. 92, p. 8843–8858.

Bindschadler, R. A., Stephenson, S. N., MacAyeal, D. R., and Shabtaie, S., 1987, Ice dynamics at the mouth of Ice Stream B: Journal of Geophysical Research, v. 92, p. 8885–8894.

Blankenship, D. D., Bentley, C. R., and Rooney, S. T., 1986, Seismic measurements reveal a saturated porous layer beneath an active Antarctic ice stream: Nature, v. 322, p. 54–57.

Boulton, G. S., Smith, G. D., Jones, A. S., and Newsome, J., 1985, Glacial geology and glaciology of the last mid-latitude ice sheets: Journal of Geological Society of London, v. 142, p. 447–474.

Clark, P. U., 1992, Surface form of the southern Laurentide Ice Sheet and its implications to ice-sheet dynamics: Geological Society of America Bulletin, v. 104, p. 595–605.

Clark, J. A., Hendriks, M., Timmermans, T. J., Struck, C., and Hilverda, K. J., 1994, Glacial isostatic deformation of the Great Lakes region: Geological Society of America Bulletin, v. 106, p. 1931.

Clark, P. U., Licciardi, J. M., MacAyeal, D. R., and Jensen, J. W., 1996, Numerical reconstruction of a soft-bedded Laurentide Ice Sheet during the last glacial maximum: Geology, v. 24, p.679–682.

Clayton, L., Attig, J. W., Mickelson, D. M., and Johnson, M. D., 1992, Glaciation of Wisconsin: Wisconsin Geological Natural History Survey, Educational Series no. 36, 4 p.

Colgan, P. M., 1996, The Green Bay and Des Moines Lobes of the Laurentide Ice Sheet: Evidence for stable and unstable glacier dynamics 18,000 to 12,000 BP [Ph.D. thesis]: Madison, University of Wisconsin, 293 p.

Colgan, P. M., and Mickelson, D. M., 1997, Genesis of landforms and flow history of the Green Bay Lobe, Wisconsin, USA: Sedimentary Geology, v. 111, p. 119.

Cuffey, K. M., Clow, G. D., Alley, R. B., Stuvier, M., Waddington, E. D., and Saltus, R. W., 1995, Large arctic temperature change at the Wisconsin-Holocene glacial transition: Science, v. 270, p. 455–459.

Dyke, A. S., and Prest, V. K., 1987, Late Wisconsin and Holocene history of the Laurentide Ice Sheet: Geographie et Quaternaire, v. 41, p. 237–263.

Engelhardt, H. F., and Kamb, B., 1998, Basal sliding of Ice Stream B, West Antarctica: Journal of Glaciology, v. 44, p. 223–230.

Engelhardt, H., Humphrey, N., Kamb, B., and Fahnestock, M., 1990, Physical conditions at the base of a fast moving Antarctic ice stream: Science, v. 248, p. 57–59.

Fisher, D. A., Reeh, N., and Langley, K., 1985, Objective reconstructions of late Wisconsinan Laurentide Ice sheet and the significance of deformable beds: Geographie et Quaternaire, v. 39, p. 229–238.

Hooke, R. L., and Mooers, H. D., 1986, Glaciology of some mid-continent ice lobes: Problems and some possible avenues for research: American Quaternary Association, Program and Abstracts, v. 9, p. 34–36.

Kaiser, K. F., 1994, Two Creeks Interstade dated through dendrochronology and AMS: Quaternary Research, v. 42, p. 288–298.

Maher, L. J., Jr., and Mickelson, D. M., 1996, Palynological and radiocarbon evidence for deglaciation in the Green Bay Lobe, Wisconsin: Quaternary Research, v. 46, p. 251–259.

Mathews, W. H., 1974, Surface profiles of the Laurentide Ice Sheet in its marginal areas: Journal of Glaciology, v. 13, p. 37–43.

McCartney, M. C., and Mickelson, D. M., 1982, Late Woodfordian and Greatlakean history of the Green Bay Lobe, Wisconsin: Geological Society of America Bulletin, v. 93, p. 297–302.

Mickelson, D. M., Acomb, L. J., and Bentley, C. R., 1981, Possible mechanisms for rapid advance and retreat of the Lake Michigan Lobe between 13,000 and 11,000 BP: Annals of Glaciology, v. 2, p. 185–186.

Mickelson, D. M., Clayton, L., Baker, R. W., Mode, W. N., and Schneider, A. F., 1984, Pleistocene stratigraphic units of Wisconsin: Wisconsin Geological and Natural History Survey, Miscellaneous Paper no. 84-1, 97 p.

Mooers, H. D., 1989, Drumlin formation: a time transgressive model: Boreas, v. 18, p. 99–108.

Need, E. A., 1983, Pleistocene geology of Brown County, Wisconsin: Wisconsin Geological and Natural History Survey, Information Circular no. 48, 19 p.

Need, E. A., 1985, Pleistocene geology of Brown County: Wisconsin, Wisconsin Geological and Natural History Survey, Map #83-1, scale 1:100,000.

Paterson, W. S. B., 1981, The physics of glaciers, (second edition): New York, Pergamon Press, 380 p.

Paterson, W. S. B., 1994, The physics of glaciers, (third edition): Trowbridge, Pergamon Press, 480 p.

Raymond, C. F., 1980, Temperate valley glaciers, in Colbeck, S. C., ed., Dynamics of snow and ice masses: New York, Academic Press, p. 79–139.

Thwaites, F. T., 1943, Pleistocene of part of northeastern Wisconsin: Geological Society of America Bulletin, v. 54, p. 87–144.

Thwaites, F. T., and Bertrand, K., 1957, Pleistocene geology of the Door Peninsula, Wisconsin: Geological Society of America Bulletin, v. 68, p. 831–879.

Trotta, L. C., and Cotter, R. D, 1973, Depth to bedrock in Wisconsin: Wisconsin Geological and Natural History Survey, scale 1:1,000,000.

Wright, H. E., 1971, Retreat of the Laurentide Ice Sheet from 14,000 to 9,000 years ago: Quaternary Research, v. 1, p. 316–330.

Wright, H. E., 1972, Quaternary history of Minnesota, in Sims, P. K., and Morey, G. B., eds., Geology of Minnesota: A centennial volume: St. Paul, Minnesota, Minnesota Geological Survey, p. 515–546.

MANUSCRIPT ACCEPTED BY THE SOCIETY OCTOBER 8, 1998

Ice sliding over weak, fine-grained tills: Dependence of ice-till interactions on till granulometry

Slawek Tulaczyk*
Division of Geological and Planetary Sciences, California Institute of Technology, Pasadena, California 91125

ABSTRACT

Two fundamental aspects of ice-till interactions, the strength of the ice-till coupling and the vertical distribution of deformation in till, may be strongly dependent on till granulometry. In particular, results of theoretical analysis of several physical processes involved in such interactions suggest the following hypotheses: (1) fine-grained tills facilitate ice sliding with ploughing and little distributed deformation, and (2) coarse-grained tills facilitate strong ice-till coupling and relatively deep till deformation (~0.1 m). The theoretical analysis is limited to Coulomb-plastic tills under low subglacial effective stresses (0–100 kPa). Fine-grained tills are represented in the analysis by a clay-rich till from beneath Ice Stream B (ISB), West Antarctica, and a silty Pleistocene till from Ohio. For comparison, two coarse-grained, clast-rich tills are also considered (from beneath the Trapridge Glacier, Yukon, and the Breidamerkurjökull Glacier, Iceland). The mechanical condition for ice sliding over till is defined as the situation in which the strength of the ice-till interface is lower than the strength of the till itself. Model calculations predict that this condition is more likely to be met in fine-grained rather than coarse-grained tills because of (1) lower abundance of ploughing clasts (clast fraction ~0.01 vs. ~0.1), (2) widespread submergence of fine matrix particles even by a very thin basal water film (~10^{-6} m), and (3) greater susceptibility to interface smoothing due to ice-water surface tension. In addition, the theoretical analysis of ice-till interactions considers three potential mechanisms for distribution of deformation in tills of Coulomb-plastic rheology: (1) plastic deformation of till around a ploughing clast, which may affect till to depth of c. 2.7 to c. 4.5 times the clast diameter; (2) particle/clast bridging, which is typically observed to result in a shear-zone that is 10 times greater than the characteristic clast/particle diameter; and (3) vertical shear-zone migration due to water-pressure fluctuations. Combined, these three effects may result in distribution of a significant fraction of ice motion throughout ~0.1 m thickness of a coarse-grained, clast-rich till. However, lower clast abundance and smaller hydraulic diffusivity of a fine-grained till makes it a less favorable environment for significant strain distribution (predicted shear zone thickness ~0.01 m).

---

*Present address: Department of Geological Sciences, University of Kentucky, Lexington, Kentucky 40506

Tulaczyk, S., 1999, Ice sliding over weak, fine-grained tills: Dependence of ice-till interactions on till granulometry, *in* Mickelson, D. M., and Attig, J. W., eds., Glacial Processes Past and Present: Boulder, Colorado, Geological Society of America Special Paper 337.

INTRODUCTION

The importance of ice-till interactions to glacier mechanics has been fully recognized relatively recently (Alley et al., 1986, 1987a, b, c; Beget, 1986; Boulton, 1986; Boulton and Hindmarsh, 1987; Brown et al., 1987; Clarke, 1987). Over the last decade, it became apparent that ice motion over weak tills may play a major role in controlling dynamics of ice masses and in formation of geologic record of glaciations (Alley, 1989a, b, 1991; Boulton, 1996a, b; Clark, 1992; Clark and Walder, 1994; Engelhardt et al., 1990; Kamb, 1991; MacAyeal, 1992). In order to understand the function of till in evolution of ice masses and glacial geologic sequences, it is necessary to identify and quantify the physical processes that determine the ice-till interactions.

Significant advancements in this direction have already been made by a number of research groups studying modern subglacial zones (Boulton and Hindmarsh, 1987; Blake, 1992; Blake et al., 1994; Engelhardt et al., 1978; Engelhardt and Kamb, 1997, 1998; Engelhardt et al., 1990; Fisher and Clarke, 1994; Hooke et al., 1997; Iverson et al., 1994, 1995). Observations beneath mountain glaciers have shown mostly strong ice-till coupling and distribution of some till deformation to a depth of 0.1–0.6 m (Boulton and Hindmarsh, 1987; Blake et al., 1994; Engelhardt et al., 1978; Hooke et al., 1997). However, a recent borehole experiment of Engelhardt and Kamb (1998) suggests that the fast motion of a West Antarctic ice stream over a fine-grained till is accommodated predominantly by sliding with possible shallow deformation. This apparent contrast in the character of ice interactions with tills beneath mountain glaciers and the till beneath Ice Stream B may, potentially, be related to the contrasting granulometry of these distinctly different tills.

For logistical reasons, it is usually difficult to collect all the field data that are necessary to build full physical description of the individual processes involved in ice-till interactions. Theoretical and laboratory research is needed to supplement and generalize field observations (e.g., Iverson et al., 1994; Kamb, 1991). In this work, I use theoretical constraints from mechanics of plastic granular media to show that two fundamental aspects of an ice-till system, the strength of ice-till coupling and the depth of till deformation, may significantly depend on till granulometry. The results suggest that sliding, with little distributed deformation, may be characteristic for ice motion over fine-grained tills. On the other hand, coarse, clast-rich tills may promote stronger ice-till coupling and relatively deeper distribution of till deformation. Throughout this work, the emphasis is on soft-bedded subglacial conditions in which tills are weak and deformable because they are under low subglacial effective stresses (<100 kPa; Brown et al., 1987). The term "till" will be used interchangeably with the terms "granular medium" or "soil" (in engineering sense).

INTRODUCTORY CONCEPTS

Till rheology

The most important decision that must be made at the beginning of a theoretical study of ice-till interactions is the choice of the rheologic models for both phases. It is widely accepted that deforming ice behaves as a power-law fluid with exponent of about 3 (Patterson, 1994, Chapter 5). However, the rheology of till, a complex mixture of solids, water, and sometimes gas, is less firmly constrained. Till is treated commonly as a material of either roughly linearly viscous or nearly Coulomb-plastic rheology (e.g., Boulton and Hindmarsh, 1987, versus Kamb, 1991). There are fundamental differences between these two alternatives. For instance, ice motion over Coulomb-plastic till is more likely to be unstable than ice motion over viscous till (Kamb, 1991). I believe that the preponderance of the existing evidence supports the Coulomb-plastic model for till rheology, and this model will be assumed in this paper. The only unequivocal support for the idea that till may have a viscous or Bingham-type rheology comes from the stress and strain-rate data presented for the subglacial zone of Breidamerkurjökull by Boulton and Hindmarsh (1987, Fig. 7). However, the reliability of the Breidamerkurjökull dataset is unclear since the source of the highly variable shear-stresses estimates has never been explained (Hooke et al., 1997, p. 173). Extensive studies on two other mountain glaciers overriding till, the Trapridge glacier and the Storglaciären, failed to confirm the viscous till model (Blake, 1992, p. 62; Hooke et al., 1997). In addition, the data from Storglaciären support the Coulomb-plastic model. Two sets of extensive laboratory shear box and ring shear tests on three different tills provide additional backing for the latter model (Iverson et al., 1998; Kamb, 1991). Readers interested in the viscous representation of till rheology are encouraged to explore the extensive literature on this subject (Alley, 1989a, b; Alley et al., 1986, 1987a, b, c; Boulton, 1996a, b; Boulton and Hindmarsh, 1987; Clark, 1991, 1992; Clark and Walder, 1994; Clark et al., 1996; Jenson et al., 1995, 1996; MacAyeal, 1992).

In the Coulomb-plastic model, till is idealized as a material with yield strength given by (Terzaghi et al., 1996, eq. 17.4):

$$\tau_f = c + p' \tan\phi \tag{1a}$$

where c is the cohesion, ϕ is the angle of internal friction, and p' is the effective pressure typically expressed as:

$$p' = P - p_w \tag{1b}$$

where P is the total load, and p_w is the pore water pressure (see also Table 1 for notations). If a shear stress lower than the yield strength is applied to such material, small or no deformation takes place (Fig. 1). Thus, large-strain or continuous deforma-

TABLE 1. LIST OF SYMBOLS

Symbol	Meaning	Dimension*
A	Area of ice-water interface	L^2
P	Total load	FL^{-2}
R_j	Median particle radius in the j-th size class	L
SSA	Specific surface area	L^{-1}
T	Period of water-pressure fluctuations	T
U_i	Total ice velocity	LT^{-1}
U_s	Velocity component due to sliding	LT^{-1}
U_t	Velocity component due to till deformation	LT^{-1}
V	Volume of an ice crystal/protrusion	L^3
Z	Depth of till deformation	L
a	Depth to which a ploughing clast protrudes into till	L
c	Cohesion	FL^{-2}
c_v	Hydraulic diffusivity (coefficient of consolidation)	$L^2 T^{-1}$
d_c	Clast diameter	L
d_{ch}	Characteristic particle diameter	L
d_w	Basal water film thickness	L
f_c	Fractional area covered by clasts	-
f_{im}	Unsubmerged fractional area of ice-matrix interface	-
f_j	Fractional area of ice contact with particles in the j-th size range	-
g	Acceleration of gravity	LT^{-2}
k_c	Till strength:critical stress proportionality factor	-
k_{im}	Ice-matrix coupling factor	-
k_p	Deformation depth:particle size proportionality factor	-
k_r	Shear zone thickness:particle size proportionality factor	-
k_s	Particle shape factor	-
n	Porosity	-
p'	Effective pressure	FL^{-2}
p_c'	Critical effective pressure	FL^{-2}
p_i	Ice pressure	FL^{-2}
p_w	Pore water pressure	FL^{-2}
r_c	Size boundary between clasts and matrix particles	L
r_j	Characteristic particle radius in the j-th size range	L
u	Excess pore pressure over hydrostatic pressure	FL^{-2}
u_o	Time-averaged excess pore pressure at the top of the till	FL^{-2}
t	Time	T
w_c	Dry clast weight fraction	-
w_j	Dry weight fraction of the j-th particle size range	-
z	Depth in till ($z = 0$ at the ice-till interface)	L
$\Delta p'$	Hydrostatic vertical effective pressure gradient	FL^{-3}
Δu	Magnitude of water-pressure fluctuations	FL^{-2}
δ	Characteristic depthscale	L^{-1}
ϕ	Internal friction angle	°
γ	Trailing angle of the cavity behind a ploughing clast	°
φ	Leading angle of a ploughing clas	°
μ	Total Gibbs free-energy	FL
μ_i	Gibbs free energy of bulk ice	FL
μ_{im}	Coefficient of ice-matrix friction	-
$\mu_{(\phi)}$	Coefficient of internal friction ($\equiv \tan\phi$)	-
v_i	Molar volume of ice	L^3
ρ_b	Buoyant till density	ML^{-3}
ρ_w	Water density	ML^{-3}
σ_{iw}	Specific surface energy of ice-water interface	FL^{-1}
τ_c	Critical stress for ploughing	FL^{-2}
τ_f	Till failure strength	FL^{-2}
τ_i	Ice-till interface strength	FL^{-2}
τ_{ic}	Ploughing component of interface strength	FL^{-2}
τ_{im}	Ice-matrix interface strength	FL^{-2}
ω	Frequency of water-pressure fluctuations	T^{-1}
ψ	Reciprocal of the depthscale for water-pressure fluctuations	L^{-1}
~	A number of the order of ...	N/A

*In addition to the usual dimensions of F, force (MLT^{-2}), L, length, M, mass, T, time, I use '°' for degrees and '-' to denote a nondimensional variable.

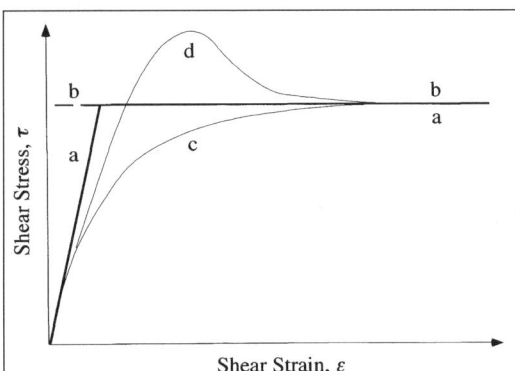

Figure 1. a, Elastic-plastic, and b, rigid-plastic model for stress-strain behavior of granular materials compared to typical stress-strain curves from laboratory tests on normally consolidated (c) and overconsolidated (d) soils (modified from Scott, 1963, Figs. 8-8 and 9-1).

tion is possible only when the yield strength of the till is reached. Unlike in the Bingham-model of Boulton and Hindmarsh (1987, eq. 1), shear stresses in excess of the yield strength cannot be applied to the Coulomb-plastic till. In addition, strain rates in this till are not explicitly determined by shear stresses but rather by other factors (e.g., rate of motion of the ice base that is applying the stresses to the till). Coulomb-plastic behavior of granular materials has long been accepted in soil mechanics because it has proven itself to adequately represent exhaustive field and laboratory data (reviews in Kamb, 1991; Mitchell, 1993; Scott, 1963; Terzaghi et al., 1996). Application of the Coulomb-plastic model to tills makes it possible to utilize existing solutions from soil mechanics. However, this approximation of till rheology does neglect some second-order effects that occur during deformation of granular media (e.g., the slight strain-rate dependence of shear strength; Kamb 1991). The simplifying assumptions made here are justified by the goal of this work, which is focused on providing useful theoretical insights into first-order aspects of ice-till interactions.

Clast ploughing and breakdown of hard-bed sliding theory

In the case of ice sliding over bedrock, basal resistance to ice motion arises from ice regelation and plastic flow around obstacles of different sizes (Kamb, 1970; Llibutry, 1979; Nye, 1969; Weertman, 1957). During this motion, relatively high stresses concentrate on bedrock obstacles (megapascals, MPa) but it is assumed that the obstacles are capable of withstanding these high stresses without being moved or destroyed. Such an assumption is reasonable for typical bedrock since strength of common rock lithologies is very high (~100 MPa; Jaeger and Cook, 1969, p. 146). When ice moves over unconsolidated sediments, ice velocity may be accommodated to some extent through deformation of the underlying sediments (e.g., Alley et al., 1986; Boulton and Hindmarsh, 1987). To avoid terminological confusion, I would like to clarify that the term "basal sliding" is used here to refer to this component of ice movement which is accommodated at the ice-till interface rather than within the deformable bed or the ice (Piotrowski and Tulaczyk, 1999).

Brown et al. (1987, p. 8991) have recognized that when ice moves over till, the strength of the latter imposes a strict limit on how much stress can be applied by ice to any clast protruding from till into ice. This limit stems from the fact that when the force applied by ice to a clast exceeds the force necessary to produce a local sediment failure around the clast, the clast will start to plough through the till. To estimate the magnitude of this limiting force, Brown et al. (1987, p. 8991) made a simplifying assumption that the local failure surface surrounding a clast has a hemiconical shape. In reality, a protrusion ploughing through a till matrix must result in a more complex pattern of deformation because of the requirement of till (near) incompressibility. To satisfy this requirement, the failing matrix must deform around the moving clast away from its stoss side and into its lee side (Fig. 2). This deformation pattern increases significantly the force required to produce local failure around the clast.

Here, I estimate the stresses required to move a clast horizontally through perfectly plastic matrix (Fig. 2). The till matrix is assumed to be spatially homogeneous, incompressible, and deforming plastically under yield stress equal to its undrained shear strength (i.e., strength at no volume change; Terzaghi et al., 1996, p. 259). The two latter assumptions are reasonable as long as the rate of clast displacement in the matrix exceeds the rate of pore pressure dissipation around the clast. From dimensional analysis this condition is expressed by:

$$U_s > \frac{c_v}{d_c} \qquad (2)$$

where U_s is the sliding velocity, c_v is the hydraulic diffusion coefficient (coefficient of consolidation), and d_c is the clast diameter. The existing limited data suggest that the hydraulic diffusion coefficient is small for fine-grained tills and significantly greater for coarse tills, ~0.1 m² y⁻¹ for Ice Stream B till and the Two Rivers Till, Wisconsin, versus ~10 m² y⁻¹ for Storglaciären till (Iverson et al., 1998; Engelhardt, unpublished data). The above condition holds easily for any clast with $d_c >$ 10⁻² m and for sliding velocity >10 m y⁻¹ in the case of fine-grained tills but it breaks down for coarse tills when $d_c < 10^{-1}$ and $U_p < 100$ m y⁻¹. Upon breakdown of condition (2), strengthening of the till matrix should occur in the zone of compression in front of the clast. However, more detailed analysis shows that this strengthening is relatively small and may be neglected in the following order-of-magnitude estimates (Appendix A). Use of the undrained till strength makes treatment of different aspects of ice-till interactions much simpler and it frequently facilitates application of existing solutions from soil mechanics and the theory of plasticity.

Clasts at an ice-till interface are typically approximated in theoretical analysis as spheres submerged halfway in till and halfway in ice (Fig. 3; Brown et al., 1987; Alley, 1989b). Unfortunately, solutions for motion of rigid hemispheres through a plastic matrix do not seem to be available or easily derivable

(Johnson, 1970, p. 481–482). However, soil resistance to indentation by a protrusion is not sensitive to the exact shape of the protrusion (Baligh, 1972, p. 67; Johnson, 1970, p. 481–482). By approximating the portion of a clast submerged in a till as a tilted cube (Fig. 2), I can take advantage of the analytical solution of Baligh (1972, eqs. 5 & 8) for a rough wedge moving through a homogeneous plastic matrix. The resulting equation shows that the ratio of the horizontal force necessary to move the clast to the horizontal area of the clast (critical stress, τ_c) exceeds the shear strength of the matrix (τ_f) by a small constant (k_c):

$$\tau_c = k_c \tau_f \approx 4.7 \tau_f. \quad (3)$$

Thus, ice needs to act on a ploughing clast with a stress that is roughly five times the yield strength of the matrix. Comparable values of k_c have been obtained for a similar problem of penetration of a flat punch into a plastic soil (4.83, 5.14, 5.2 to 5.7; Johnson, 1970, p. 481–482). The result given in equation (3) is also generally consistent with the treatments used previously by other workers in analysis of an indentor ploughing through till (Brown et al., 1987; Fischer and Clarke, 1994; Humphrey et al., 1993; Iverson et al., 1994).

Calculations of shear stress exerted by sliding ice on a hemispherical particle (Lliboutry, 1979, eq. 46; Brown et al., 1987, eq. 2) show that for any reasonable ice sliding speeds all particles larger than $\sim 10^{-4}$ m should plough if the particles are embedded in a weak till (Fig. 4). Thus, in spite of the fact that clasts in a till do offer increased resistance as compared to the bulk till, as shown in equation (3), they still cannot provide nearly as significant retardation to ice motion as obstacles on a rigid bed. Therefore, the formulations used in the hard-bed sliding theory cannot be used for calculating the strength of the ice-till interface or the ice-sliding velocities. New expressions must be developed for ice motion over till.

STRENGTH OF ICE-TILL COUPLING

Influence of clasts

A model of an ice-till interface will be developed for the simplest case and then complicated by introduction of additional physical processes that should have important, first-order effect on the interface strength. Figure 4 suggests that an uncomplicated two-phase model for till (clasts + matrix) is a useful approximation of reality in theoretical analysis of an ice-till interface. It is clear from previous arguments that clasts trapped on the ice-till interface will plough and contribute a stress τ_{ic} to the total interface strength:

$$\tau_{ic} = f_c \tau_c = f_c k_c \tau_f \quad (4)$$

where f_c is the fractional area of the interface covered by clasts. This expression is a simple generalization of the previously discussed case of a single ploughing clast, as shown in equation (3).

The question that now arises is how to treat adequately the direct contact between till matrix and ice. Formulation of such expression is dependent on the choice of the predominant mechanism of ice motion over till matrix. At first sight (Fig. 4), regelation past small particles seems a logical choice since it can occur at relatively low stresses for particles of small diameter (~ 1–10 kPa for $<10^{-5}$ m). However, both observations and theory suggest that surface-tension effects will retard formation of ice in small void spaces between small particles and, thus, hinder or prevent regelation past these particles (Alley et al.,

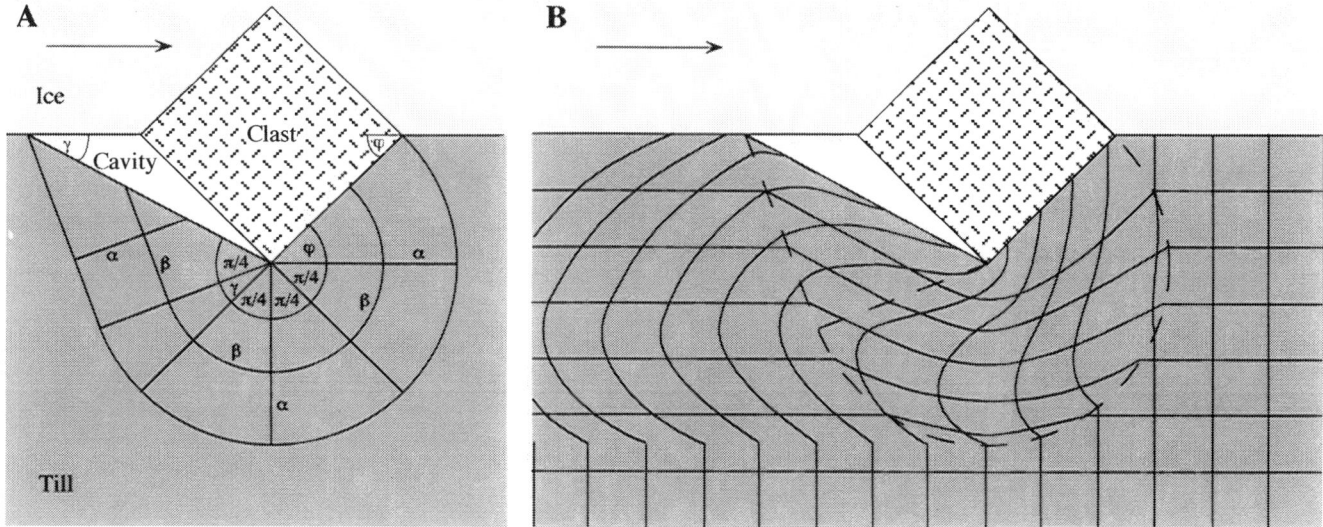

Figure 2. A, Physical plane for the two-dimensional problem of a square clast ploughing perfectly rigid-plastic till. α and β denote two families of slip-lines; φ and γ are the leading angle of the ploughing clast and the trailing angle of the cavity developed behind the clast ($\gamma = \arcsin[\sin\varphi/\sqrt{2}]$), respectively. B, Deformation of a square grid produced by migration of the ploughing clast from A. Both figures adapted from Baligh's (1972, Figs. II-4 and II-11) work on the plane-strain problem of a wedge indenting homogenous, perfectly rigid-plastic soil.

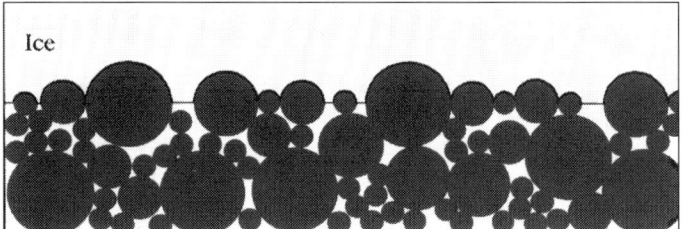

Figure 3. Illustration of the assumption that ice-till interface consists of spherical till particles halfway submerged in ice. Roughness of such interface is controlled at all scales by grain size distribution of the till.

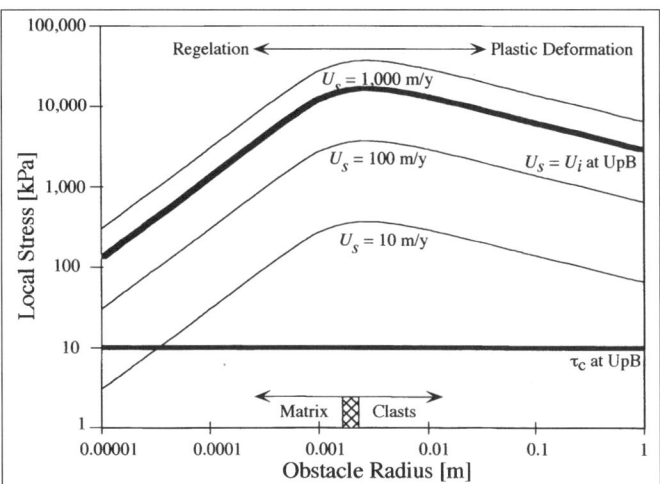

Figure 4. Local shear stress on a spherical particle as a function of the particle radius and ice sliding velocity (Brown et al., 1987, eq. 2; Lliboutry, 1979, eq. 46). The critical obstacle size is in the range of 0.001 to 0.01 m of particle radius. Ice motion over smaller particles is accommodated predominantly by regulation and over larger particles predominantly by plastic deformation. This fact gives a glaciological justification for treatment of till as a mixture of matrix and clasts. The crosshatched rectangle indicates the selected size boundary between these two phases ($r_c = 2 \times 10^{-3}$ m). The thick gray lines show the stress-radius relationship for sliding velocity (U_s) equal to the velocity (U_i) of Ice Stream B at the camp UpB, West Antarctica (c. 440 m y^{-1}, Whillans and van der Veen, 1993), and the magnitude of the critical stress (τ_c) necessary for the particle to plough the weak till beneath Ice Stream B (this study, eq. [3] with $\tau_f \approx 2$ kPa; Kamb, 1991).

1998; Everett, 1961; surface-tension effect will be later introduced into the interface-strength formulations in a slightly different context). If regulation is neglected, the problem may be simplified to the case of soil interaction with a rigid solid body. Data from soil mechanics suggest that the interface shearing stress (τ_{im}) between a solid and a granular medium separated by a macroscopically flat boundary can be expressed as (Baligh, 1972, p. 68–69):

$$\tau_{im} = (1 - f_c) k_{im} \tau_f, \qquad (5)$$

where $0 < k_{im} < 1.0$. From a physical standpoint, the value of this constant tends towards 1.0 with the roughness of the solid-soil boundary approaching the roughness of intra-soil (till) failure planes. Thus, the interface shearing stress may be at most equal to the strength of the till. On the other hand, k_{im} should tend to zero when solid boundary becomes smooth. In general, k_{im} is a function of the ratio of the coefficient of internal friction ($\mu_{(\phi)} = \tan\phi$ in eq. [1a]) and the coefficient of interface friction (μ_{im}). However, lack of experimental data for ice-till interfaces prevents introduction of this more physical expression into equation (5). Soil mechanics investigations of soil-structure interactions suggest that the value of k_{im} lies frequently in the range of 0.5–1.0 (Scott and Schoustra, 1968, p. 205). For the time being, however, the usual assumption will be made that the roughness of the ice-till interface is governed at all scales by grain size (Fig. 3, particles are half-submerged hemispheric bumps; Brown et al., 1987; Alley, 1989a, b). This postulate prompts the use of a conservative k_{im} value of 1.0. Toward the end of this chapter, surface-tension effects will be introduced to argue against universal applicability of this assumption to fine-grained tills.

Combination of equations (4) and (5) estimates the strength of an idealized ice-till interface as a simple function of one variable, the fractional area of clasts (f_c):

$$\tau_i = (1 - f_c) k_{im} \tau_f + f_c k_c \tau_f. \qquad (6)$$

Under this condition the strength of the interface is necessarily equal to ($\tau_i = \tau_f$ for $f_c = 0$) or greater than the intrinsic strength of the till. Thus, it can be expected that in this very simplified case there should be no tendency for the ice to slide along the interface because deformation on shear planes within the till is mechanically more favorable. However, even this very simple model does already suggest that interface strength should be greater for clast-rich and smaller for clast-poor tills.

Basal water film

The presence of a basal water film is likely to have an important influence on the interface shearing strength because such a film may separate the ice base from the underlying till over relatively large areas of the bed. Where such separation occurs, the interface strength goes to practically zero. Even if a channelized water system provides an important means of water drainage in a given ice-till system (e.g., Walder and Fowler, 1994), the presence of a relatively widespread water film can be expected because basal meltwater production has a distributed character and some form of a distributed drainage is needed to deliver the water to the channels/canals. Following Alley (1989a, b), I assume that a water film of thickness d_w submerges all particles whose radii are equal to or smaller than d_w. Thus the fractional area of the bed submerged by the water film scales with the grain-size distribution of the till matrix. The assumption is made here that a basal water film is not thick enough to submerge clasts. This is a sound assumption because water films are expected to have thickness of the order of millimeters or less (Weertman, 1972, Table 1).

Grain-size distribution is typically given for tills in the form of weight fractions of particles occurring in discrete size ranges. Therefore, it is useful to cast the mathematical expressions of interface shearing strength in a way that accounts for this discretization of till granulometry. The fractional area of ice-clast contact can be calculated from the dry clast weight fraction (w_c) through (Brown et al., 1987):

$$f_c = (1 - n)w_c \quad (7a)$$

and the contribution of the j-th matrix size range to the ice-matrix contact area is obtained from the weight fraction of this size range (w_j):

$$f_j = w_j/(1 - f_c), \quad (7b)$$

where n is the till porosity. Note that equation (7b) assumes that pore spaces are part of the till matrix and that their contribution to the ice-matrix contact area scales in the same way as the contribution of the different particle size ranges. Since grain-size and pore-size distributions are related entities this assumption is reasonable.

The new formulation for the ice-till interface strength accounts for the influence of water film thickness in the following way:

$$\tau_i(d_w) = \tau_{ic} + \tau_{im}(d_w) = \\ (k_c f_c + k_{im} f_{im}(d_w))\tau_f = (f_c k_c + k_{im} \Sigma f_j)\tau_f, \quad (8)$$

where $f_{im}(d_w) = \Sigma f_j$ denotes summation of equation (7b) over the particle size ranges that are not submerged by the water film and are smaller than the minimum clast size (r_c; i.e., $d_w < r_j < r_c$). In order to apply equation (8) to real tills, it is necessary to first specify the size-boundary between clasts and matrix particles (r_c). In sedimentology, the lower size boundary for pebbles is frequently taken to be 2×10^{-3} m (Pettijohn, 1975, p. 28). Figure 4 also supports a choice of r_c in the size range of the critical obstacles (10^{-3} to 10^{-2} m). In soil mechanics tests, which provide in practice the basis for estimation of till matrix strength (τ_f), particles greater than this are typically not included. Therefore, the sedimentological definition of clasts will be used here ($r_c > 2 \times 10^{-3}$ m). This value should be treated only as an approximate clast-matrix boundary. However, tills are typically poorly sorted with only several percent of weight fraction falling into each size interval. This is why the main features of my further analysis are not likely to be significantly affected by the uncertainties in r_c.

To examine the sensitivity of the interface shear strength to water-film thickness for tills of different granulometry, equation (8) is applied to two examples of fine-grained tills and two examples of coarse-grained tills (Fig. 5). The fine-grained tills are represented by the clay-rich Ice Stream B till (henceforth the ISB till; Tulaczyk et al., 1998, Fig. 3) and an average of the silt-rich Tazewell and Cary tills from northeast Ohio (henceforth the Ohio till; Shepps, 1953, Table 1). The two coarse-grained tills are the Breidamerkurjökull till (Boulton and Hindmarsh, 1987, Fig. 3), and the Trapridge till (Clarke, 1987, Fig. 4). The two latter have been selected for this analysis because in situ measurements have documented that some deformation takes place in these two tills to a depth of several decimeters. The ISB till is the proposed deforming bed of Ice Stream B (Alley et al., 1986, 1987a, b) and the Ohio till exemplifies the matrix-dominated southern Laurentide Pleistocene tills for which deforming-bed origin has been also advocated (Alley, 1991).

Figure 6 gives the results of application of equation (8) to these four tills. It plots the ratio of the interface shear strength to the matrix shear strength (τ_i/τ_f) versus the water film thickness (d_w). This ratio is used not only for the convenience arising from its dimensionless character, but also because of the special importance that is associated with the critical value of $\tau_i/\tau_f = 1.0$. When the interface is weaker than the till ($\tau_i/\tau_f < 1.0$) sliding along the interface is the mechanically favorable way to accommodate ice motion. For the opposite condition ($\tau_i/\tau_f > 1.0$), shear within the till should take place. When the two strengths are equal, either sliding or shear may accommodate the ice motion. It is important to remember that this discussion concentrates on the simple case of a till whose strength does not change with depth. This is physically equivalent to assuming a lithostatic pore pressure distribution with depth. Another important case of a hydrostatic pore pressure distribution would tend to favor sliding or shear localized near the ice-till interface more than the lithostatic case. This is because till strength increases with depth under hydrostatic conditions (Alley, 1989b, p. 123).

There is a significant difference in the impact of a water film of a given thickness on the interface strength for ice contact with coarse- and fine-grained tills (Fig. 6). For the two fine-grained tills the interface may be weaker than the till matrix even in the presence of an extremely thin water film ($< 10^{-6}$ m). This is espe-

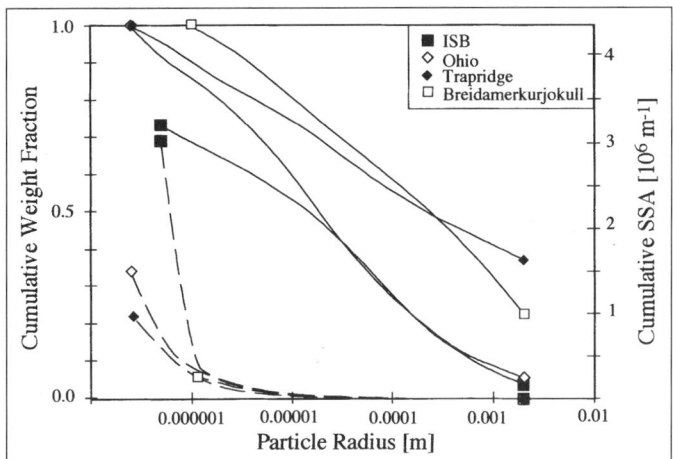

Figure 5. Cumulative weight fraction (solid lines, scale on the left) and cumulative specific surface area (*SSA*, dashed lines, scale on the right) of the four selected tills as a function of particle radius.

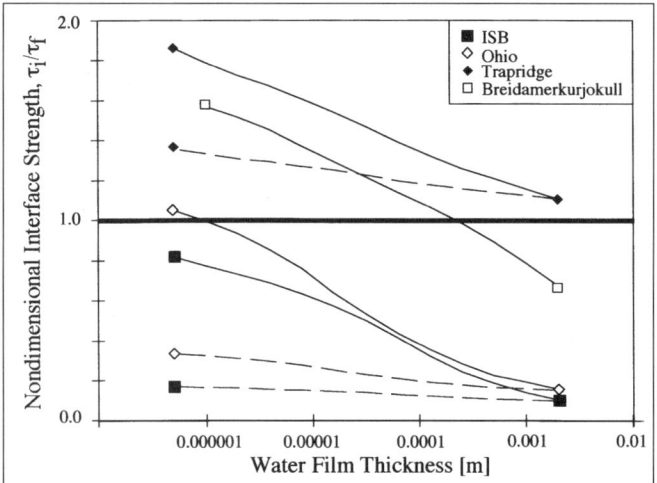

Figure 6. Dependence of the nondimensional interface strength on water film thickness for the four selected tills calculated using equation (8). The case of perfect ice-matrix coupling ($k_{im} = 1.0$) is given by the solid lines and the case of ice-matrix coupling determined by surface tension at 10 kPa subglacial effective pressure (k_{im} from eq. [13]) is given by dashed lines. The thick gray line marks the sliding criterion $\tau_i/\tau_f = 1$.

cially true for the ISB till that has one-third of its material in particles smaller than 10^{-6} m. In addition, both of the fine-grained tills have a relatively low content of clasts (Fig. 6) and the ice-till interface cannot be significantly strengthened by ploughing. On the other hand, the high abundance of clasts in the two coarse-grained tills more than makes up for the weakening caused by the presence of a water film. Thus, the interface strength for the coarse tills exceeds the bulk strength of the till matrix for almost any reasonable thickness of the basal water film (Fig. 6).

Surface-tension effect

Additional support for the proposed weakness of ice coupling with fine-grained tills is provided by another grain-size-dependent physical effect, the surface-tension effect. This phenomenon stems from the existence of ice-water capillary forces that hinder infiltration of ice into small pore spaces (Alley et al., 1998; Everett, 1961). The introduction of the surface-tension effect into the model of ice-till interface is used here to argue against the previous assumption that the roughness of such interface is determined on all scales by particle size only (Fig. 2). In this assumption, particles of all sizes form hemispherical bumps at the top of the till surface and the ice surface conforms to them by invading the pore spaces between the particles. However, capillary forces may prevent this invasion of ice into small pore spaces and, thus, may make the ice base much smoother than the top of the till or any intra-till shear planes. This smoothing effect will act to decrease the strength of the interface by reducing the value of the coefficient of ice-matrix coupling, k_{im} in equations (5), (6), and (8), below its previously assumed maximum value of one.

Both theory and observations indicate that ice-water surface tension hinders infiltration and growth of ice into small pore spaces (Alley et al., 1998; Everett, 1961; Hallet et al., 1991). As a result, a high effective pressure may be necessary to make the geometry of an ice base comply perfectly with the roughness of a fine-grained till. Everett (1961) derived the mathematical expression that accounts for the surface-tension effect in growth of small ice crystals and protrusions. To better illustrate the physical basis of this phenomenon, parts of his derivation are reproduced here. The equation for Gibbs free energy of a small ice crystal/protrusion growing in contact with water under pressure of p_w is given by (Everett, 1961, eq. 2):

$$\mu = \mu_i(p_w) + v_i \sigma_{iw} dA/dV, \qquad (9)$$

where μ is the total Gibbs free-energy of the small ice crystal/protrusion, μ_i is the Gibbs free energy of bulk ice (function of pressure in the adjoining water, p_w), v_i is the molar volume of ice, σ_{iw} is the ice-water surface energy (0.034 J m^{-2}, Ketcham and Hobbs, 1969), A is the area of the ice-water interface, and V is the total volume of the ice crystal/protrusion. The second term in equation (9) represents the excess free energy that results from increasing the ice-water contact area when an ice crystal or protrusion experiences growth. Formally, the magnitude by which the energy of the small crystal/protrusion exceeds the free energy of bulk ice (i.e., $\mu - \mu_i(p_w)$) can be ascribed to increased pressure (p_i) within the crystal/protrusion:

$$\mu(p_i) - \mu_i(p_w) = v_i(p_i - p_w). \qquad (10a)$$

Comparison of equations (10a) and (9), shows that the pressure difference between the ice and water (i.e., the effective pressure, p') is uniquely related to the curvature of the ice-water interface:

$$p' = p_i - p_w = \sigma_{iw} dA/dV. \qquad (10b)$$

In turn, if the effective pressure is treated as the independent variable, equation (10b) implies that the curvature of ice-water interface (dA/dV) changes in such a way as to always satisfy this relationship.

The implication of equation (10b) for ice-base geometry is that it can no longer be assumed that the ice base simply complies at all scales with the geometry of the top of the till. At the microscale, the geometry of the ice-till interface is now determined by a combination of till granulometry and the magnitude of the subglacial effective pressure (Fig. 7). When the subglacial effective pressure is zero, the ice base remains absolutely flat ($dA/dV = 0$) and, thus, the roughness of the ice-till interface is vanishingly small. Only when a critical value of subglacial effective pressure (p_c') is reached will the ice perfectly comply with the curvature of the top of the till, and in this case the previously applied assumption will be valid. The measure of the curvature at the top of the till is provided by the specific surface area of the till (SSA), which is an intrinsic till

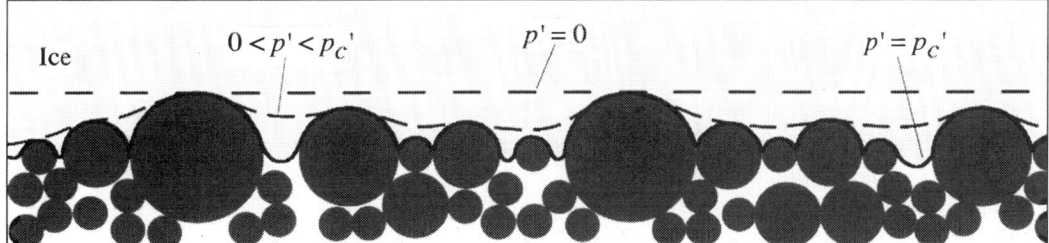

Figure 7. Microscopic geometry of the ice-till interface as a function of subglacial effective pressure. Ice can intrude into all pore spaces as illustrated previously in Figure 3 only when effective pressure is at or above its critical value (p_c', given by eq. [12]). Smaller value of the subglacial effective pressure ($0 < p' < p_c'$) means that the maximum curvature of the ice base is limited by ice-water surface tension following equation (10b). Finally, at the effective pressure of zero, the ice base is constricted to be flat.

property determined by grain size distribution (i.e., for critical p_c', $dA/dV = SSA$, both with units of area per volume; Parks, 1990). Clearly, the subglacial effective pressure can take on a whole range of values between 0 and p_c' and for this range, the curvature of the ice base lies between 0 and SSA. When subglacial effective pressure exceeds the critical value, ice is free to regelate into the till following the physical law verified empirically by Iverson (1993) and Iverson and Semmens (1995; also Alley et al., 1998).

For a particular till, the magnitude of p_c' can be calculated from a modification of equation (10b):

$$p_c' = \sigma_{iw} SSA \qquad (11)$$

The specific surface area, SSA, can be estimated from grain-size distribution (e.g., Parks, 1990, p. 133–135) by applying the following summation over the discrete size ranges:

$$SSA = (1 - n)\Sigma(w_j k_s / R_j), \qquad (12)$$

where k_s is the particle shape factor (3 for a sphere, Parks, 1990, p. 133–134), and R_j is the characteristic particle radius in the j-th size class (chosen hereafter to be the midrange for each class). Equations (11) and (12) are used to calculate p_c' and SSA for the four examples of coarse- and fine-grained tills. The results of these calculations are presented in Figures 8 and 5, respectively. These results illustrate well that the surface-tension effect is much less significant for the coarse tills than for the fine-grained tills. For instance, full compliance of the ice-surface geometry to the till-surface geometry is possible at p_c' of only a few kPa for the Breidamerkurjökull till but of 104 kPa for the ISB till.

Explicit introduction of the surface-tension effect into the mathematical model of ice-till interface strength will require modification of the coefficient k_{im}, which provides a parametric measure of ice-base roughness and ice coupling with till matrix. From equation (10b) it follows that this coefficient should be dependent on the subglacial effective stress that controls the microscale ice-base roughness for $0 \leq p' \leq p_c'$. The exact form of this dependence is difficult to constrain because no relevant observational data are available. Nevertheless, it is reasonable to expect that the following two trends should hold in general: when $p' \to p_c'$ then $dA/dV \to SSA$ and $k_{im} \to 1.0$; and when $p' \to 0$ then $dA/dV \to 0$ and $k_{im} \to 0$. For the intermediate values, the simplest linear relationship is assumed here giving the following expressions:

$$k_{im}(p') = p'/p_c' = p'/(\sigma_{iw} SSA) \qquad \text{for } 0 \leq p' \leq p_c', \qquad (13a)$$

$$k_{im}(p') = 1.0 \qquad \text{for } p' > p_c', \text{ and} \qquad (13b)$$

$$k_{im}(p') = 0.0 \qquad \text{for } p' < 0, \qquad (13c)$$

where all of the terms have been explained previously.

Since the condition in equation (13b) is equivalent to the previous assumption of the maximum value of k_{im} and the condition equation (13c) simply implies no strength along the ice-till-matrix contact, the only case for which the equation for ice-till interface strength has to be modified is equation (13a). This new expression for the interface strength (with no water film) as a function of effective stress is derived by adding equation (5) to equation (6) and substituting equation (13a):

$$\tau_i = [f_c k_c + (1 - f_c) p'/(\sigma_{iw} SSA)]\tau_f = [f_c k_c + (1 - f_c) p'/(\sigma_{iw} SSA)] p' \tan\phi, \qquad (14)$$

where the second part is cast in terms of subglacial effective pressure by substituting $\tau_f = p' \tan\phi$ (till shear strength without cohesion, eq. [1a]).

Equation (14) indicates that, in absolute terms, strength of the interface is quite sensitive to the effective stress because it increases proportionally to the second power of p'. As before, however, it is most useful to treat the strength of the interface in relative terms as the nondimensional ratio τ_i/τ_f. Application of equation (14) to the two fine-grained tills considered here shows clearly that the surface-tension effect incorporated into this mathematical model causes significant weakening of the ice-till interface. The nondimensional strength ratio is below its critical value of one for almost the whole range of subglacial effective pressures relevant to the soft-bed conditions (0–100 kPa; Fig. 8).

Therefore, ice sliding with ploughing should be the mechanically preferred mode of ice motion for these fine-grained tills. In contrast, the influence of the surface-tension effect on the coupling of ice with the two coarse tills is not enough to make the interface weaker than the till, except for the Breidamerkurjökull till at effective pressures very near zero (Fig. 8).

In general, the decreased geometric coupling of ice and till due to ice-water tension has a very similar impact on the ice-till interface strength as the presence of a widespread basal water film (compare Figs. 6 and 8). In nature, these two effects are likely to act together and reinforce each other. To work through an example of such combined influence, it is assumed here that water-film thickness and subglacial effective pressure are mutually independent. A value of k_{im} calculated for each till (eq. [13a]) for a selected effective pressure (10 kPa) is plugged into equation (8). The calculated interface strengths (dashed lines in Fig. 6) are extremely low for the fine-grained tills.

This section demonstrated relatively simple but insightful ways of calculating the ice-till interface strength with incorporation of three physical effects that have the potential of being the main controllers of ice-till interactions at and near the interface. The exact form of the mathematical formulations chosen to represent these physical effects is not very well constrained by observational data. However, a consistent, robust feature displayed by the interface models examined here is the significant dependence of ice-till interface strength on till granulometry.

DISTRIBUTION OF DEFORMATION

Distribution of till deformation with depth represents another extremely important aspect of ice-till interactions. Understanding of the individual processes that may distribute shear in tills is necessary to properly interpret field observations and to generate reliable models of coupled ice-till flow. In the viscous model of subglacial bed deformation it is assumed that distributed shear in tills results from the strain-rate dependence of till strength, typical for viscous materials (Alley, 1993). This simple effect is, however, no longer applicable if till is a material of Coulomb-plastic rheology that has no such dependence. The latter rheology is assumed here. In general, shear strain rates and strains are not uniquely determined by shear stresses in plastic materials (Salencon, 1977).

Influence of clast ploughing

As it has been mentioned in one of the previous sections, clast ploughing requires plastic flow of the till material from the stoss to the lee side of the clast (Fig. 2). This flow distributes the deformation associated with the passing clast downwards to depths well below the ones that come directly in contact with the clast. Baligh (1972) conducted theoretical and experimental studies of plain-strain patterns of deformation around wedges of different shapes indenting homogenous soil of plastic rheology. Figure 2B shows adaptation of Baligh's theoretical results

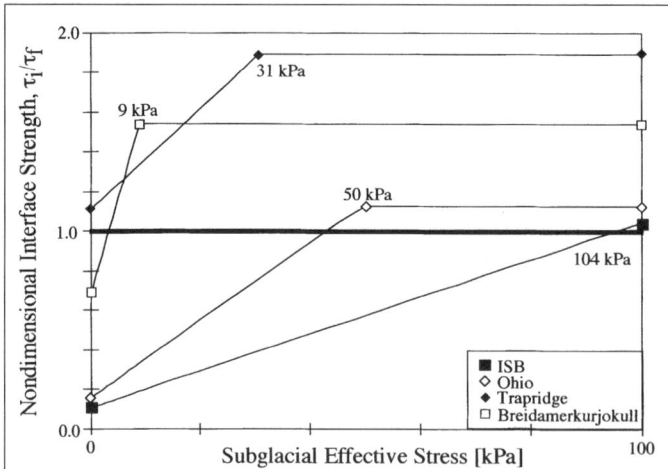

Figure 8. Dependence of the nondimensional interface strength on subglacial effective pressure in the case of surface-tension-controlled roughness of the ice-till interface (eq. [14]). Four selected tills shown by thin solid lines and the sliding criterion shown by the thick gray line. Values of the critical subglacial effective pressure calculated from equation (12) are also given for each of the four tills.

to a ploughing clast. The patterns of deformation for an initially square grid are predicted from theory of perfect plasticity. This solution has a well-defined zone of distributed deformation, separated from the surrounding matrix by a sharp discontinuity and extending to the depth of:

$$Z = k_p a = (1 + \sqrt{2})a, \qquad (15)$$

where a is the maximum vertical dimension of the clast protruding into the till, and k_p is a constant. After the ploughing clast passes, horizontal markers come back to their original position but vertical markers remain permanently deformed (Fig. 2B). Experimental results are broadly consistent with the theory but differ in a few important details (Fig. 9). The experimental zone of deformation has a much more diffused character and extends to significantly greater depth, $k_p \approx 4.5$. In addition, the ultimate shape of vertical markers is somewhat different because they indicate permanent strain only in the direction of ploughing. The differences between the theoretical and experimental results must result from the fact that the experimental clay matrix, like soils in general, does not behave as the perfectly rigid-plastic material assumed in the theory.

In nature, an ice-till interface can contain abundant ploughing clasts. Superposition of numerous ploughing events like the one illustrated by Figures 2 and 9 will cause distributed deformation that corresponds to some fraction of the total relative ice-till motion. The thickness of the resulting shear zone should be several times greater than the depth to which ploughing clasts protrude from the ice base, a in equation (15). The transport distance at the top of the till due to one passage of a ploughing clast is equal to about a (Figs. 2 and 9). If clast spacing is not much greater than two times a, then nearly all of the

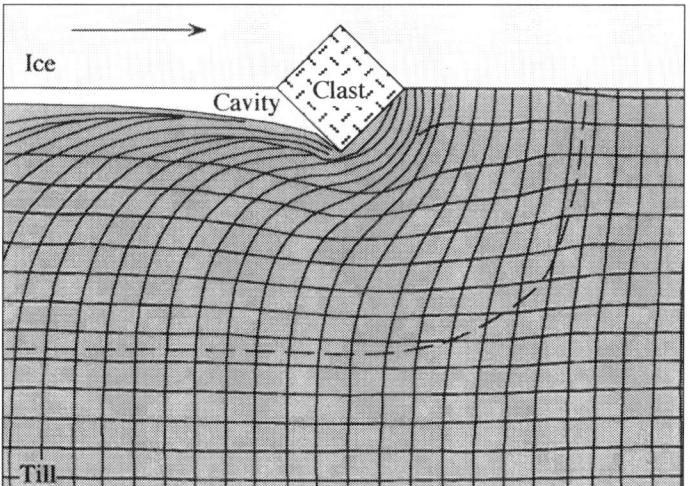

Figure 9. Deformation of a square grid produced by a clast ploughing homogeneous till. The dashed line marks approximate extent of the zone of deformation around the clast. The marker lines near the top of the domain are locally too disturbed to recognize their location. Markers traced from a photograph by Baligh (1972, Fig. II-15) after an experiment in which a wedge with a 45° leading angle ploughed through a homogeneous layer of clay. Ice base drawn along the axis of symmetry of Baligh's experimental wedge. Figure 2, this study, presents the theoretical solution to the same problem.

ice motion can be transferred to the underlying till via the deformation zones surrounding ploughing clasts. This mechanism may provide a very efficient way of distributing deformation in coarse tills in which clast spacing can easily be of that order. On the other hand, fine-grained tills have typically small clasts and low clast abundance (Fig. 5) and the ploughing-related deformation should be relatively insignificant in their case.

Grain bridging

The mechanism of shear distribution discussed in the previous section concentrated on the case of a homogeneous, very fine-grained matrix being ploughed by clasts that are many times greater than the matrix grain-size (i.e., micron-size clays). Tills, however, typically contain a variety of different size particles that will interact with each other when ice motion is accommodated either at the ice-till interface or on shear zones within the till. It has been inferred from observational data that the thickness of a granular shear zone is controlled by the characteristic grain size of particles contained in the sheared material (Mulhaus and Vardoulakis, 1987; Roscoe, 1970). In geotechnical practice, it is commonly assumed that shear strain should distribute over a thickness Z given by a simple expression analogous to equation (15):

$$Z = k_r d_{ch} \approx 10 d_{ch} \qquad (16)$$

where d_{ch} is the characteristic grain diameter, and k_r is a constant. The value of the constant is usually assumed to be c. 10 but some researchers put it at as much as 15–50 (Maltman, 1992, p. 270). Tests performed as a part of this study on the Ottawa sand (c. 0.25 mm in diameter) sheared in a ring shear device are consistent with $k_p \approx 10$ (Fig. 10). At the basic level, the micromechanism of strain distribution in granular shear zones has probably to do with the formation and failure of grain bridges and grain networks (Hooke and Iverson, 1995; Iverson et al., 1996).

The soil mechanics data that have led to formulation of the simple relationship expressed in equation (16) were collected for well-sorted materials containing similar-size particles. Application of the same rule to poorly sorted materials, such as tills, is greatly complicated by the uncertainty in selection of the characteristic grain diameter. Intuitively, d_{ch} should represent the largest particles, which during shear come frequently in contact with similar size particles. For matrix-dominated tills, these would be the matrix-size particles ($d_{ch} < 2$ mm). As the tills coarsen, the importance of clast interactions in distributing strain should increase. At the coarse extreme of clast-supported tills d_{ch} is of the order of the typical clast size (~0.01 to ~0.1 m). Thus, distribution of strain in tills due to particle interactions may vary from a zone whose thickness is of the order of several microns for clay-rich tills to a zone that is decimeters in thickness for clast-rich tills.

Fluctuations of water pressure

The two mechanisms of strain distribution discussed above have to do with the granular character of tills composed of rigid particles of different sizes. However, this inhomogeneous nature of tills is frequently neglected, leading to a theoretical

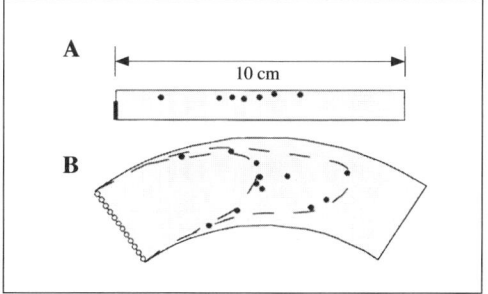

Figure 10. Strain distribution in Ottawa sand (c. 0.25 mm diameter) deformed in a ring shear device to a total displacement of c. 8 cm at 6 kPa normal effective stress. The figure shows postshear distribution of marker beads (black dots) that were placed at the beginning of the shear experiment (A) in a vertical pile in the center of the sample chamber (short thick line at the left) and (B) in a straight line across the sample chamber (empty circles). Dashed lines in B show an envelope encircling the final positions of all twelve marker beads. Both results (A and B) illustrate the fact that some distribution of strain takes place in the Ottawa sand, in spite of the fact that the material has not shown a significant strain-rate dependence of strength in a series of strain-rate-controlled tests which were also performed as a part of this study (data not shown).

approximation of a till as a fluidlike continuum (Boulton and Hindmarsh, 1987; Alley et al., 1986, 1987a, b, c). Under such an assumption, it has been argued that concentrated deformation should be characteristic for perfectly plastic tills and that distributed deformation observed beneath some glaciers provides an evidence for nearly linearly viscous rheology of tills (Alley, 1993, p. 205). This is a sound argument because strain-rate dependence of strength typical for viscous materials forces distribution of strain, but when strength is strain-rate independent deformation may collapse towards a single plane (Turcotte and Schubert, 1982, p. 318). However, for a Coulomb-plastic till changes in strength and distribution of strain may be caused by changes in effective pressure with depth and time. If these changes can force a vertical migration of the shear zone that accommodates the deformation, the time-integrated effect of this process will be to create a diffused, pseudo-viscous zone of deformation. This will happen even though at any instant in time the deformation will take place on a discrete zone in accordance with the intrinsic Coulomb-plastic rheology of a till. In other words, the effective-pressure dependence of strength in a Coulomb-plastic till may play a similar function with regard to strain distribution as the strain-rate dependence does for a viscous material. Therefore, even if the "grainy," inhomogeneous nature of tills is neglected there is still a nonviscous mechanism that may lead to strain distribution.

To verify whether the conjecture of Coulomb-plastic strain distribution is plausible in the context of subglacial physical conditions, I solve a problem of time-dependent distribution of effective pressure driven by a periodic (daily) variation of water pressure in the basal water system. Diurnal fluctuations of basal water pressure are very common because of changes in meltwater supply (especially for mountain glaciers) and tidal forcing occurring on these timescales (Blake, 1992; Boulton and Hindmarsh, 1987; Engelhardt and Kamb, 1997; Hooke et al., 1997; Iverson et al., 1995). Assuming a constant total load (ice overburden pressure) and infinitesimal strains due to consolidation, the time-variable distribution of excess pore water pressure in a vertical column of till can be described by the diffusion equation (Scott, 1963, eq. 5-34):

$$c_v u_{zz} = u_t, \quad (17)$$

where c_v is the hydraulic diffusion coefficient (coefficient of consolidation), u is the excess pore pressure (total water pressure less hydrostatic pressure), and u_{zz}, and u_t denote the second derivative of u with respect to depth and the first derivative of u with respect to time. A commonly encountered analytical solution to equation (17) exists when the periodic boundary condition applied to the top of the till has the form:

$$u(0,t) = u_o + \Delta u \cos(\omega t), \quad (18)$$

where u_o is the time-averaged excess pore pressure at the top of the till, Δu is the magnitude of water pressure fluctuations, and $\omega = \sqrt{(2\pi/T)}$, where T is the period of the fluctuation. The solution is then given by (Turcotte and Schubert, 1982, p. 155–157):

$$u(z,t) = u_o + \Delta u \exp(-\psi z)\cos(\omega t - \psi z), \quad (19)$$

where $\psi = \sqrt{[\pi/(c_v T)]}$. It follows from equation (19) that pore water pressure fluctuations decay quickly with depth and are very small already at the characteristic depth $2\delta = 2\sqrt{(c_v T)}$. Thus, the depth to which water-pressure fluctuations may influence strength properties of till is dependent on the hydraulic diffusivity of till, c_v. Hydraulic diffusivity depends sensitively on the content and mineralogy of clays in soil (Mitchell, 1993, p. 180). Data for tills are sparse but they suggest that clay-rich tills have $c_v \sim 10^{-8}$ m^2 s^{-1} and coarse tills $c_v \sim 10^{-6}$ m^2 s^{-1} (Engelhardt, unpublished data; Iverson et al., 1998, Table 1; Sauer et al., 1993, Table 2). It is obvious already from this dimensional analysis that any effects of pore-pressure fluctuations on strain distribution will propagate significantly deeper (roughly ~10 times deeper) in coarse tills than in fine-grained tills.

To illustrate how pore water fluctuations may cause a time-dependent vertical migration of a shear zone, I work through an example. Changes in the subglacial effective pressure can be calculated from equation (19) using:

$$p'(z,t) = \Delta p' z - u(z,t), \quad (20)$$

where $\Delta p' = (\rho_b - \rho_w)g$ is the hydrostatic effective pressure gradient ($\Delta p' \approx 10$ kPa m^{-1} for ρ_b, the buoyant till density of c. 1,000 kg m^{-3}; ρ_w, the water density of c. 1,000 kg m^{-3}, and g, the acceleration of gravity of 9.81 m s^{-2}). The sign convention used here assumes that pore pressures below the overburden ice pressure (P in eq. [1b]) are negative. This expression is used to calculate changes in effective pressure for a set of assumptions that is intended to emulate a 0.6-m-thick layer of coarse till with hydraulic diffusivity of 10^{-6} m^2 s^{-1}. The system is subjected to a basal pore water fluctuation of amplitude $\Delta u = \pm 10$ kPa of period $T = 24$ hrs. = 86,400 s around an average basal excess pore water pressure of $u_o = -50$ kPa. Figure 11A shows the effective pressure timelines at one-hour intervals throughout the whole 24-hour cycle.

A model of till for which an assumption is made that till strength at any given depth is always a simple linear function of the current effective pressure (p' in eq. [1a]), irrespective of previous effective pressures, can be called a "perfectly remolded till model." For this model, at any point in time the motion of overlying ice is accommodated on the weakest shear zone whose depth is determined by the depth of the minimum effective pressure, p_{min}'. One can use equation (20) to track the depth and record the magnitude of the minimum effective pressure through time (Fig. 11B). For the considered example, the minimum effective pressure migrates downward to the depth of c. 0.3 m ($\approx \delta$) for half of the water-pressure cycle and remains at the ice-till interface ($z = 0$) for the other half of the cycle (Fig. 11B). If the "perfectly remolded till model" and a constant

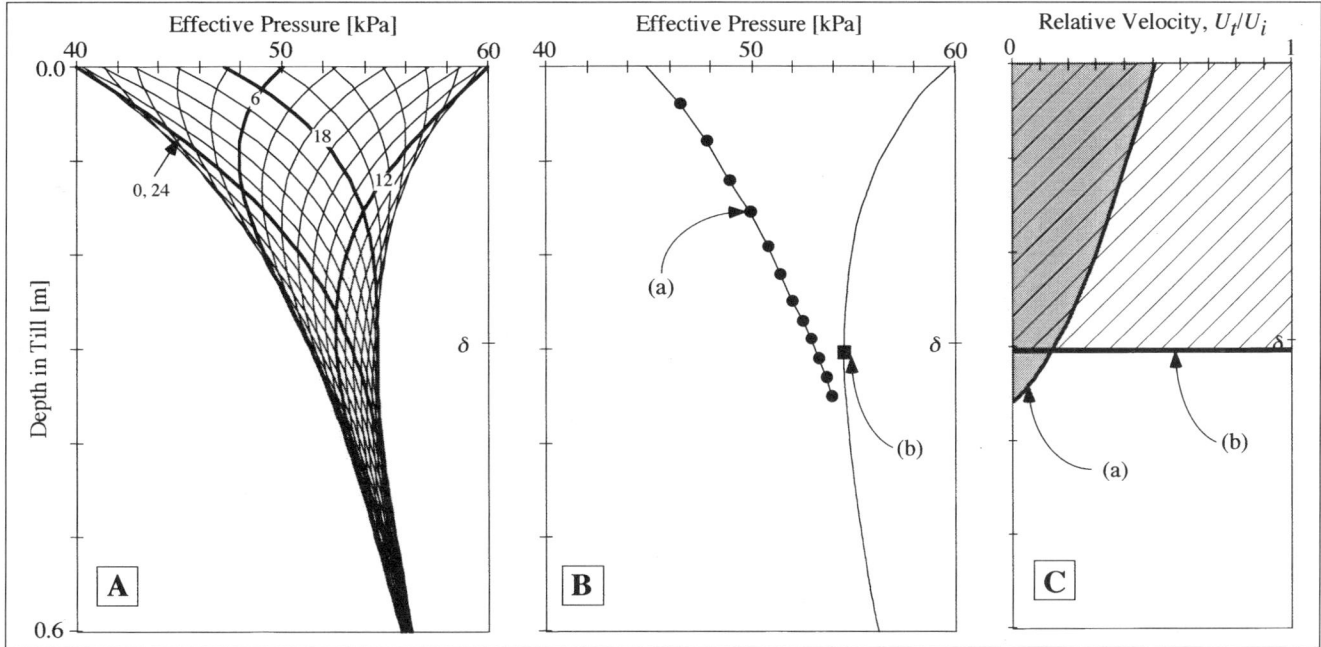

Figure 11. A, 1-hour timelines showing the distribution of effective pressure with depth during a 24-hours cycle of pore-water pressure fluctuations. A time-averaged hydrostatic pore pressure distribution is assumed (eq. [20]). B, (a) Migration of the minimum effective pressure with depth during 12 hours of the water-pressure cycle (line with solid circles). During the other half of the cycle, the minimum effective pressure is located at the top of the till (depth $z = 0$) and it is assumed that ice is then sliding over the top of the till. The second solid line (b) plots the values of the maximum effective pressure experienced at each depth throughout the water-pressure cycle. The square symbol marks the minimum of the latter function. C, Strain distribution in till (a) for the case of an intra-till shear zone that follows the migration of the minimum water pressure in B ("perfectly remolded till model"). Plug-flow of till atop an intratill shear zone (b) that is fixed at the location of the least maximum effective pressure shown in B ("perfectly overconsolidated till model"). In both cases the relative velocity of till, U_t, with depth is expressed as a fraction of the ice velocity, U_i, which is assumed to be constant throughout the water-pressure cycle. The position of the characteristic depthscale, $\delta = \sqrt{(c_v T)} \approx 0.3$ m, is shown in all three plots (A, B, and C).

velocity of ice motion, U_i, are assumed and ice moves at a constant velocity throughout the cycle, till deformation will be distributed in a manner shown in Figure 11C.

The assumption of the "perfectly remolded till" represents, however, an idealization. In nature, it is observed that granular materials subjected first to a higher effective pressure and then sheared under a lower effective pressure (e.g., p_{min}') show a transient peak in strength (case d in Fig. 1). This additional stress threshold will hinder the "perfectly-remolded" strain distribution. Here, the other end-member assumption is considered: that the till behaves as a "perfectly overconsolidated" material whose strength is always determined (following equ. [1a]) by the maximum effective pressure that this material has ever experienced (p_{max}'). Under this assumption, all of the ice motion must be accommodated at the depth in till where the maximum effective pressure during the whole water-pressure cycle is smaller than the maximum effective pressure at all other depths (Fig. 11B). For the considered example, this condition is met at the depth of c. 0.3 m ($\approx \delta$) resulting in a plug flow of till trapped between this shear zone and the ice base (Fig. 11C).

Real granular materials show a behavior that is somewhere between the perfectly remolded and perfectly overconsolidated approximations (Scott, 1963). Thus, the shear stress threshold due to overconsolidation (Fig. 1) may be at times small enough to be overcome. Then the strain in the till can be accommodated on planes characterized by the minimum effective pressure at any given time. In addition, for till to remain overconsolidated, no physical remolding of the till structure acquired at higher effective pressures can take place. However, clasts dragged by the ice base may provide an efficient agent for remolding of till matrix and removing the effects of overconsolidation, making the material behave more like the "perfectly remolded" end-member. Behavior of a "real" till probably falls between the two end-members shown in Figure 11C.

My calculation of the fluctuating subglacial effective stress (Fig. 11A) also illustrates the fact that the assumption of temporally and spatially constant till strength made in the preceding discussion of ice-till interface represents a relatively simple case. Ploughing clasts may interact with a till whose strength changes with depth and time and these clasts themselves are likely to influence the distribution of till strength through remolding. Future detailed and case-specific analyses of ice-till interactions should take these effects into account.

There is no simple way to verify whether water-pressure fluctuations are to any degree responsible for the distributed deformation occurring in <0.6-m-thick shear zones beneath some mountain glaciers (Blake, 1992; Boulton and Hindmarsh, 1987; Hooke et al., 1997; Iverson et al., 1995). However, the physical conditions that have been observed beneath these glaciers (diurnal fluctuations of water pressure, relatively coarse tills) are consistent with the ones assumed here. The presented model suggests that strain distribution due to water-pressure fluctuations will occur over depths that should be roughly an order of magnitude greater for coarse tills ($Z \sim 0.1$ m) than for fine-grained tills ($Z \sim 0.01$ m).

DISCUSSION

There is a dearth of physical models describing ice-till coupling and distribution of deformation in till (Hooke, 1997). The exception is the viscous-till model for which distributed till deformation is a result of the observed and assumed viscous till rheology (Alley, 1989b; Boulton and Hindmarsh, 1987). The question of whether there are significant differences in subglacial behavior of different types of till has not been explored enough, even though till has one of the broadest ranges of properties of sediment types. An initial step in this direction was made by Boulton (1974) who inferred that the strength of the ice-till coupling may be dependent on till granulometry, with coarser tills favoring strong coupling. The different physical mechanisms of ice-till interactions reviewed here support and refine the inference of Boulton (1974) and, in addition, suggest that the depth of till deformation may also be sensitive to till granulometry.

The great variability in till granulometry is well illustrated by a review of some of the published data on grain-size distribution in tills. The clast content reported from coarse, sand-rich tills can be relatively high (25–50%) and commonly clasts are of a diameter of ~0.1 m (Boulton and Hindmarsh, 1987, Fig. 3; Clarke, 1987, Fig. 4; Dreimanis and Vagners, 1972, Fig. 8; Holmes, 1960, Fig. 2; Jiao et al., 1989). On the other side of the spectrum, fine-grained tills have typically 0–15% clasts and their matrix is dominated by silt- and clay-size material (Johnson, 1983; Kemmis, 1981, p. 148; Shepps, 1953, Table 1; Tulaczyk et al., 1998, Fig. 3). The distribution of fine-grained and coarse-grained till is mostly controlled by the character of source material and the distance of glacial transport (Dreimanis and Vagners, 1972). In general, coarse-grained till results from direct glacial erosion of lithified bedrock and fine-grained till is incorporated material from preexisting fine sediment, clay-rich sedimentary rocks and far-traveled, highly comminuted bedrock (Mickelson et al., 1983; Sladen and Wrigley, 1983).

Up-to-date, direct observations of ice-till interactions were mainly made beneath mountain glaciers that are underlain by coarse-grained till (Boulton and Hindmarsh, 1987; Blake, 1992; Blake et al., 1994; Engelhardt et al., 1978; Fischer and Clarke, 1994; Hooke et al., 1997; Iverson et al., 1995). Most of these studies reveal complex, space- and time-variable patterns of ice-till coupling and widely fluctuating, positive and negative strain rates. Despite the very rough nature of at least some of these tills (e.g., Blue Glacier till of Engelhardt et al., 1978), complete decoupling of ice from till occurs when water pressures are near flotation level ($p' \approx 0$; Engelhardt et al., 1978; Hooke et al., 1997; Iverson et al., 1995). Engelhardt et al. (1978) have made a qualitative inference from their direct borehole observations that the ability of ice to intrude into till pore spaces is the main factor that controls the strength of the ice-till coupling. This qualitative statement is reflected in a physical way by the surface-tension effect whose importance was inferred here from basic thermodynamic laws of ice interactions with granular media.

Beneath mountain glaciers where extensive studies of till deformation were conducted, the thickness of subglacial shear zones (Z) was observed/estimated to be 0.1 m (Blue Glacier, Engelhardt et al., 1978), 0.15–0.5 m (Trapridge, Blake et al., 1994; Fischer and Clarke, 1994), 0.3 m (Storglaciären; Hooke et al., 1997; Iverson et al., 1995), and 0.6 m (Breidamerkurjökull; Boulton and Hindmarsh, 1987). Grain-size distribution data available for Trapridge and Breidamerkurjökull (Fig. 5) as well as the visual observations of Engelhardt et al. (1978) beneath Blue Glacier indicate that these coarse tills contain a significant fraction of clasts whose diameter is comparable to the thickness of the deforming zone in the sense of equation (16; i.e., $0.1Z < d < 1.0Z$). Beneath the Blue Glacier, the only site where real-time borehole observations of till deformation were made, larger clasts were often spanning the whole distance between the moving ice and the subtill bedrock. These clasts were accommodating ice motion by rolling in a ball-bearing fashion (Engelhardt et al., 1978). It is difficult to assess just from the published data how important clasts are in distributing strain in coarse till, but there is the possibility that, in accordance with equations (15) and (16), they represent one of the main factors leading to creation of distributed shear zones in plastic tills.

However, not all of the features of the existing strain-rate records from tills of mountain glaciers can be attributed to the influence of clasts (Blake et al., 1994). For instance, strain-rate fluctuations appear to be generally correlated with water-pressure fluctuations (Blake et al., 1994; Hooke et al., 1997; Iverson et al., 1995). This suggests that changes in effective pressure may distribute till deformation. There are a number of potential mechanisms that can explain the linkage between strain-rates and effective pressure. The shear-zone migration model described above represents one of them. Other possibilities that have been considered in the literature include till thinning/thickening and elastic response to cyclic loading (Blake, 1992; Hooke et al., 1997, p. 177).

Data on ice interactions with fine-grained tills is very sparse. The only direct investigation of such modern subglacial system is the study of Ice Stream B, West Antarctica (Engelhardt et al., 1990; Kamb, 1991; Engelhardt and Kamb, 1997, 1998). The presence of a weak, continuous layer of till beneath this ice stream was inferred first from active seismic data (Blankenship et al., 1986, 1987; Rooney et al., 1987, 1991) and then confirmed through drilling and sampling of the bed material (Engelhardt et al., 1990). This material is composed of glacially recycled, Tertiary glacimarine sediments whose clay-rich, clast-poor character predetermines the very fine-grained nature of the ISB till (Tulaczyk et al., 1998).

Borehole experiments suggest that bulk of the ice stream motion is accommodated by basal sliding, which may involve also till deformation in a thin shear zone (~centimeters; Engelhardt and Kamb, 1998). This concentration of sliding/shearing near the ice base is fully consistent with the results of the theoretical analysis presented here (e.g., Figs. 6 and 8). These results predict that the low clast content, susceptibility of the fine-till matrix to submergence by very thin water films, and strong surface effects, should combine beneath ISB to make the ice-till interface significantly weaker than the till itself. This situation favors a mechanical decoupling at the interface even though physically the till and the ice are separated only by a very thin water film (<0.1 mm, Engelhardt and Kamb, 1998). In the case of a perfectly flat ice base pure sliding would be possible. However, the ice base is likely to contain some clasts and/or ice protrusions that will locally plough and deform the underlying till to depths scaling with their amplitude. The apparent variations in sliding velocity obtained by Engelhardt and Kamb (1998) may be due to such irregularities interfering with the tethered stake that was placed just beneath the ice base to measure the sliding. Water pressure beneath the ice stream fluctuates over about ±10 kPa with a predominantly diurnal period (Engelhardt and Kamb, 1997). These fluctuations cannot, however, cause significant distribution of deformation with depth because of the very low hydraulic diffusion coefficient of this clay-rich material ($c_v \sim 10^{-8}$ m^2 s^{-1}; Engelhardt, unpublished data).

Pleistocene fine-grained tills are quite common, but interpretation of their shear strain history from their sedimentological properties is very difficult (Alley, 1991, versus Clayton et al., 1989). Beget (1986, p. 238–239) hypothesized that sliding may have been a common mode of ice motion over the matrix-dominated tills of the Lake Michigan lobe. He proposed also that this sliding should be associated with localized deformation extending to a depth controlled by the size of ice and rocks protruding from the base. There is some sedimentological evidence that supports occurrence of ice sliding associated with clast ploughing and shallow deformation (Clark and Hansel, 1989; Ehlers and Stephan, 1979). Ice-till decoupling is very difficult to document for Pleistocene tills, unless it was associated with a water film thick enough to produce extensive stringers of sorted sediments (Alley, 1991, p. 72). In some localities, these stringers are common enough to provide strong evidence for widespread and long-lasting decoupling of ice from the underlying till (Brown et al., 1987; Piotrowski and Tulaczyk, 1998).

It has been previously proposed that only tills of viscous rheology are capable of deforming in a distributed manner and, thus, transporting debris subglacially (e.g., Alley, 1993). This proposition stems from the assumption that till is a fluid-like continuum in which strain distribution may be caused by one and only one mechanism, the strain-rate dependence of strength. However, simple physical arguments presented here suggest that distributed deformation is not inherently inconsistent with Coulomb-plastic till. The reasons that may make such distributed deformation possible in such till include (1) the fact that the till and the ice-till interface are not perfectly fluidlike or smooth at the scale of subglacial shear zones (i.e., the granular character of the till cannot be neglected), and (2) the presence of mechanisms other than strain-rate strengthening that may prevent concentration of shear strain on one failure plane (e.g., changing strength distribution due to changes in effective pressure). Unless it can be demonstrated that these influences are unimportant for a given till where distributed deformation is observed, the mere fact of distributed deformation cannot be used as conclusive proof of the viscous character of this till. Such unequivocal proof is provided only by data that show a viscous-type relationship between subglacial strain rates and stresses (e.g., Boulton and Hindmarsh, 1987).

Ice motion over till of plastic rheology can result in relatively low, but nonzero, subglacial transport and erosion rates. Taking the shear-zone-migration model as an example (Fig. 11), one can estimate that the depth-averaged till velocity represents a significant fraction of ice velocity, 0.3 and 1.0. These values are broadly consistent with the ones calculated by Alley (0.1–0.5; 1989b, Fig. 3) for viscously behaving tills, even though the assumed mechanism of strain distribution is greatly different. The depth over which the shear-zone migration distributes deformation is likely to be only of the order of a few centimeters for fine-grained tills and a few decimeters for coarse tills ($Z \approx \delta \equiv \sqrt{(c_v/T)}$ with $c_v = 10^{-8}$ m^2 s^{-1} for fine-grained tills and 10^{-6} m^2 s^{-1} for coarse tills). This is between one and two orders of magnitude less than the commonly postulated thickness of several meters for an actively deforming viscous till (Alley, 1991; Alley et al., 1986, 1987a, b, c; Boulton, 1996a, b; Clark et al., 1996; Jenson et al., 1995, 1996). To remain in a steady state, these thick viscous till beds are thought to require a relatively fast rate of debris delivery by subtill erosion and/or deposition from basal debris-laden ice (~0.1 mm/y; Cuffey and Alley, 1996). Debris flux for a plastic, mobile till could be roughly one or more orders of magnitude smaller. Thus, such a mobile till bed needs to be resupplied with debris at a rate of only <~0.01 mm/y. These low till evacuation rates can be matched and exceeded by deposition of debris from basal ice melting at a rate ~1 mm/y and containing a reasonable volumetric concentration of debris, <~0.1% (Kirkbride, 1995, Fig. 8.3).

CONCLUSIONS

In this paper, I present quantitative analysis of several physical mechanisms of potentially great importance to ice interactions with tills of Coulomb-plastic rheology. The nature of these mechanisms suggests that till granulometry has an important control over the strength of ice-till coupling and over the depth to which till deformation is distributed. With all other factors being equal, coarse, clast-rich tills should provide stronger ice-till coupling and thicker subglacial shear zones than fine-grained tills. The latter should favor ice decoupling and concentration of deformation in the vicinity of the ice base. A weak interface is predicted in the case of ice contact with fine-grained tills because (1) support provided by the relatively scarce ploughing clasts is small, (2) large areas of the interface are submerged even by a thin water film (~10^{-6} m), and (3) surface-tension effects hinder coupling of the ice base with the top of the till. Within the range of low subglacial effective pressures applicable to the soft-bed conditions (<100 kPa), fine-grained tills are likely to have interface strength significantly lower than the strength of the till itself ($\tau_i < \tau_f$). As long as the latter condition is met, sliding with ploughing is mechanically more advantageous than pervasive deformation in the underlying till. The strength of ice coupling with coarse tills is dominated by the effect of clast ploughing and is much less sensitive to the presence of a thin basal water film and to the surface-tension effect. As a result, ice-till decoupling and sliding may occur only over a relatively narrow range of effective pressures that are very near zero.

I propose that there are at least three mechanisms that may cause greater depth distribution of strain in coarse, clast-rich tills than in fine-grained tills: (1) matrix deformation around ploughing clasts, (2) particle bridging, and (3) vertical shear-zone migration due to effective pressure fluctuations. For the first two, the thickness of a till shear zone should scale as a small multiple of the characteristic clast/particle diameter. These two mechanisms may result in relatively thick shear zones (~0.1 m) when large clasts (~0.01 m to ~0.1 m) are abundant. However, their influence should be negligible for clast-poor, fine-grained tills. In the case of the third mechanism, the thickness over which till deformation is distributed decreases with the square root of the hydraulic diffusivity of till. The existing measurements of the latter suggest that this thickness may, again, vary from ~0.1 m for coarse tills to ~0.01 m for fine-grained tills. The fact that there are physically viable mechanisms for distributing deformation in tills of Coulomb-plastic rheology indicates that, by themselves, observations of distributed deformation in modern subglacial zones do not provide an unequivocal evidence for viscous rheology of these tills.

The inference that behavior of a weak subglacial till in contact with moving ice may depend fundamentally on till granulometry is consistent with the existing set of observations from modern subglacial zones. Most of the previous interpretations of subglacial till deformation emphasized the controlling role of other physical conditions in determining till behavior. Moreover, till itself was largely treated within a framework of continuum mechanics and as a material whose basic properties do not differ significantly from one location to another. However, till is a very broad term, which encompasses a variety of sediments with widely differing characteristics. To formulate accurate physical laws that govern the mechanics of ice motion over till and till generation by subglacial shear, it is necessary to account for the variable properties of real till. The relative importance of the physical aspects of ice-till interactions discussed here can be further verified through direct field observations and laboratory experiments.

ACKNOWLEDGMENTS

This work was supported by the NSF grant OPP-9219279 to Barclay Kamb and Hermann Engelhardt as a part of research on the role of subglacial till in ice stream mechanics. Additional support to S. T. was provided by the Henry and Grazyna Bauer Fellowship. Sincere thanks go to Barclay Kamb, Ronald F. Scott, and Hermann Engelhardt of the California Institute of Technology for the many discussions that have continuously stimulated this research. However, the author bears sole responsibility for the content of this paper. Neal Iverson and Richard Alley have gratefully provided copies of their manuscripts in press. I thank Richard Alley and Jan Piotrowski for helpful criticisms, some of which are not answered here.

APPENDIX A. CLAST PLOUGHING IN DRAINED AND UNDRAINED CONDITIONS

In the analysis of clast ploughing presented in the main body of this paper I have used undrained till strength to calculate ploughing resistance. However, the assumption of undrained conditions during ploughing will break down for relatively slow clast motion and for tills with high hydraulic diffusivity (as per eq. [2]). It is important to verify whether the breakdown of this assumption may change in a fundamental way my conclusion regarding the high sensitivity of ice-till interactions to till granulometry. In this appendix I estimate the difference in till resistance to ploughing for drained and undrained conditions.

Relative motion of a clast through till produces a zone of compression in front of the clast and a zone of extension behind it (Fig. 2). Under drained conditions, the strength of the till in both of these zones will be different, higher in compression (τ_c) and lower in extension (τ_e) (see also notations in Table A1). The exact stress distribution and the geometry of slip lines accommodating till deformation around the clast are likely to be complex. However, the approximate symmetry of the shear zone surrounding the clast suggests that the drained till strength can be reasonably treated as an average of τ_c and τ_e (Fig. 2):

$$\tau_d = 0.5(\tau_c + \tau_e). \qquad (21)$$

To estimate the values of τ_c and τ_e I consider a two-dimensional till experiencing ploughing by a single clast. Representative stress paths for the zones of compression and extension are illustrated in the Mohr diagram in Figure 12A. Since the considered system is perfectly drained there is no buildup of pore water pressures and total stresses are equal to effective stresses. Initially, the horizontal and vertical effective stresses are assumed to have the same magnitude ($\sigma_{ho}' = \sigma_{vo}' = p_o'$). During ploughing, the vertical stress remains constant at p_o' but the horizontal stress increases in front of the clast and decreases behind it. These changes continue until failure is reached. At failure, the two Mohr circles defined by the constant vertical stress and the two horizontal stresses (σ_{he}' and σ_{hc}') are tangential to the failure envelope (Fig. 12A). The intersections of these circles with this envelope give τ_c and τ_e.

Analytical solutions for the stresses in extension and compression (Fig. 12A) represent a classical problem in soil mechanics solved orig-

TABLE A1. LIST OF SYMBOLS

Symbol	Meaning	Dimension*
A_f	Pore pressure parameter at failure	-
p_o'	Initial effective pressure	FL^{-2}
u	Excess pore pressure	FL^{-2}
ϕ	Internal friction angle	°
σ_1, σ_2	Principal stresses	FL^{-2}
σ_D	Deviatoric stress ($\sigma_1 - \sigma_2$)	FL^{-2}
σ_{ho}'	Initial horizontal effective stress	FL^{-2}
σ_{hc}'	Horizontal effective stress in compression	FL^{-2}
σ_{he}'	Horizontal effective stress in extension	FL^{-2}
σ_{vo}'	Initial vertical effective stress	FL^{-2}
σ_{vc}'	Vertical effective stress in compression	FL^{-2}
σ_{ve}'	Vertical effective stress in extension	FL^{-2}
τ_c	Till strength in drained compression	FL^{-2}
τ_d	Overall drained till strength	FL^{-2}
τ_e	Till strength in drained extension	FL^{-2}
τ_f	Till strength	FL^{-2}
τ_u	Undrained till strength	FL^{-2}

*In addition to the usual dimensions of F, force (MLT^{-2}), L, length, M, mass, T, time, I use '°' for degrees and '-' for nondimensional variables.

inally by Rankine (1857). It can be shown that the magnitudes of horizontal stresses are uniquely related to the constant vertical stress through the following expressions:

$$\sigma_{he}' = \frac{(1-\sin\phi)}{(1+\sin\phi)} p_o', \text{ and} \quad (22a)$$

$$\sigma_{he}' = \frac{(1+\sin\phi)}{(1-\sin\phi)} p_o' \quad (2b)$$

From the Mohr diagram, the failure shear strength can be expressed as a function of the deviatoric stress:

$$\tau = 0.5\sigma_d\cos\phi \quad (23)$$

where σ_d is the deviatoric stress (i.e., the principal stress difference, $\sigma_1 - \sigma_2$). In the considered simple case the horizontal and vertical stresses are the principal stresses (i.e., $\sigma_d = \sigma_{hc} - \sigma_{vc}$ in compression and $\sigma_d = \sigma_{ve} - \sigma_{he}$ in extension; Fig. 12A). Using this fact one can combine equations (22ab) and (23) to obtain the till shear strength in extension and compression:

$$\tau_e = \frac{\sin\phi\cos\phi}{(1+\sin\phi)} p_o', \text{ and} \quad (24a)$$

$$\tau_c = \frac{\sin\phi\cos\phi}{(1-\sin\phi)} p_o'. \quad (24b)$$

Finally, equation (21) combined with equation (24ab) gives the overall drained till strength:

$$\tau_d = \frac{\sin\phi}{\cos\phi} p_o' = \tan\phi \, p_o'. \quad (25)$$

Since I want to compare the drained and undrained till strength during ploughing, it is now necessary to express the undrained strength in terms of the same variables, p_o' and ϕ. To do so, let us start from the same initial isotropic stress state (Fig. 12B, $\sigma_{ho}' = \sigma_{vo}' = p_o'$). At this initial state the total stresses are equal to the effective stresses (i.e., $u = 0$). However, this changes once the motion of the ploughing clasts induces a deviatoric stress that leads to a buildup in pore pressure, which at failure will be given by (Scott, 1963, p. 272):

$$u = 0.5(\sigma_1 + \sigma_3) + A_f(\sigma_1 - \sigma_3) \quad (26)$$

where A_f is the value of the pore pressure parameter at failure. For a normally consolidated soil a reasonable and a computationally convenient value of A_f is 1.0 (Scott, 1963, Fig. 6–11). Having this constraint one can define the effective failure stresses for the zone of compression in front of the ploughing clast and the zone of extension behind it (Fig. 12B). Both of these stress states are described by the same Mohr circle and the strength of the till in the undrained state is everywhere the same:

$$\tau_u = [\sin\phi\cos\phi/(1 + 2\sin\phi)]p_o'. \quad (27)$$

Comparison of the equations (25) and (27) shows that in the drained case till strength is greater than the undrained till strength by a small factor whose exact value depends on the internal friction angle, (1 + 2sinϕ)(cosϕ)$^{-2}$. The value of ϕ for tills lies typically within the range of 20°–40° (Patterson, 1994, Table 8.1). Therefore, in perfectly drained conditions a stress on a ploughing clast should be only 2.1–3.9 times greater than the equivalent stress in perfectly undrained conditions. This is a relatively small effect that does not change significantly the order-of-magnitude calculations shown in the main body of this paper. Moreover, the influence of this effect is generally consistent with my main proposition that ice coupling with coarse tills should be stronger than ice coupling with fine-grained tills. Fine-grained tills are expected to have low hydraulic diffusivity and they should experience drained conditions less frequently than the coarse tills (see eq. [2]). In addition, the

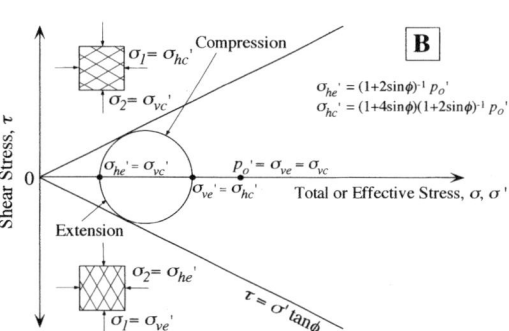

Figure 12. Mohr diagrams showing two-dimensional effective stress states at failure in a pure horizontal extension and a pure horizontal compression under (A) drained, and (B) undrained, conditions. Horizontal and vertical stresses are the principal stresses. Stresses in the undrained case (B) are calculated assuming that at failure the excess pore pressure parameter, A_f, equals to one. In both examples, (A) and (B), the internal friction angle ϕ is 26.5°. Approximate orientation of slip lines is shown in the square boxes. In extension, slip lines make an angle of 45° + 0.5ϕ with the horizontal and in compression this angle is 45° – 0.5ϕ.

coarse-grained tills have greater angle of internal friction than the fine-grained tills (e.g., for the ISB till $\phi = 24°$ and $\tau_d/\tau_u = 1.7$ whereas for the Breidamerkurjökull till $\phi = 32°$ and $\tau_d/\tau_u = 2.1$; Boulton and Hindmarsh, 1987; Tulaczyk, unpublished data). With all other factors being equal, this will make the ice-till interface stronger for coarse tills.

The analysis of pore-pressure-dependence of ploughing stress presented in this appendix clearly shows that the assumption of undrained conditions made before does not limit significantly the generality of my conclusions.

REFERENCES CITED

Alley, R. B., 1989a, Water-pressure coupling of sliding and bed deformation 1. Water system: Journal of Glaciology, v. 35, p. 108–118.

Alley, R. B., 1989b, Water-pressure coupling of sliding and bed deformation 2. Velocity-depth profiles: Journal of Glaciology, v. 35, p. 119–129.

Alley, R. B., 1991, Deforming-bed origin for southern Laurentide till sheets?: Journal of Glaciology, v. 37, p. 67–76.

Alley, R. B., 1993, How can low-pressure channels and deforming tills coexist subglacially: Journal of Glaciology, v. 38, p. 200–207.

Alley, R. B., Blankenship, D. D., Bentley, C. R., and Rooney, S. T., 1986, Deformation of till beneath Ice Stream B, West Antarctica: Nature, v. 322, p. 57–59.

Alley, R. B., Blankenship, D. D., Bentley, C. R., and Rooney, S. T., 1987a, Till beneath Ice Stream B, 3. Till deformation: evidence and implications: Journal of Geophysical Research, v. 92, p. 8921–8929.

Alley, R. B., Blankenship, D. D., Bentley, C. R., and Rooney, S. T., 1987b, Till beneath Ice Stream B, 4. A coupled ice-till flow model: Journal of Geophysical Research, v. 92, p. 8921–8929.

Alley, R. B., Blankenship, D. D., Rooney, S. T., and Bentley, C. R., 1987c, Continuous till deformation beneath ice sheets: Publication of the International Association of Hydrological Sciences, no. 170, p. 81–90.

Alley, R. B., Cuffey, K. M., Evenson, E. B., Strasser, J. C., Lawson, D. E., and Larson, G. J., 1998, How glaciers entrain and transport basal sediments: physical constraints: Quaternary Science Reviews, v. 16, p. 1017–1038.

Baligh, M. M., 1972, Applications of plasticity theory to selected problems in soil mechanics [Ph.D. thesis]: Pasadena, California Institute of Technology, 228 p.

Beget, J., 1986, Influence of till rheology on Pleistocene glacier flow in the southern Great Lakes area, U.S.A.: Journal of Glaciology, v. 32, p. 235–241.

Blake, E. W., 1992, The deforming bed beneath a surge-type glacier: Measurement of mechanical and electrical properties [Ph.D. thesis]: Vancouver, University of British Columbia, 179 p.

Blake, E. W., Fischer, U. H., and Clarke, G. K. C., 1994, Direct measurements of sliding at the glacier bed: Journal of Glaciology, v. 40, p. 595–599.

Blankenship, D. D., Bentley, C. R., Rooney, S. T., and Alley, R. B., 1986, Seismic measurements reveal a saturated porous layer beneath an active Antarctic ice stream: Nature, v. 322, p. 54–57.

Blankenship, D. D., Bentley, C. R., Rooney, S. T., and Alley, R. B., 1987, Till beneath Ice Stream B, 1. Properties derived from seismic travel times: Journal of Geophysical Research, v. 92, p. 8903–8911.

Boulton, G. S., 1974, Processes and patterns of glacial erosion, in Coates, D. R., ed., Glacial geomorphology: Binghamton, New York, State University of New York, p. 41–87.

Boulton, G. S., 1986, Geophysics—a paradigm shift in glaciology: Nature, v. 322, p. 18.

Boulton, G. S., 1996a, The origin of till sequences by subglacial sediment deformation beneath mid-latitude ice sheets: Annals of Glaciology, v. 22, p. 75–84.

Boulton, G. S., 1996b, Theory of glacial erosion, transport and deposition as a consequence of subglacial sediment deformation: Journal of Glacioogy, v. 42, p. 43–62.

Boulton, G. S., and Hindmarsh, R. C. A., 1987, Sediment deformation beneath glaciers—rheology and geological consequences: Journal of Geophysical Research, v. 92, p. 9059–9082.

Brown, N. E., Hallet, B., and Booth, D. B., 1987, Rapid soft bed sliding of the Puget glacial lobe: Journal of Geophysical Research, v. 92, p. 8985–8997.

Clark, P. U., 1991, Striated clast pavements: Products of deforming bed?: Geology, v. 19, p. 530–533.

Clark, P. U., 1992, Surface form of the southern Laurentide Ice Sheet and its implications to ice-sheet dynamics: Geological Society of America Bulletin, v. 104, p. 595–605.

Clark, P. U., and Hansel, A. K., 1989, Clast ploughing, lodgement, and glacier sliding over a soft glacier bed: Boreas, v. 18, p. 201–207.

Clark, P. U., and Walder, J. S., 1994, Subglacial drainage, eskers, and deforming beds beneath the Laurentide and Eurasian ice sheets: Geological Society of America Bulletin, v. 106, p. 304–314.

Clark, P. U., Licciardi, J. M., MacAyeal, D. R., and Jenson, J. W., 1996, Numerical reconstruction of a soft-bedded Laurentide ice sheet during the last glacial maximum: Geology, v. 24, p. 679–682.

Clarke, G. K. C., 1987, Subglacial till: a physical framework for its properties and processes: Journal of Geophysical Research, v. 92, p. 9023–9036.

Clayton, L., Mickelson, D. M., and Attig, J. W., 1989, Evidence against pervasively deformed bed material beneath rapidly moving lobes of the southern Laurentide Ice Sheet: Sedimentary Geology, v. 62, p. 203–208.

Cuffey, K., and Alley, R. B., 1996, Is erosion by deforming subglacial sediments significant? (Toward till continuity): Annals of Glaciology, v. 22, p. 17–24.

Dreimanis, A., and Vagners, U. J., 1972, Bimodal distribution of rock and mineral fragments in basal tills, in Goldthwait, R. P., ed., Till; a symposium: Columbus, Ohio State University Press, p. 237–250.

Ehlers, J., and Stephan, H.-J., 1979, Forms at the base of till strata as indicators of ice movement: Journal of Glaciology, v. 22, p. 345–355.

Engelhardt, H. F., and Kamb, B., 1997, Basal hydraulic system of a West Antarctic ice stream: constraints from borehole observations: Journal of Glaciology, v. 43, p. 207–230.

Engelhardt, H. F., and Kamb, B., 1998, Sliding velocity of Ice Stream B: Journal of Glaciology, v. 44, p. 207–230.

Engelhardt, H. F., Harrison, W. D., and Kamb, B., 1978, Basal sliding and conditions at the glacier bed as revealed by borehole photography: Journal of Glaciology, v. 20, p. 469–508.

Engelhardt, H. F., Humphrey, N., Kamb, B., and Fahnestock, M., 1990, Physical conditions at the base of a fast moving Antarctic ice stream: Science, v. 248, p. 57–59.

Everett, D. H., 1961, The thermodynamics of frost damage to porous solids: Transactions of the Faraday Society, v. 57, p. 1541–1551.

Fischer, U. H., and Clarke, G. K. C., 1994, Plowing of subglacial sediment: Journal of Glaciology, v. 40, p. 97–106.

Hallet, B., Walder, J. S., and Stubbs, C. W., 1991, Weathering by segregation ice growth in microcracks at sustained sub-zero temperatures, verification from an experimental study using acoustic emissions: Permafrost and Periglacial Processes, v. 2, p. 283–300.

Holmes, C. D., 1960, Evolution of till-stone shapes, central New York: Geological Society of America, v. 71, p. 1645–1660.

Hooke, R. L., 1997, Principles of glacier mechanics: Prentice Hall, London, 248 p.

Hooke, R. L., and Iverson, N. R., 1995, Grain-size distribution in deforming subglacial tills: role of grain fracture: Geology, v. 23, p. 57–60.

Hooke, R. L., Hanson, B., Iverson, N. R., Jansson, P., and Fischer, U. H., 1997, Rheology of till beneath Storglaciären, Sweden: Journal of Glaciology, v. 43, p. 172–179.

Humphrey, N., Kamb, B., Fahnestock, M., and Engelhardt, H., 1993, Characteristics of the bed of the lower Columbia Glacier, Alaska: Journal of Geophysical Research, v. 98, p. 837–846.

Iverson, N. R., 1993, Regelation of ice through debris at glacier beds: implications for sediment transport: Geology, v. 21, p. 559–562.

Iverson, N. R., and Semmens, D. J., 1995, Intrusion of ice into porous media by regelation: a mechanism of sediment entrainment by glaciers: Journal of Geophysical Research, v. 100, p. 10219–10230.

Iverson, N. R., Jansson, P., and Hooke, R. L., 1994, In-situ measurements of the strength of deforming subglacial till: Journal of Glaciology, v. 40, p. 497–503.

Iverson, N. R., Hanson, B., Hooke, R. L., and Jansson, P., 1995, Flow mechanics of glaciers on soft beds: Science, v. 267, p. 80–81.

Iverson, N. R., Hooyer, T., and Hooke, R. L., 1996, A laboratory study of sediment deformation, stress heterogeneity and grain-size evolution: Annals of Glaciology, v. 22, p. 167–175.

Iverson, N. R., Baker, R. W., and Hooyer, T. S., 1998, A ring shear device for the study of till deformation: Tests on tills with contrasting clay content: Quaternary Science Reviews, v. 16, p. 1057–1066.

Jaeger, J. C., and Cook, N. G. W., 1969, Fundamentals of rock mechanics: London, Metheuen, 513 p.

Jenson, J. W., Clark, P. U., MacAyeal, D. R., Ho, C., and Vela, J. C., 1995, Numerical modeling of advective transport of saturated deforming sediment beneath the Lake Michigan lobe, Laurentide ice sheet: Geomorphology, v. 14, p. 157–166.

Jenson, J. W., MacAyeal, D. R., Clark, P. U., Ho, C. L., and Vela, J. C., 1996, Numerical modeling of subglacial sediment deformation—implications for the behavior of the Lake-Michigan lobe, Laurentide Ice Sheet: Journal of Geophysical Research, v. 101, p. 8717–8728.

Jiao, K., Zheng, B., and Ma, Q., 1989, Particle composition of glacial deposits in the West Kunlun Mts.: Bulletin of Glacier Research, v. 7, p. 153–159.

Johnson, A. M., 1970, Physical processes in geology: San Francisco, Freeman and Cooper, 577 p.

Johnson, M. D., 1983, The origin and microfabric of Lake Superior red clay: Journal of Sedimentary Petrology, v. 53, p. 859–873.

Kamb, B., 1970, Sliding motion of glaciers: theory and observations: Reviews of Geophysics and Space Physics, v. 8, p. 673–728.

Kamb, B., 1991, Rheological nonlinearity and flow instability in the deforming-bed mechanism of ice stream motion: Journal of Geophysical Research, v. 96, p. 16585–16595.

Kemmis, T. J., 1981, Importance of the regelation process to certain properties of basal tills deposited by the Laurentide ice sheet in Iowa and Illinois, USA: Annals of Glaciology, v. 2, p. 147–152.

Ketcham, W. M., and Hobbs, P. V., 1969, An experimental determination of the surface energies of ice: Philosophical Magazine, v. 19, p. 1161–1173.

Kirkbride, M. P., 1995, Processes of transportation, in Menzies, J., ed., Modern glacial environments: Oxford, United Kingdom, Butterworth Heinemann, p. 261–292.

Llibourty, L., 1979, Local friction laws for glaciers: a critical review and new openings: Journal of Glaciology, v. 23, p. 67–95.

MacAyeal, D. R., 1992, Irregular oscillations of the West Antarctic ice sheet: Nature, v. 359, p. 29–32.

Maltman, A., 1992, Deformation structures preserved in rocks, in Maltman, A., ed., The geologic deformation of sediments: London, Chapman and Hall, p. 261–308.

Mickelson, D. M., Clayton, L., Fullerton, D. S., and Borns, H. W., 1983, The Late Wisconsin glacial record of the Laurentide ice sheet in the United States, in Wright, H. E., and Porter, S. C., eds., Late Quaternary environments of the United States: Minneapolis, University of Minnesota Press, p. 3–37.

Mitchell, J. K., 1993, Fundamentals of soil behavior: New York, John Wiley & Sons, 437 p.

Mulhaus, H. B., and Vardoulakis, I., 1987, The thickness of shear bands in granular materials: Geotechnique, v. 37, p. 271–283.

Nye, J. F., 1969, The calculation of sliding of ice over a wavy surface using a Newtonian viscous approximation: Proceedings of the Royal Society of London, v. 311, p. 445–467.

Parks, G. A., 1990, Surface energy and adsorption at mineral-water interfaces: an introduction, in Hochella, M. F., and White, A. F., eds., Mineral-water interface geochemistry: Washington, D.C., Mineralogical Society of America, Reviews in Mineralogy, v. 23, p. 133–175.

Patterson, W. S. B., 1994, The physics of glaciers: Oxford, United Kingdom, Pergamon, 480 p.

Piotrowski, J. A., and Tulaczyk, S., 1999, Subglacial conditions under the last ice sheet in northwest Germany: ice-bed separation and enhanced basal sliding: Quaternary Science Reviews, (in press).

Rankine, W. J. M., 1857, On the stability of loose earth: Philosophical Transactions of the Royal Society of London, v. 147, p. 9–27.

Rooney, S. T., Blankenship, D. D., Alley, R. B., and Bentley, C. R., 1987, Till beneath Ice Stream B, 2. Structure and continuity: Journal of Geophysical Research, v. 92, p. 8913–8920.

Rooney, S. T., Blankenship, D. D., Alley, R. B., and Bentley, C. R., 1991, Seismic reflection profiling of a sediment-filled graben beneath Ice Stream B, West Antarctica, in Thomson, M. R., Crame, J. A., and Thomson, J. W., eds., Geological evolution of Antarctica: Cambridge, United Kingdom, British Antarctic Survey, p. 261–265.

Roscoe, K. H., 1970, The influence of strains in soil mechanics: Geotechnique, v. 20, p. 129–170.

Salencon, J., 1977, Applications of the theory of plasticity in soil mechanics: New York, John Wiley & Sons, 158 p.

Sauer, E. K., Egeland, A. K., and Christiansen, E. A., 1993, Compression characteristics and index properties of tills and intertill clays in southern Saskatchewan, Canada: Canadian Geotechnical Journal, v. 30, p. 257–275.

Scott, R. F., 1963, Principles of soil mechanics: Reading, Massachusetts, Addison-Wesley Publishing Company, 550 p.

Scott, R. F., and Schoustra, J. J., 1968, Soil: mechanics and engineering: New York, McGraw-Hill, 314 p.

Shepps, V. C., 1953, Correlation of the tills of northeastern Ohio by size analysis: Journal of Sedimentary Petrology, v. 23, p. 34–48.

Sladen, J. A., and Wrigley, W., 1983, Geotechnical properties of lodgement till—a review, in Eyles, N., ed., Glacial geology, an introduction for engineers and earth scientists: Oxford, United Kingdom, Pergamon Press, p. 184–212.

Terzaghi, K., Peck, R. B., and Mesri, G., 1996, Soil mechanics in engineering practice: New York, John Wiley & Sons, 549 p.

Tulaczyk, S., Kamb, B., Scherer, R., and Engelhardt, H. F., 1998, Sedimentary processes at the base of a West Antarctic ice stream: constraints from textural and compositional properties of subglacial debris: Journal of Sedimentary Research, v. 68, p. 487–496.

Turcotte, D. L., and Schubert, G., 1982, Geodynamics; Applications of continuum physics to geological problems: New York, John Wiley and Sons, 450 p.

Walder, J. S., and Fowler, A., 1994, Channelized subglacial drainage over a deformable bed: Journal of Glaciology, v. 40, p. 3–15.

Weertman, J., 1957, On sliding of glaciers: Journal of Glaciology, v. 3, p. 33–38.

Weertman, J., 1972, General theory of water flow at the base of a glacier or ice sheet: Reviews of Geophysics and Space Physics, v. 10, p. 287–333.

Whillans, I. M., and van der Veen, C. J., 1993, New and improved determinations of velocity of Ice Stream B and C, West Antarctica: Journal of Glaciology, v. 39, p. 483–490.

MANUSCRIPT ACCEPTED BY THE SOCIETY OCTOBER 8, 1998

Quaternary glacial deposits and landforms of the north Timan region, Russia—a possible center of local glaciation

Andrei V. Matoshko
National Academy of Sciences of Ukraine, Institute of Geological Sciences, Gonchara Str.55b, Kiev, 252054, Ukraine; matoshko@radiogeo.kiev.ua

ABSTRACT

During the Quaternary the north Timan region of northern Russia experienced repeated episodes of glaciation and marine transgression. The distribution of Quaternary materials and landforms is concentric around the central axis of the north Timan plateau. The central axis of the plateau is characterized by ablation moraines, presence of a weathered mantle of residual deposits, and little trace of glacial erosion. The slopes around the periphery of the plateau are characterized by basal till, meltwater deposits, thrust moraines, marine deposits, and erosional landforms of different scales. Radially projecting glacial valleys infilled by glacial and marine deposits, dissect the plateau slopes. The areal distribution of glacial materials and landforms, their composition, and other features indicate that during the expansion of Quaternary ice sheets the north Timan plateau was a local center of glacial outflow. A small ice dome is hypothesized to have formed there at the beginning of glacial expansion. The spreading of short outlet glaciers from the dome initiated the formation of the glacial valleys within the plateau slopes. During the glacial maximum, the Eurasian ice sheet overrode the dome. During the retreat stage, the plateau was a massif of dead ice on the plateau separated from active oscillations of the main ice sheet. Centers of local glaciation may have formed on similar elevated plateaus (eastern part of Kola Peninsula, most northern part of Kanin Peninsula) and along the periphery of high-latitude ice sheets, initiating local glacial erosion patterns and affecting the distribution of glacial materials and landforms.

INTRODUCTION

The north Timan plateau (Timansky kryazh) reaches elevations of 200–300 m a.s.l., between latitudes 66° and 68° north (Fig. 1). It is located in the northern part of the Arhangelsk oblast (Russian Federation). The north Timan plateau extends from the shore of the Barents Sea for more than 200 km in a southeast direction. The width of the plateau increases from 30 km in the north to 150 km in the south. Tectonically the plateau is an arched part of the East-European platform, which is composed of Riphean, Paleozoic, and Mezozoic igneous, metamorphic, and sedimentary rocks in its axial part. These rock units are overlain by Quaternary deposits.

The north Timan region is within the area glaciated during the Valdaian (Weichselian) glaciation. The first studies of the glacial geology of this region were those of Grigoriev (1924), Ramsay (1931, in Grosvald, 1983), Lavrov (1947, in Grosvald, 1983), Chernov (1947), and others. These workers delineated the positions of end moraines, and directions of erratic debris transport. Lavrov (1977) defined the general position of the retreat stages of the Valdai glaciation on the northern part of the East European Plain, including the north Timan region, and described geomorphological features of the ice marginal deposits in the Malozemelskaya Tundra area (including the northeastern part of the north Timan region). Gornostai (1990) established the existence of huge glaciodislocations on the western slopes of the north Timan plateau. As a whole, problems of the north Timan region glaciation in relation to the main Valdai ice sheet is con-

Matoshko, A. V., 1999, Quaternary glacial deposits and landforms of the north Timan region, Russia—a possible center of local glaciation, *in* Mickelson, D. M., and Attig, J. W., eds., Glacial Processes Past and Present: Boulder, Colorado, Geological Society of America Special Paper 337.

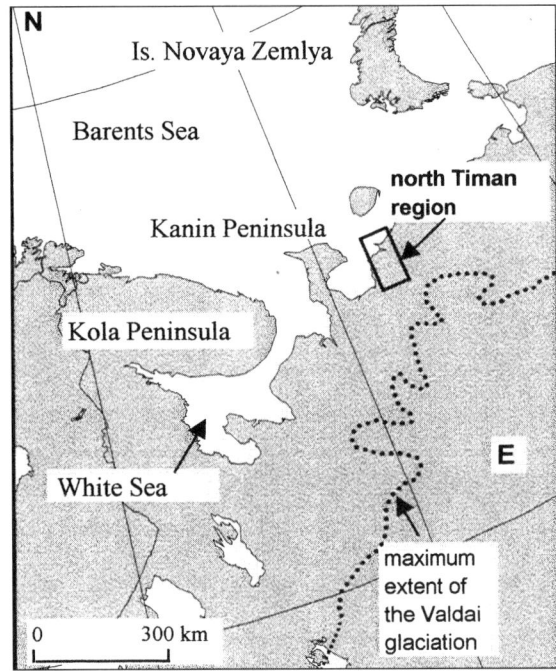

Figure 1. General location map of the investigated area.

sidered in many works, the main ideas of which are summarized in publications by Chebotareva (1977), Velichko and Hodackov (1981), and Grosvald (1983).

Despite this, the glacial materials and landforms of the north Timan region and ice sheet dynamics of the area have been only scantily studied. All other modern information about Timan glacial deposits is contained in the materials of geological survey and unpublished prospecting reports of Timan surveys (Naryan Mar, Arhangelskaya oblast, Russian Federation). These materials, as well as other literature sources, were evaluated in the course of investigations of glacial deposits and landforms accomplished while the author prospected for diamonds between 1988 and 1993. Maps of glacial landforms and Quaternary deposits, a contour map of their base, a map of their thickness in the scale 1:200,000, and tens of geological profiles were compiled and analyzed. The investigation results gave an opportunity to evaluate the glacial landforms, materials, and structure of the north Timan region. This evaluation provides some insight into important problems of glaciation and dynamics of ice sheets in high latitudes.

DATA

General nature of the Quaternary cover

The total amplitude of the north Timan region sub-Quaternary surface reaches 370 m (some elements of this map are shown in Fig. 2). As a whole, this surface corresponds to the modern surface of the territory and may be divided in two main parts: a plateau proper, and its slopes. The plateau is intensively dissected with numerous hills and ridges. Short valleys with orientations

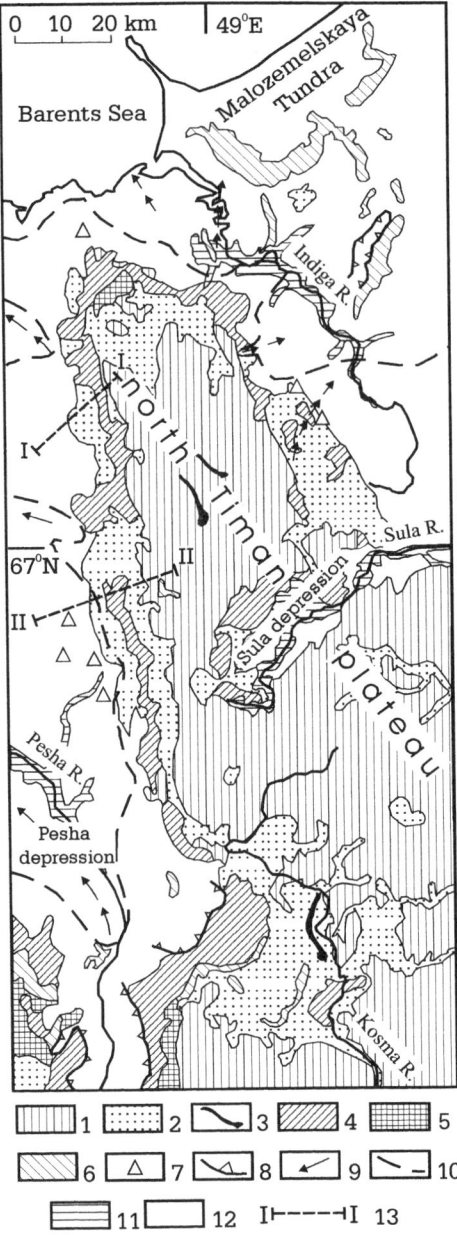

Figure 2. Distribution of the upper Pleistocene–early Holocene glacial formations in the north Timan region. 1, fragmentary cover of ablation tills (a plateau); 2, meltwater deposits; 3, great eskers; 4, cover of basal tills (plateau slopes); 5, massifs of hummocky moraines; 6, separate great bodies of thrust moraines; 7, glacial rafts and glaciodislocations; 8, boundaries of subglacial depressions reflected in the modern relief; 9, thalwegs of buried Quaternary valleys and their tilt direction shown by arrows; 10, contour of 0 m altitude of the pre-Quaternary surface; 11, areas of the Holocene alluvial deposits distribution; 12, areas of the Holocene marine deposits distribution; 13, position of the geological profiles (see Fig. 3).

northwest-southeast and northeast-southwest divide them. This orientation coincides with the direction of the principal structural elements of the north Timan uplift. Negative landforms, such as the river valleys, are present near the plateau slopes only.

The plateau is surrounded on the north, west, and possibly on the east by a large ringlike depression that reaches 100 m below sea level. Probably the depression has a continuation within the limits of the Barents Sea shelf. Grosvald (1983) noted that the surface of the Barents-Kara shelf is inclined towards the coast in many places. In the south part of the territory this main depression intrudes as a narrow bay along the Pesha River to the central Timan zone (Pesha depression). The Sula depression cleaves the Timan plateau across its strike (Fig. 2). The plateau slopes toward the depression have different profiles (Fig. 3). The thickness of Quaternary deposits relates to the sub-Quaternary surface. On the plateau proper, it varies from 0 to 15 m, thick rarely reaching 20 m. Towards the slopes, the thickness of the Quaternary deposits increases to about 60 m.

Late Pleistocene–early Holocene glacial formations and erosional landforms

Glacial and residual deposits are dominant within the plateau proper (160–270 m a.s.l.). Till deposits are typically several meters thick and cover most areas. They are absent on the top of some bedrock hills, ridges, and steep valleys sides. Till thickness increases in the valleys to as much as 8–9 m. Till deposits of the plateau lie on the weathered mantle of residual deposits or on the unweathered bedrock. It should be mentioned that the largest remnants of Neogene waste mantle in the territory of Kola Peninsula (see Fig. 1) are associated with its central part where the position of the main ice divide has been reconstructed (Evzerov and Koshechkin, 1981).

The plateau tills are brown (of different tints) loose diamictions without shearing or folded contacts, lack stratification, and contain a high percentage of coarse clasts in addition to a large clay fraction. Unfortunately, there are no data about petrographic composition of coarse glacial clasts in the relation to their primary sources. The clayey composition of till leads to mass wasting and covering of exposures. As a whole, variations in the occurrence and composition of the plateau deposits occur because most are ablation till or other supraglacial sediment. It should be added that except for minor exceptions, traces of glacial erosion or pressure on the glacier bed are not found.

The upper part of the plateau slopes (100–160 m a.s.l.) is covered by Valdai basal till. It lies on older till or marine deposits and also on bedrock. Ablation till and solifluction mantles occur at the top of the sequence locally. The zone of basal till cover is usually narrow (some kilometers), but in the upper reaches of the Pesha River it expands to 8–10 km and is present down to the level of 80–90 m a.s.l. (see Fig. 2). Maximum till thickness (20–25 m) is present along the middle part of the plateau slopes and at lower elevations it gradually decreases (see Fig. 3). Below altitudes of 80–90 m, the basal till is overlain by meltwater

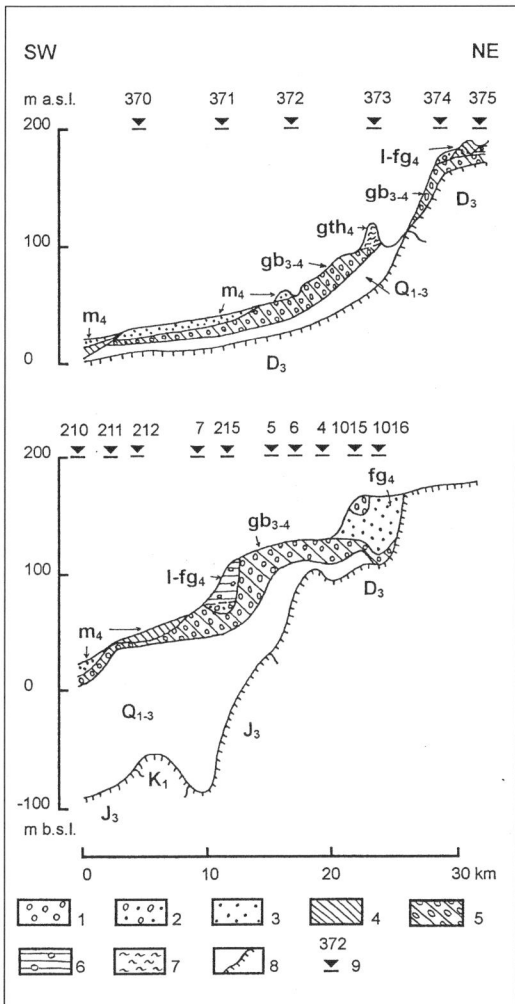

Figure 3. Schematic geological profiles across western slopes of the north Timan plateau. Upper profile: the slope of common simple structure; I-I in Figure 2. Lower profile: the slope that is complicated by Quaternary erosional depression; II-II in Figure 2. 1, coarse debris (gravel, pebble, cobble); 2, sands with high percentage of coarse debris; 3, sands; 4, aleurites; 5, diamictite (till); 6, clays with coarse debris; 7, dislocated clays of thrust moraines; 8, top of the bedrock; 9, bore holes and their numbers. Genetical indices and indices of age: Holocene deposits: l-fg$_4$, meltwater; fg$_4$, fluvioglacial; gth$_4$, thrust moraines; m$_4$, marine; gb$_{3-4}$, upper Pleistocene–Holocene basal tills; Q$_{1-3}$, nonstratified lower Quaternary–upper Pleistocene marine and glacial deposits; D$_3$, Upper Devonian; J$_3$, Upper Jurassic; K$_1$, Lower Cretaceous.

deposits of the same age, and below 60–70 m (maximum level of the greatest late Pleistocene marine transgression, the "Boreal transgression") marine and solifluction deposits of early Holocene rest on the basal till.

According to the Lavrushin (1976) classification, most of the basal till belongs to the monolithic facies, but the basal till imbricate facies has been recorded at two points. Single glacial rafts (they are considered also as a facies of basal till) of Paleozoic and Mezozoic sandstones and limestones are visible in bore holes on the east slopes of the Timan plateau (see Fig. 2). Their thickness is not known to exceed 10 m. Monolithic facies of basal till are characterized by a brecciated glaciodynamic microstructure (Fig. 4). Platy microstructures are found in one place (basin of the Pesha River, upper reaches) only. Basal tills have high density, predominantly gray or dark gray color (Figs. 4, 5), and a high percentage of clay fraction. These last two characteristics suggest a local origin of the till, which is enriched by the material picked up by the glacier from the underlying gray and black, marine, late Pleistocene clays. Therefore, the till contains numerous remnants of marine organisms, especially foraminifera. The percentage of pebbles, cobbles, and boulders in the basal till is distinctly less than in the ablation till.

Marginal thrust moraines (ice-pushed moraines), which are concentric ridges and hills, surround the north Timan plateau. Thrust moraines are the prominent landforms of the area (as much as 50 km in length, 6 km in width, and 60 m in height). They form broken chains, which join together with other marginal formations to define the outer position of the ice sheet margin during the Marhida and Velt retreat phases of the last glaciation (Lavrov, 1977). They are identified by their morphological expression. Information about their internal structure is still not available. The thrust moraines are associated with areas of hummocky moraine (see Fig. 2) in some places on the western plateau slopes and have an irregular combination of small hills and with ridges of different orientation. They are composed of poorly sorted sand and gravel.

The samples of peat and wood from lacustrine deposits underlying the uppermost till horizon in the lower reaches of Pechora River near the Village of Marhida are dated by C14 (Arslanov et al., 1975). According to this dating their age is 9,000 B.P. or younger. But there is one more reach of marginal formations northwards near the sea shore, Velt (the retreat phase "Velt"), and its age must be younger (Grosvald, 1983). Formations of this reach are also correlated with the northernmost thrust moraines of the investigated area. Thus, the marginal formations were deposited early in the Holocene. They are the youngest ones in the East-European Plain.

Very important indicators of the glacial flow history are the orientation of glacial striations, grooves, and furrows on the abraded surfaces of igneous bedrock protrusions. Single measurements of some of these forms in the northwestern part of the territory were interpreted to show a southeast direction of ice flow. This interpretation is used for the glaciodynamic scheme of Lavrov (1977). There is also an indication of glacial transport of local Timan rocks from the Barmin Cape (Timan coast) to the tops of its plateau indicating flow in the south-southeast and southeast direction in the work by Grigoriev (1924). However, the latest measurements in the same area carried out by geologists of the Timan expedition (materials of geological survey) and by this author clearly indicate two directions: 250°–260° and 170°–180° in azimuth (see Fig. 6).

Meltwater deposits occur irregularly throughout area. Meltwater drainage systems of restricted extent in valleys as well as separate areas of meltwater deposits in the structural depressions of the bedrock surface are characteristic of the plateau below 200 m a.s.l. (see Fig. 2). They are flat surfaces amidst the dissected plateau topography. There are also single groups of kames, fluvioglacial fans, and eskers.

An extensive zone of meltwater deposits (as much as 12 km^2) occupies two levels (140–170 m and 90–120 m a.s.l.) surrounding the plateau (see Fig. 2). Areas of valley sandar with fluvioglacial terraces are divided by zones of basal till and by scarps and are most abundant on the west plateau slopes (see Fig. 3). The sandar consist predominantly of sand and pebble gravel. Cobble and boulder beds are present in the upstream part of the sandur sequence as well. There are a lot of modern lakes

Figure 4. Brecciated microstructure of the local basal till. The exposure is on the right bank of the Volonga River valley.

Figure 5. Exposure of the local basal till enriched of dark gray and black marine Pleistocene clays. D_3, Upper Devonian sandstones; gb_{3-4}, upper Pleistocene–Holocene local basal till diamictite; ga_3, Holocene ablation moraine.

within the limits of the area. The author speculates that these lakes are the remnants of the large intraglacial and proglacial basins, where glacial lacustrine sediments accumulated.

Buried early Pleistocene–late Pleistocene glacial formations of the slopes area

In summary, the Pleistocene sequence of the north Timan plateau slopes area is characterized by an alternation of glacial, marine-glacial, and marine deposits, where the glacial complex dominates. Foraminiferal and diatomaceous fauna data (geological survey materials) indicate the Odintsovo (middle Pleistocene), the Mikulino (Eemian–upper Pleistocene), and the Holocene marine horizons are present in several bore holes. The two former ones divide three glacial and marine-glacial horizons: the Dnieper, the Moscow (middle Pleistocene), and the Valdai or the late Valdai (upper Pleistocene). The glacial deposits also contain faunal remnants but they are mixed. Pre-Quaternary microfossils prevail there.

This relatively simple picture is substantially complicated in many other bore holes and sections, where as much as four

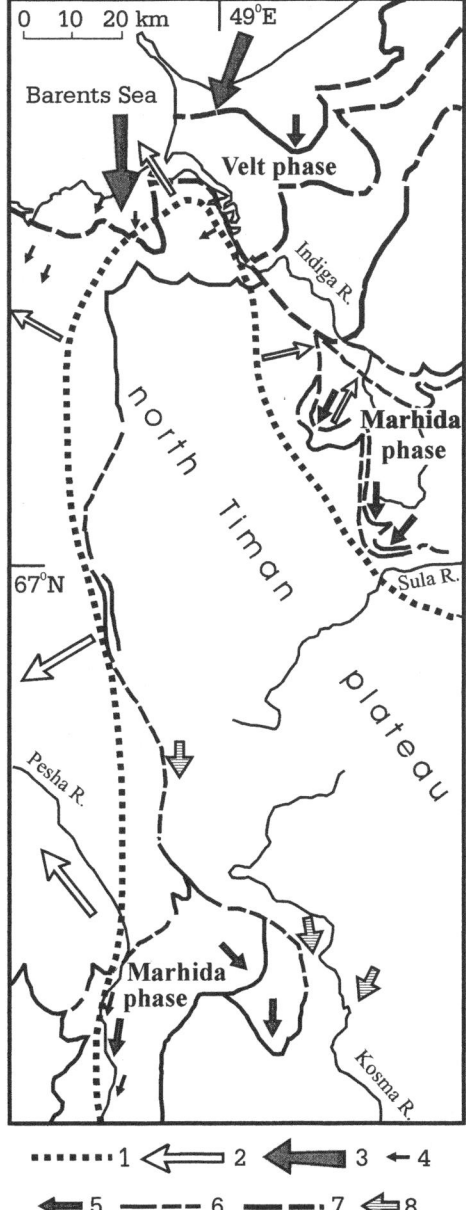

Figure 6. Schematic dynamics map of the north Timan region glaciation. 1, hypothetical position of the rudimentary north Timan ice dome; 2, directions of ice dome tongues flow; 3, main directions of ice sheet advance; 4, orientation of glacial furrows, grooves, and gravel-pebble clasts in the basal tills; 5, directions of ice tongues readvance during the retreat stage; 6, position of the Marhida retreat phase marginal zones; 7, position of the Velt retreat phase marginal zones; 8, main directions of the regional melt water runoff.

marine and five glacial horizons can be distinguished by lithological signs and mode of occurrence. It is especially characteristic for boreholes of the buried radial valleys that cut the plateau slopes (Fig. 2, Fig. 3, lower profile). Some of them are narrow with steep slopes, others have a flat bed, but all of them lack branches like those of river valleys. The deposits mentioned above occur with a variety of regional and local uncomformities. Their lithological composition and thickness change over short distances, and they are glacially disturbed in many places. Therefore, the correlation of bore holes is not possible in most cases. This explains the simplified drawing of geological profiles shown in Figure 3.

The complex of buried valleys contains the following major types of glacial deposits: basal tills, glacial rafts, and glacial-marine deposits, including glacial-marine turbidites. In many bore holes till deposits cover the floor of the depressions. It should be noted that alluvial deposits are not present there. As a whole, basal tills beds are thicker than areas beyond the valleys. On the western slopes of the Timan plateau, glaciodislocations of bedrock removed from the depressions (predominantly of Late Jurassic and Early Cretaceous age) has been found by Gornostai (1990; see Fig. 2). All features of the buried depressions mentioned above are evidence of their glacial origin, and their radial orientation indicates the existence of concentrated local ice outflow from the plateau. Therefore, they are considered glacial valleys of short outlet glaciers from the plateau.

DEVELOPMENT OF NORTH TIMAN REGION GLACIATION; AN HYPOTHESIS

In summary, the following points concerning the north Timan region lead to the following hypothesis. This region has undergone repeated glaciations and marine transgressions during the Quaternary. The youngest glacial formations that are reflected in the modern surface are dated to the late Pleistocene–Holocene epochs. The distribution of Quaternary glacial materials and landforms is concentric about the plateau. An area characterized by ablation moraines, little evidence of glacial erosion, weathered residual deposits, and marine deposits is located in the center of the region. This is the plateau proper, with only traces of the impact of glaciation.

In contrast, there are numerous and different types of evidence of active glaciers and of glacier/bed interaction on the plateau slopes. Slopes are occupied by upper Pleistocene and Holocene glacial marginal deposits, and also by alternating glacial, glacial-marine, and marine deposits, of early, middle, and upper Pleistocene age. The latter are most fully represented in the buried radial glacial valleys that dissected the plateau slopes. Different forms of glacial erosion (striations, grooves, furrows, depressions, rafts of local rocks, local moraines) and glacial dislocations are also characteristic of the plateau slopes.

The facts cited above can be used to test general ideas of the development of Quaternary and especially the last European or Eurasian ice sheets development. People have suggested ice sheet initiation from mountain regions (Scandinavian Peninsula, Hibiny Mountains, Spitsbergen, Novaya Zemlya, Urals), from the Barents and the Kara Sea shelves, or a combination. These ideas are analyzed in the works of Velichko and Hodakov (1981), Grosvald (1983), and some others. The concepts of one common or several separate ice sheets during the advance stage are not in contradiction with the possibility of small ice domes developing in the peripheral area of the expanding ice sheet. Some elevated plateaus (above about 200 m a.s.l.) of the eastern part of Kola Peninsula, the most northern part of Kanin Peninsula, and the north Timan plateau were appropriate for that. The firn line in the adjacent arctic regions of present island glaciation (Spitsbergen, Novaya Zemlya, Severnaya Zemlya) lies in the range 200–600 m a.s.l. (Koryakin, 1988). Grosvald (1983) concluded that the favorable climatic conditions for firn and ice accumulation existed in most parts of the Arctic during Wurm glaciation. According to Denton and Hughes (1981), calculated mean annual air temperatures decreased to –17 °C at latitude 70° north during this time.

I suggest that during Quaternary glacial epochs, the north Timan plateau was a local center of glaciation. There was a small glacial dome on it at the beginning of each epoch (Fig. 7 G.1., see also Fig. 6). It looked like the modern ice caps on the plateaus of Spitsbergen and Novaya Zemlya. These ice caps are described by Koryakin (1988). The radial spreading of the glacial tongues (outlet glaciers) from the dome used preexisting valleys and started deepening the glacial valleys on the Timan plateau slopes and also glacially modified the Pesha and Sula tectonic depressions.

There are several points of view on the position of ice sheet centers and their influence on the history of the northeastern part of the East European Plain. The possible centers are four: the Scandinavian center, a center in the flank of the Barents Sea shelf, the Novaya Zemlya center, and the Kara Sea shelf center (Chebotareva, 1977; Grosvald, 1983). The observations in this paper about the glacial erosional forms and glacial marginal deposits and their orientation document the influence of the latter three. The main glacial stream or common glacial front of the ice sheet moved from the north and northwest until it joined with the Timan dome and overrode it. It can be assumed that the upper parts of the dome ice were deformed and carried along with general ice sheet movement (Fig. 7 G.2.). But the lowest part of the ice, frozen to the glacier bed, was immobile under the deforming glacial ice. In contrast to the north Timan plateau, its periphery was subjected to intensive glacial erosion and glacial pressure, accompanied by the accumulation of basal till.

During the retreat stages of the last late Pleistocene–Holocene glaciation (Valdai ice sheet) the residual glacial dome of the north Timan served as the ice divide for the glacial lobes. Along the contacts between the lobes and the dome a great glaciotectonic tension zone with crevasses formed (Fig. 7 G.3.). It was completed by the Marhida phase of the last glaciation, when the residual dome of the north Timan plateau separated (see Fig. 6) from the ice sheet. After that, small oscillations of its west margin followed, but the dome had disintegrated into two massifs of dead ice along the Sula depression. Its further disintegration triggered accumulation of meltwater deposits in crevasses and debris accumulation on the dome ice. The final phase of glacial dome stagnation was associated with the inversion of dead ice landforms

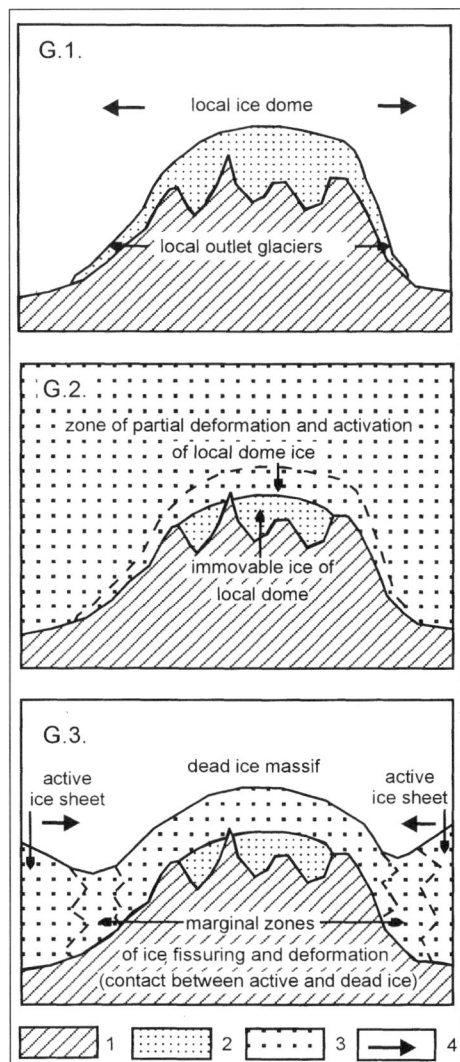

Figure 7. Hypothetical evolution of the north Timan region glaciation shown in schematic cross sections. 1, north Timan plateau (geological substrate); 2, ice of the local dome; 3, ice of the main sheet; 4, directions of ice flow. G.1., G.2., G.3., principal phases of the north Timan glaciation: G.1., phase of the local plateau glaciation at the beginning of ice age; G.2., maximum phase of ice sheet expansion; G.3., phase of active ice sheet retreat.

(eskers, kames), redeposition of tills from the melting dome ice surface of the plateau top as ablation moraines, accompanied by local systems of melt water runoff and glacial lake formation. As a whole, outlines of the marginal domes (see Fig. 6) are similar to those shown by Lavrov (1977).

At the same time, the active retreat of the Valdai ice sheet continued. The Marhida phase had about three oscillations around the north Timan plateau, resulting in marginal landforms and deposits. Further retreat of the ice sheet was interrupted by a new phase of readvance, the Velt phase, which included three more oscillations.

Other centers of local glaciation similar to that of the north Timan region could have originated around the periphery of ice sheets in high latitudes causing local and regional changes in the glaciation dynamics and geological activity. This hypothesis should be taken into account in the course of paleoglaciological reconstructions.

ACKNOWLEDGMENTS

The author thanks Y. Golubev, S. Koldaev, M. Palanski, and B. Mackarov for the valuable comments and the assistance in realization of the field investigations.

REFERENCES CITED

Arslanov, K. A., Lavrov, A. S., and Nikiforova, L. D., 1981, On stratigraphy, geochronology and climatic changes during the middle and late Pleistocene in the north-eastern region of the Russian Plain, in Velichko, A. A., and Faustova, M. A., ed., Pleistocene glaciations at the East European Plain: Moscow, Nauka, p. 37–52.

Chebotareva, N. S., editor, 1977, The structure and dynamics of the last ice sheet of Europe: Moscow, Nauka, 144 p.

Chernov, G. A., 1947, New data about Quaternary history of the Bolshezemelskaya tundra area: Bulletin of the Commission on the Quaternary Study, no. 9, p. 61–74.

Denton, G. H., and Hughes, T. J., 1981, The Arctic Ice Sheet: an outrageous hypothesis, in Denton, G. H., and Hughes, T. J., ed., The last great ice sheets: New York, Wiley-Interscience, p. 437–467.

Evzerov, V., and Koshechkin, B., 1981, Glaciation flow stages in the western part of the Kola Peninsula, in Glacial deposits and glacial history of Eastern Fennoskandia: Apatity, p. 38–47.

Gornostai, B. A., 1990, Glaciodislocations adjacent to the Timan region: Bulletin of the Commission on the Quaternary Study, no. 59, p. 152–155.

Grigoriev, A. A., 1924, Geology and relief of the Bolshezemelskaya tundra area and related problems: Transactions of the Northern Scientific-Producers' Expedition of USSR SSNE (Supreme Soviet of the National Economy), v. 43, no. 22, 63 p.

Grosvald, M. G., 1983, Ice sheets of the continental shelves: Moscow, Nauka, 216 p.

Koryakin, V. S., 1988, Arctic glaciers: Moscow, Nauka, 160 p.

Lavrov, A. S., 1977, The Barents Sea—Pechora ice stream, in Chebotareva, N. S., ed., The structure and dynamics of the last ice sheet of Europe: Moscow, Nauka, p. 89–95.

Lavrushin, Yu. A., 1976, Structure and formation of the basal tills of continental glaciations: Moscow, Nauka, Transactions of the Geological Institute of the Academy of Sciences, USSR, no. 288, 237 p.

Velitchko, A. A. and Hodakov, V. G., 1981, On the base of the specific data and paleoglaciological modeling: Transactions of the Institute of Geology and Geophysics, Siberian Branch of Academy of Sciences of USSR, no. 494, p. 24–34.

MANUSCRIPT ACCEPTED BY THE SOCIETY OCTOBER 8, 1998

Discussion of the observed asymmetrical distribution of landforms of the southeastern sector of the Scandinavian Ice Sheet

Reet Karukäpp
Institute of Geology, EE-00001 Tallinn, Estonia

ABSTRACT

Reconstruction of the flow pattern in the southeastern sector of the Scandinavian Ice Sheet indicates that the westerly oriented parts of glacial lobes were more active than other parts of the glacier. This is reflected in the asymmetry of eroded bedrock forms at the margins of glacial lobe depressions and in the form of drumlins within lobe depressions. For discussion it is proposed that the preferred flow of the glacier towards the right during its movement was influenced by the Coriolis force.

INTRODUCTION

The flow pattern of the Scandinavian Ice Sheet during the Weichselian Glaciation is reflected in the glacier-bed topography most notably by alternation of glacial lobe depressions and interlobate formations (Fig. 1). Investigations in the southeastern sector of the Scandinavian Ice Sheet have provided a great amount of data for detailed reconstruction of local glacier flow patterns. The reconstruction of the Gotiglacial stage of deglaciation (Fig. 1) is based on the author's investigations and published data (Āboltiņš et al., 1976; Aseev, 1974; Znamenskaya et al., 1977; Ekman et al., 1981; Isachenkov, 1981; Punkari, 1993; Dreimanis and Zelčs, 1995). Deglaciation is time transgressive: the age of the distal part is older than the proximal part and reflects everywhere the last active stage of the glacier lobe (tongue) in the area of its location. The flow lines pattern (Fig. 1) shows evidence of flow to the right from theoretical lines starting from the glaciation center in the Gulf of Botnia.

THE GOTIGLACIAL TIME OF DEGLACIATION

Gotiglacial time of deglaciation (the term first used by De Geer, 1940) marks the intermediate stage between the maximum extent of the glacier (Daniglacial) and the final deglaciation in early Holocene (Finiglacial). It is characterized by active movement and rapid changes in the flow patterns of the glacier streams and lobes. The pattern of alternated lobe depressions and interlobate formations (as plinth type of uplands of glacial erosion and insular accumulative heights) was typical for the Gotiglacial.

Against the background of the flat surface topography with small absolute and relative (generally less than 100 m) heights of the East European Plain, the glacial accumulative islandlike (insular) heights have a remarkable position. They cover thousands of square kilometers and form submeridional systems of 2–4 heights in each. They have typical hummocky dissected topography, are separated from each other by lowlands, and project 80–200 m above their surroundings. Usually the heights have a Quaternary cover 100–150 m thick, composed of several till beds with related intermorainic deposits. The heights have developed during several glaciations in the course of their subglacial, englacial, marginal formation, and stagnant ice stages (Āboltiņš et al., 1976).

In general, the bifurcation of the glacial streams took place in their right side and towards the west. For example (Fig. 1B), the Eastern Latvian (E) and Võru-Hargla (V) lobes branched off from the Peipsi glacial stream (P) and formed distinct, radially oriented glacial relief (Karukäpp, 1975). The western wing of the Central Lativan glacier lobe (C) deviated from its main direction by as much as 140° and moved around the Eastern Kursa Upland towards another lobe from the depression of the Baltic Sea (Dreimanis and Zelčs, 1995). About the same time (13,000–13,500 BP), the western wing of the Baltic ice stream (Fig. 1A) formed a westward-moving ice lobe (Low Baltic readvance) in southern Sweden and on the Danish Islands (Ringberg, 1988, 1989).

BEDROCK TOPOGRAPHY

The level of the bedrock surface of the area (Fig. 2) varies from the present sea level to 100–150 m a.s.l. (typically –50 to

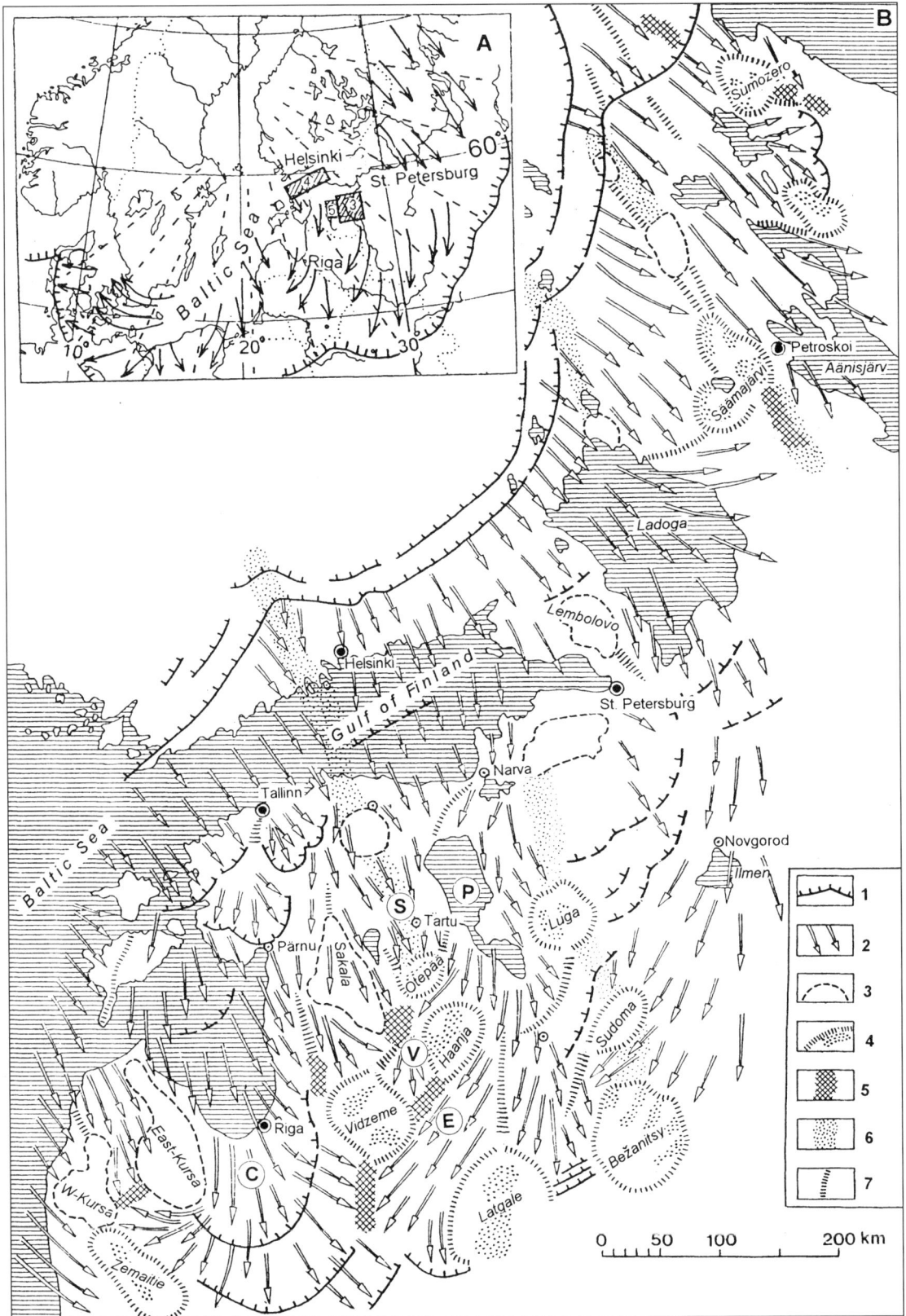

Figure 1. A, Sketch of dynamics of southern and southeastern parts of the late Weischelian glacier based on the published data. Dashed lines mark radial orientation from the center of the glaciation, arrows show real direction of ice lobe movement. Cross-hatched areas refer to locations of Figures 3–5. B, Glacial dynamics and morphogenesis in the southeastern sector of the Scandinavian glaciation in Gotiglacial time. Compiled by R. Karukäpp using published data by O. Āboltiņš, J. Straume, I. Ekman, M. Punkari, and V. Isachenkov. 1, glacier margin; 2, ice flow direction in the final stage of its activity; 3, plinth-type bedrock upland of glacier erosion; 4, accumulative insular heights; 5, interlobate complex of landforms; 6, ice shed between glacier flows; 7, local ice shed. The letters in circles mark the lobe depressions mentioned in the text: (E), Eastern Latvian; (V), Võru-Hargla; (P), Peipsi; (C), Central Latvian; and (S), Saadjärve Drumlin Field.

100 m). The lithological composition of bedrock is different from area to area: those most resistant to glacial erosion are limestone and dolomite; those most easily eroded are Paleozoic or Mesozoic soft sandstones and shales. The difference of the composition is slightly reflected in the bedrock topography. Every lobe depression discussed below (Peipsi Lake and Central Latvian; Fig. 1) is characterized by homogeneous bedrock conditions.

The peripheral cover of the continental ice sheet was considered to have been less than 500 m thick (Aseev, 1974) and, in all likelihood, the rate of its movement varied from tens to several hundred meters per year. With the thinning of the glacier ice the subglacial surface came to exert an ever-growing effect on the direction of the ice movement. Nevertheless, the detailed investigations in Skåne (Lidmark-Bergström et al., 1991) proved that even big differences (about 20 m) in altitudes of the bedrock, oriented perpendicular to the ice movement, did not prevent the ice lobe (Low Baltic Ice) from declining to the right by 140°. In analyzing the morphology of modern bedrock relief, the preconditions for glacier movement cannot be often distinguished from the results of the glacier movement.

LOBE DEPRESSIONS MORPHOLOGY

Analysis of the morphology of the lobe depressions in the southeastern sector of the Scandinavian Ice Sheet shows that declination to the right from the geometrical (theoretical) radial lines (Fig. 1A) of the ice sheet was typical of the ice streams and lobes in that area. This right declination is clearly revealed by the northeast/southwest orientation of the local ice divide zones (interlobate formations; Fig. 1). If the energy of the glacier's movement had been equal radially in all directions, the interlobate formations would have mostly northwest/southeast orientations.

If the glacial streams and lobes tended to deviate to the right, the effect of the glacier (erosion, transport, accumulation) on the lobe depression would have been asymmetric. Let us consider an example of the Peipsi depression (Fig. 3). As a macroform of glacial relief the Peipsi depression was relatively small but there is more or less reliable coring data available on its bedrock relief obtained from the land (Tavast and Raukas, 1982) and on the basis of seismic profiling data from the western part of Lake Peipsi (Noormets et al., 1998). The north-south-trending lobe depression in Middle and Upper Devonian sandstones and siltstones measures 50–60 km in width and 30–40 m in depth. Three cross sections across the depression clearly demonstrate slope asymmetry (Fig. 3b), which suggests that erosion by glacial streams and lobes was more intensive on the west side.

Analysis of the topography of the thoroughly studied Pandivere and Ahtme bedrock uplands of Ordovician and Silurian limestones and dolomites (Tavast and Raukas, 1982) shows that their east slopes are steeper than the west slopes, which also confirms the above-described asymmetry. The same asymmetry is evident in the cross section of the Riga lobe depression (C, Fig. 1B) cut into the Devonian sandstones.

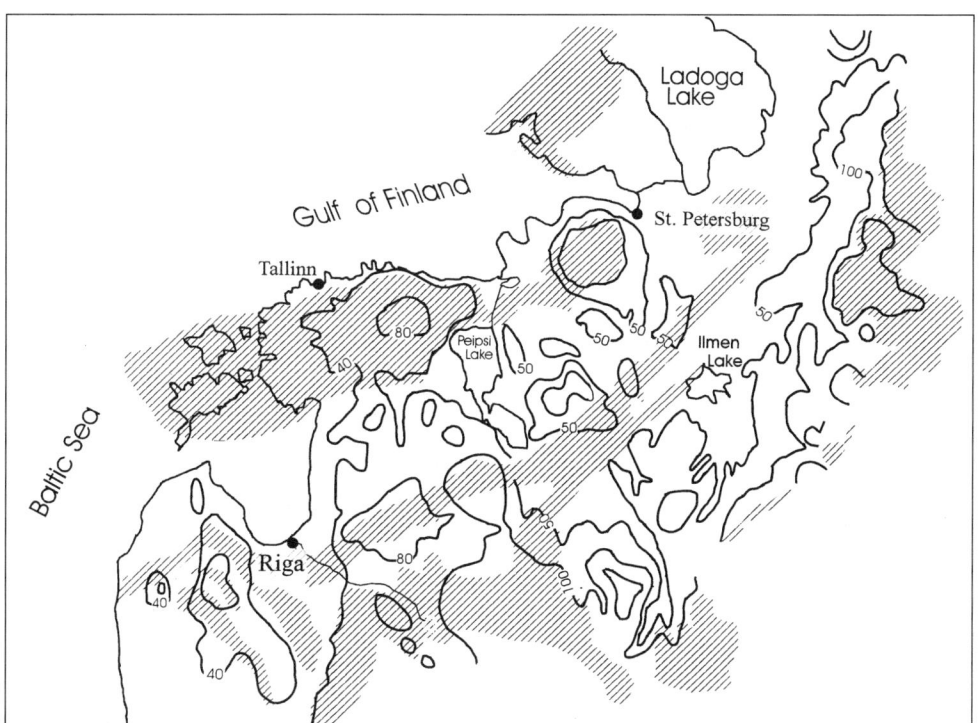

Figure 2. Generalized bedrock topography. Shaded areas mark the domination of the varieties of the bedrock relatively resistant to the glacial erosion. Compiled by the author using published data (Āboltiņš, et al., 1976; Ekman et al., 1981; Isachenkov, 1981; Tavast and Raukas, 1982).

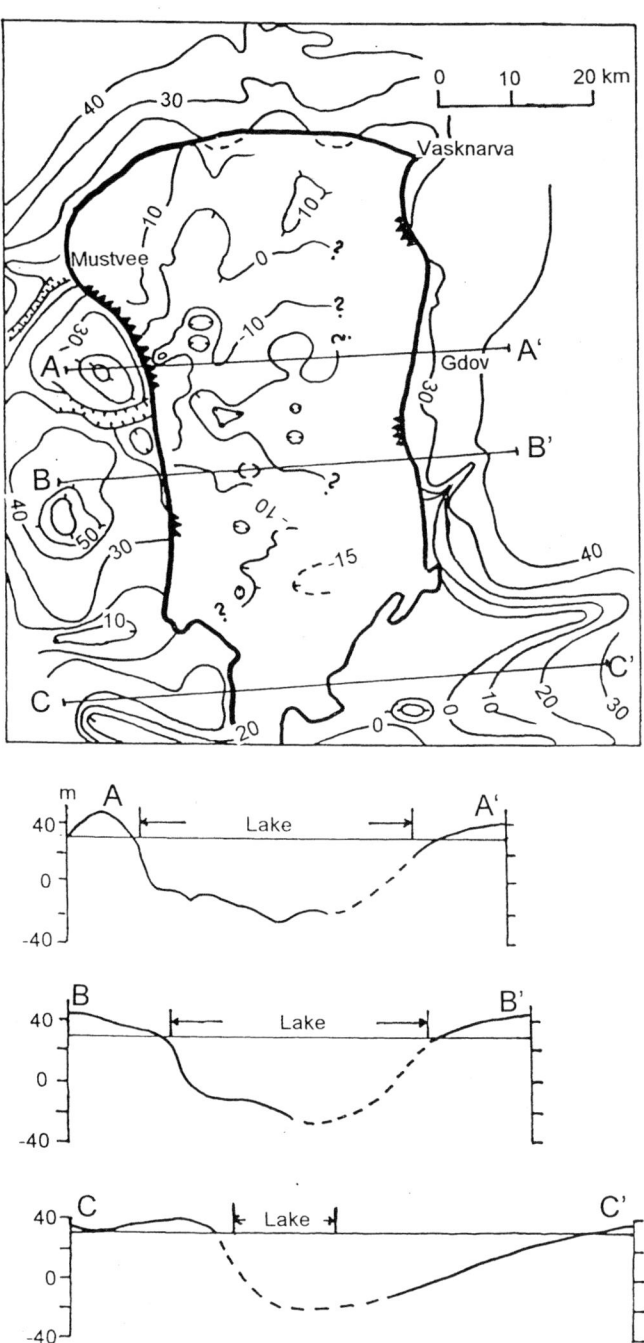

Figure 3. Bedrock relief of Peipsi Lake depression. Compiled by R. Karukäpp using data by Tavast and Raukas (1982) and seismic profiling data (Noormets et al., 1998). Isohypses mark height from sea level. Dip of bedrock structure is from north to south. For location see Figure 1A.

The given examples demonstrate the asymmetry of the dynamics of the glacial stream (lobe) depression for the whole macroformation. Let us analyze now the data on the glacier movement within the glacial stream (lobe) itself, choosing for that purpose the depression of the Gulf of Finland on the east margin of the Baltic glacial stream (Fig. 4). The glacier flow direction is revealed by large drumlinlike forms, as much as 90 m high and 10 km long. Corings on Pragli Island and indirect (seismic profiling) data concerning the inner structure allow these forms to be classified as predominantly erosional. The cores of the drumlins contain lower Paleozoic sedimentary bedrock, sediments of earlier glaciations (till), and interglacial deposits (Kajak et al., 1976; Karukäpp and Vassiljev, 1992). In the area under consideration, the northeastern slopes of these huge drumlins are somewhat steeper than the southwestern slopes (Fig. 4).

For comparison, the topography of a classical drumlin field from eastern Estonia (S, Fig. 1B)) was analyzed. These drumlins are well-developed equilibrium forms between glacial erosion and accumulation. The slopes of the drumlins of the central part of the Saadjärve drumlin field were measured (Fig. 5A). As the depressions between drumlins are occupied by glaciolacustrine deposits, lakes, or mires, only upper parts of the slopes are available for investigations. Nevertheless, the tendency of asymmetry of the slopes is traceable at the eastern Estonia example but not so evident as in the Gulf of Finland: the southwest slopes of the interdrumlin depressions are slightly steeper than northwest ones (Fig. 5B). This seems to support the statement that any moving portion of the glacier was more active to the right of its axis.

DISCUSSION AND CONCLUSIONS

The curving of the ice lobes to the right in the southeastern section of the Scandinavian glaciation is evident. Ehlers (1990, p. 81), who described the earlier (Pommeranian) stage of deglaciation in northwestern Germany, called attention to the tendency of clockwise rotation of the flow lines, explaining the phenomenon with increasing control by the shape of the Baltic Sea depression. Matoshko and Chugunny (1993, p. 145) declared with conviction that glaciers of the Dnepr Glaciation (middle Pleistocene) tended to curve to the right due to Coriolis force effect.

The author of this paper also proposed for discussion (Karukäpp, 1996) the idea that the declination (curving) of the glacier towards the right during its movement was due to the Coriolis force.

The Coriolis force is an apparent force acting on a moving object due to the rotation of the coordinate system in which the object's velocity is measured. The force is directed perpendicular to the velocity: to the right in the Northern Hemisphere, and to the left in the Southern Hemisphere. The Coriolis effect must be considered in a great variety of phenomena in which motion over the surface of the Earth is involved. It includes, in addition to rivers, air movement and streams in oceans. Gaspard Gustave de Coriolis (1792–1843), the French civil engineer, did not apply his theory to any natural process.

Karl Ernst von Baer (1792–1876), the founder of embryology, was the first to publish his observations on the asymmetry of the river erosion in Russia in 1854 (see Müürsepp, 1984). At that time he had not the faintest notion of Coriolis' theory. Jacques Babinet (1794–1872) was the first to formulate (1859) correctly the regularity

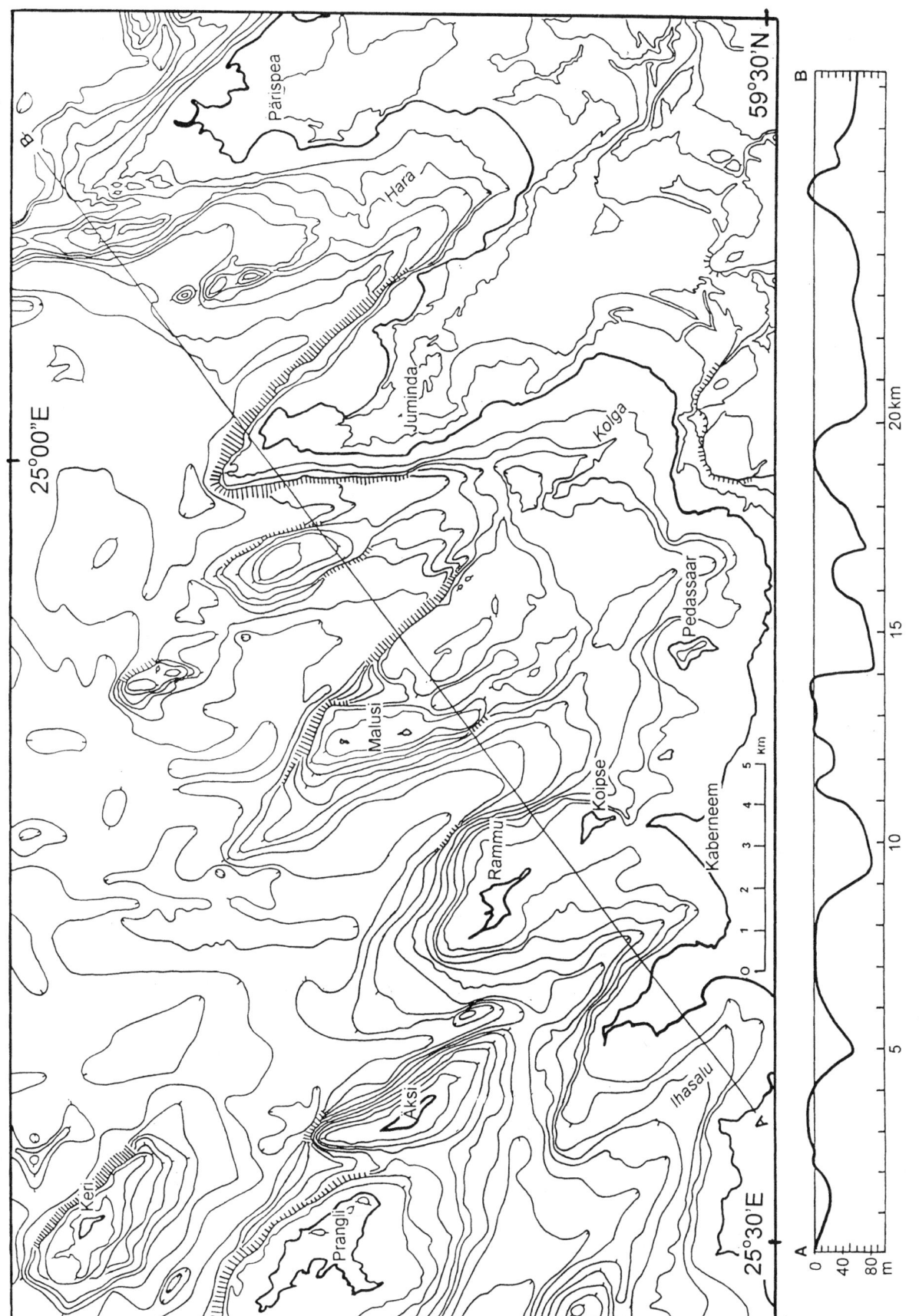

Figure 4. Example of the bottom relief of the Gulf of Finland compiled by R. Karukäpp and J. Vassiljev. For location see Figure 1A. The profile A–B is oriented perpendicular to the main direction of ice flow.

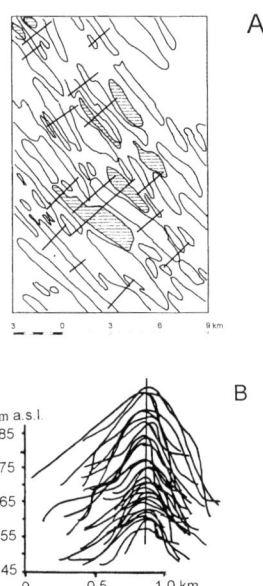

Figure 5. Slope inclination of the drumlins in the central part of the Saadjärve drumlin field. For location see Figure 1A. A, location of the cross sections; B, comparison of the slope profiles.

of river erosion, which was based on mathematical calculations and Coriolis' theory (see Müürsepp, 1984). In the course of long-lasting discussions, the regularity was formulated as the Baer-Babinet law.

The idea that the Coriolis effect impacted glacier flow was criticized by A. Salamatin from Kazan University (unpublished discussion) who asserted that the speed of ice movement was too slow for the Coriolis force to have a considerable effect.

The glacier streams and lobes of the Gotiglacial age of deglaciation were active and rapidly moving. The basal temperatures were probably balancing around the melting point, and subglacial water might be an important agent in the glacier dynamics. As the Coriolis force affects any movement, it must have influenced subglacial meltwater flow and thus, in turn, the dynamics of the glacial lobes. The results of this influence can be observed in the asymmetry of glacial landforms. What other factors possibly influenced this process will be the subject of future investigations.

ACKNOWLEDGMENTS

The author is grateful to H. Kukk and A. Noor for linguistic help and to R. Vaher and A. Molodkov for drawings. The valuable comments and suggestions were given by reviewers. The study was partly supported by the Estonian Science Foundation (Grant No. 2058).

REFERENCES CITED

Āboltiņš, O. P., Straume, J. A., and Juškevics, V., 1976, Rel'ef i osnovnoye etapy lednikovogo morfogeneza Aluksnenskoj vozvyshennosti (Topography and the main stages of glacial morphogenesis of the Aluksne Heights), in Danilans, I., ed., Voprosy chetvertichnoj geologii, Volume 9: Riga, Zinatne, p. 79–89 (in Russian).

Aseev, A. A., 1974, Drevnie materikovye oledeneija Evropy (Ancient continental glaciations): Moscow, Nauka, 317 p. (in Russian).

De Geer, G., 1940, Geochronologia Suecica Principles: Kungliska Svenska Vetenskapsakademiens Handlingar, ser. 3, v. 18, no. 6, 360 p.

Dreimanis, A., and Zelčs, V., 1995, Pleistocene stratigraphy of Latvia, in Ehlers, J., Kozarski, S., and Gibbard, P., eds., Glacial deposits in North-east Europe: Rotterdam, A. A. Balkema, p. 105–114.

Ehlers, J., 1990, Reconstructing the dynamics of the North-west European Pleistocene ice sheets: Quaternary Science Review, v. 9, p. 71–83.

Ekman, I. M., Iljin, V. A., and Lukashov, A. D., 1981, Degradation of the Late Glacial ice sheet on the territory of Karelian ASSR, in Gorbunov, G., Koshechkin, B., and Listisyn, A., eds., Glacial deposits and glacial history in eastern Fennoskandia: Apatity, Academy of Science USSR, Kola Branch, p. 103–117.

Isachenkov, V. A., 1981, Proishozhdenie krupnykh form rel'efa Severo-Zapada Russkoy ravniny (Genesis of the large scale topography in the northwestern part of the Russian plain): Geomorfologiya, v. 4., p. 14–73 (in Russian).

Kajak, K., Kessel, H., Liivrand, E., Pirrus, R., Raukas, A., and Sarv, A., 1976, Mestnaya rabochaya stratigraficheskaya skhema chetvertichnykh otlozhenij Estonii (The local stratigraphic scheme for the Quaternary deposits of Estonia), in Stratigrafiya chetvertichnykh otlozhenij Pribaltiki: Vilnius, p. 4–52 (in Russian).

Karukäpp, R., 1975, Specific features of Pleistocene relief formation of Karula Upland: Eesti NSV Teaduste Akadeemia Toimetised. Keemia. Geloogia, v. 24, no. 2, p. 145–151 (in Russian).

Karukäpp, R., 1996, Baer-Babinet law in glacial dynamics: Proceedings Estonian Academy of Science. Geology, v. 45, no. 4, p. 216–224.

Karukäpp, R., and Vassiljev, J., 1992, Geomorphology of the gulf's floor, in Raukas, A., and Hyvarien, H., eds., The geology of the Gulf of Finland: Tallinn, Valgus, p. 72–89 (in Russian).

Lidmark-Bergström, K., Elvhage, C., and Ringberg, B., 1991, Landforms in Skåne, South Sweden. Preglacial and glacial landforms analysed from two relief maps: Geografiska Annaler, v. 73(A), no. 2, p. 61–91.

Matoshko, A. V., and Chugunnyi, J. G., 1993, Dneprovskoye oledeneniye territorii Ukrainy (The Dnieper glaciation in the Ukraine), in Shelkoplyas, V. N., ed.: Kiev, Naukova Dumka, 190 p. (in Russian).

Müürsepp, P., 1984, Miks jõed uuristavad paremat kallast? (Why the rivers erode the right-side bank?): Tallinn, Valgus, 40 p. (in Estonian).

Noormets, R., Floden, T., Hang, T., Miidel, A., and Bjerkeus, M., 1998, Bedrock topography in the lake Peipsi depression: results of a seismic reflection survey: Geologiska Föreningens i Stockholm Förhandlingar, v. 120, p. 47–52.

Punkari, M., 1993, Modelling of the dynamics of the Scandinavian Ice Sheet using remote sensing and GIS methods, in Aber, J., ed., Glaciotectonics and mapping glacial deposits. Proceedings of the INQUA Commission on Formation and Properties of Glacial Deposits: University of Regina, Canadian Plains Research Center, p. 232–250.

Ringberg, B., 1988, Late Weichselian geology of southernmost Sweden: Boreas, p. 17, 243–263.

Ringberg, B., 1989, Upper late Weichselian lithostratigraphy in western Skåne, southernmost Sweden: Geologiska Föreninens i Stockholm Förhandlingar, v. 111, no. 4, p. 319–337.

Tavast, E., and Raukas, A., 1982, Bedrock relief of Estonia: Tallinn, Valgus, 193 p. (in Russian).

Znamenskya, O. M., Faustova, M. A., and Chebotareva, N. S., 1977, Ladozhskij lednikovyi potok (The Ladoga glacial lobe), in Chebotareva, N. S., ed., Struktura I dinamika poslednego lednikovogo pokrova Evropy: Moscow, Nauka, p. 54–66 (in Russian).

MANUSCRIPT ACCEPTED BY THE SOCIETY OCTOBER 8, 1998

Role of climate oscillations in determining ice-margin position: Hypothesis, examples, and implications

Thomas V. Lowell, Rosalyn K. Hayward
Department of Geology, University of Cincinnati, Cincinnati, Ohio 45221-0013
George H. Denton
Institute for Quaternary Studies, University of Maine, Orono, Maine 04469

ABSTRACT

A chronological comparison between a paleoclimate proxy and ice-sheet expansion, as recorded by end moraines, can elucidate the interplay of climate and ice-sheet behavior. Geologic evidence from Lake Gribben, Michigan, Two Creeks, Wisconsin, Des Moines, Iowa, and Chillocothe, Cincinnati, Cuba, and Todd Fork, Ohio, record ice-sheet expansions at 10,030 ± 20; 11,850 ± 100; 13,790 ± 23; 17,490 ± 230; 19,590 ± 35; 20,360 ± 84 and 23,200 ± 85 ^{14}C yr B.P. All of these, within the resolution of time-scale conversions, occur after a prolonged cooling interval and just before or at a rapid warming as recorded in the Greenland ice cores. These expansions occur too rapidly after climate change to have been propagated from the accumulation area. Similarly, the chronological comparison suggests that changes in subglacial rheology, which vary on longer times scales of 6,000 years and are unlikely to be in phase with climate, are not responsible for the observed pattern. Thus variations in ablation in the ice-margin zone emerge as a primary mechanism to force ice-sheet expansions and contractions on millennial timescales. Glaciological models may be able to replicate complex ice-sheet behavior on millennial time-scales by simply varying ablation.

INTRODUCTION

The relationship between climate change and ice sheet behavior remains under debate. The behavior of an ice sheet depends on a combination of accumulation, ablation and bed conditions. However their relative contribution to ice-sheet behavior on a millennial time-scale is not known. The role of bed conditions has been extensively examined to reconstruct the behavior and geometry of the Laurentide Ice Sheet (e.g., MacAyeal, 1993; Clark et al., 1996). MacAyeal (1993) provides a creative way to change the basal water content in the central core of the Laurentide Sheet, but on time scales in excess of 6,000 years, much too long to explain observed fluctuations. Moreover, these studies assume that (1) climate conditions, and therefore accumulation and ablation, were stable, and (2) ice sheets reached steady state. Recent reports demonstrating a highly variable glacial climate (Johnsen et al., 1992; Grootes et al., 1993) show the former to be false. The latter is in doubt because modeling experiments indicate that ice sheets require about 10,000 years to establish equilibrium (Oerlemans and van der Veen, 1984) and ice cores show that the climate fluctuated much faster on millennial frequencies. The southern margin of the largest Northern Hemisphere paleo-ice sheet, the Laurentide, fluctuated on similar times scales (Mickelson, et al., 1984). The similarity of time scales suggests a linkage of climate and ice-sheet dynamics that can be explored.

The role of accumulation and ablation can be assessed using stratigraphic comparison. The quasi-regular beat of climate change during the last glaciation (Johnsen et al., 1992; Grootes et al., 1993) would start accumulation variations in different sectors of the ice sheet at the same time. However the lag time needed for these affects to be felt at the margin of the ice sheet would vary depending on individual flow-line lengths. Thus we would anticipate little similarity between the timing of climate changes and the timing of ice margin changes. In contrast ablation, starting at zero at the equilibrium line and increasing to a maximum at the glacier end, has the most influence directly at the ice margin. The phasing of ablation changes and ice-sheet response should be very small.

Our objective is to reintroduce the importance of ablation as a primary control on ice-sheet behavior especially at the millennial time-scale. We illustrate the effectiveness of this process by comparing the proxy climate record from the Greenland ice core to the chronology of well-dated moraines along the southern margin of the Laurentide Ice Sheet. Many moraines formed when climate changed from a gradual cooling interval to an abrupt warming. This phasing requires a direct, rapid link between climate and ice-margin position; a link only changes in ablation rate can provide. Thus we suggest that ablation may be the primary control on ice margin changes on millennial and perhaps shorter time-scales.

HYPOTHESIS

Ablation controls

We suggest that ablation rate, a function primarily of mean summer temperature in the ablation zone of terrestrial ice-sheet margins, forced millennial duration changes in ice-margin position. On an annual cycle, lower winter temperatures reduce melting to less than the mass influx from flow and allow the glacier margin to advance. Conversely increased summer temperatures melt more than the mass influx forcing the ice margin to retreat. Climate change on longer times scales (decades, centuries, and millennia) should likewise force the ice margin on larger spatial scales.

Some caveats

To examine this hypothesis, we make three assumptions. First, we make the assumption that the GISP2 Greenland ice core (Grootes et al., 1993) is a proxy for air temperature change near the Laurentide Ice Sheet margins. Because similar signals are being reported worldwide (e.g., Bond et al., 1993; Lowell et al., 1995), we believe the overall pattern of change also affected the southern margins of the Laurentide Ice Sheet. We consider the Greenland ice core not so much for the absolute magnitude of these climate changes, but rather for the sign and timing of these changes.

Second, we adopt the Bard et al. (1990) and Stuiver and Reimer (1993) conversion of radiocarbon (^{14}C) ages to calendar (cal) years. Acknowledging that this conversion will be improved in the future as we learn more about shorter-term changes in ^{14}C production during the last glacial interval, we use it here to make a preliminary comparison of glacial expansions with Greenland ice core results. To accomplish this, we consider multiple age analysis from several sites (Table 1) and then average them as outlined in Ward and Wilson (1978). This mean and estimated error is then converted to calendar years following Stuiver and Reimer (1993) or Bard et al. (1990; Table 2). Ideally it would be desirable to assign confidence limits to this conversion for events beyond tree-ring calibration, but such a task lies outside the scope of this paper. We make these stratigraphic comparisons to establish a phasing relationship of climate change and ice sheet behavior. As long as the relative phasing remains valid, the conclusions drawn about the glaciological processes remain valid.

Our third assumption is that the Little Ice Age provides a model for past glacial events. The Little Ice Age is so named because most glaciers worldwide expanded for several centuries during a climatic cold interval and then withdrew uniformly within the last 150 years during a climatic warming leaving spectacular moraines (Grove, 1988).

The Little Ice Age is apparently the last of several similar millennial scale fluctuations through the Holocene and last glacial cycle (Bond et al., 1997). For the Holocene, Denton and Karlén (1973) demonstrated a close correspondence between the expansion and recession of alpine glacier margins from the St. Elias Mountains with those of Swedish Lapland implying a common driving force across the Northern Hemisphere. Intervals during which glaciers expanded lasted about 900 years whereas intervals during which glaciers contracted lasted as much as 1,750 years.

Although details vary, the pattern for the last of these, the Little Ice Age, is a general expansion of glaciers for the last 1,000 yr, terminating about 150 years ago. Grove (1988) considers the consistent, worldwide pattern of glacier advance and retreat to reflect a climate-driven event. However shorter term fluctuations are superposed on this longer cycle. Denton and Karlén (1977) report on 11 different glaciers in the White River Valley and Skolai Pass in Alaska and the Yukon Territory that have lichen on adjacent moraines that cluster at >85, >60, 45, 30, 15, and 0–10 mm maximum diameters. The 0–10 mm set are interpreted to represent a series of expansions during the last 100 years. Thus these individual moraine representing distinct events, comprise a belt that represents the Little Ice Age.

That moraines relate to the peaks of these advances can be illustrated from direct observations. Messerli et al. (1978) used historical records, maps, and drawings to reconstruct the position of the Unterer Grindelwaldgletscher in Switzerland from 1600 A.D. onward. Four moraines formed about 1600, 1780, 1820, and 1860 A.D. in a narrow zone about 1,600–1,800 m beyond its 1970 position. Another two moraines formed 800 m beyond the 1970 position in 1880 and 1930. Except for 1600 A.D., when the glacier held that position for approximately 60 years, all of these moraines formed right after a small readvance and before a retreat of the ice margin. These decade-long oscillations show the same relationship to climate as the yearly cycle; glacier margins extend when temperature drops and ablation is reduced and retreat when it rises and ablation increases (Messerli et al., 1978).

One final point; Little Ice Age moraines ring most terrestrial glaciers, except in Antarctica. Grove (1988) reports typical retreat magnitudes of 1–2 km. This applies to glaciers of all sizes and in many different settings including the Greenland Ice Sheet. Since the common thread across these contrasting envi-

TABLE 1: RADIOCARBON AGES DEFINING GLACIAL EXPANSIONS

Site	Age	Laboratory Number	Description	Reference
Lake Gribben	9,895±55	A-7878	wood buried by outwash from the Grand Marais I Moraine	Lowell, unpublished
	9,910±55	A-7876	"	"
	9,965±55	A-7877	"	"
	10,040±55	A-7879	"	"
	10,040±65	A-7881	"	"
	10,050±55	A-7883	"	"
	10,075+95/-90	A-7880	"	"
	10,155±65	A-7882	"	"
	10,200±55	A-7875	"	"
Des Moines				
Saylorville Spillway	13,420±75	A-8122	wood from alluvial below till	This report
	13,460±90	A-8118	"	"
	13,605±75	A-8115	"	"
	13,615±80	A-8121	"	"
	13,750±75	A-7957	"	"
	13,760±80	A-8116	"	"
	13,885±70	A-7957.1	replicate of A-7957	"
	13,965±75	A-8117	wood from alluvial below till	"
	14,010±75	A-8120	"	"
	14,038±70	A-8119	"	"
	14,065±115	A-7958	"	"
	14,190±135	A-7958.1	replicate of A-7958	"
Chillocothe				
Bier's Run 2	18,120±180	UGa-6704	log with bark	Lowell, unpublished
	18,340±290	UGa-6703	log with branches	"
	18,520±245	UGa-6702	log with branches	"
	18,520±200	UGa-6705	wood log	"
Dry Run	17,590±210	UGa-6709	wood log	"
	18,240±180	UGa-6711	basal 2 cm of silt	"
	18,490±280	UGa-6710	upper 8 cm of silt	"
	18,520±280	UGa-6708	wood fragment	"
	18,750±260	UGa-6706	wood log	"
	18,800±290	UGa-6707	macerated wood	"
North Fork	17,490±230	UGa-6698	single log	"
Cincinnati				
Beckett Road	19,620±150	ISGS-2646	wood in basal till	Lowell and Brockman, 1994
	19,670±230	ISGS-2652	"	"
	19,780±170	ISGS-2642	"	"
	19,800±160	ISGS-2645	"	"

Dimmick Road	19,830±190	ISGS-2643	"	"
	19,410±140	ISGS-2621	wood at basal contact of till	"
	19,450±190	ISGS-2620	"	"
	19,500±270	ISGS-2624	"	"
	19,520±180	ISGS-2619	"	"
	19,640±200	ISGS-2625	"	"
	19,730±140	ISGS-2618	"	"
Creek Road	20,110±170	ISGS-2834	organic mat at top of alluvium below till	Lowell, unpublished
	20,290±160	ISGS-2828	"	"
	22,730±460	ISGS-2837	organic mat at base of alluvium above till	"
	22,800±330	ISGS-2829	"	"
Sharonville	19,200±140	PITT-0508	*Larix* stump below till just inside the Hartwell Moraine	Lowell et al., 1990
	19,310±170	PITT-0506	"	"
	19,690±150	PITT-0509	"	"
	19,960±170	PITT-0227	"	"
	20,200±140	PITT-0507	"	"
Princeton High School	19,390±180	ISGS-2614	wood enclosed in till	Lowell and Brockman, 1994
	19,480±190	ISGS-2616	"	"
	19,500±115	PITT-0229	"	"
	19,610±180	ISGS-2617	"	"
	19,690±180	ISGS-2615	"	"
	20,210±210	PITT-0232	*Picea* enclosed in till	"
	19,135±160	Beta-34385	organic debris on top of silt	N. Miller, unpublished
Oxford	20,030±140	PITT-0625	stump rooted below till, same stump as PITT-0624	Ekburg, et al., 1993
	20,620±180	PITT-0624	stump rooted below till, same sample as PITT-0625	"
	20,820±210	ISGS-2757	stump slightly transported, replicate of ISGS-2762	Lowell and Brockman, 1994
	20,800±250	ISGS-2758	stump rooted below till, replicate of PITT-0765, PITT-0764, ISGS-2763	"
	20,800±210	ISGS-2760	stump rooted below till, replicate of PITT-0624, PITT-0625, ISGS-2761	"
	20,800±200	ISGS-2761	stump rooted below till, replicate of PITT-0624, PITT-0625, ISGS-2760	"
	20,770±210	ISGS-2763	stump rooted below till, replicate of PITT-0765, PITT-0764, ISGS-2758	"
	20,850±200	ISGS-2762	stump slightly transported, replicate of ISGS-2757	"
	21,240±150	PITT-0765	stump below till, same stump as PITT-0764	Ekburg, et al., 1993
	21,390±200	PITT-0764	stump below till, same stump as PITT-0765	"

Cuba Gully	20,240±130	ISGS-3233	Picea roots and stump from the top of organic-rich silt below glaciofluvial sediments and subglacial till. Overlies alluvium and calcareous till.	Dubois, 1996
	20,420±120	ISGS-3232	"	"
	20,550±280	ISGS-3234	"	"
Todd Fork	23,160±180	ISGS-2935	wood in organic silt-till contact	Debois, 1996
	23,250±210	ISGS-2936	"	"
	23,150±180	ISGS-2939	stump in organic silt below till	"
	23,110±240	ISGS-2940	"	"
	23,230±400	ISGS-3067	"	"
	23,150±240	ISGS-3068	"	"
	23,540±300	ISGS-3066	"	"

TABLE 2: RADIOCARBON AND CALENDAR AGES OF EXPANSIONS

Site	Radiocarbon Age	Calendar Age	Comments
Lake Gribben	10,030±20	11,550±25	Average of 9 new radiocarbon analyses from Lake Gribbe distal to the Grand Marais I Moraine (Drexler et al., 1983)
Two Creeks	11,850±100	13,840±125	From Brocker and Farrand, (1963), interior to the Two Rivers Moraine (Mickelson, et al., 1984)
Des Moines	13,790±23	16,250±28	Average of 12 analyses from Saylorville Dam of Bettis et al., (1985) interior to the Bemis Moraine (Hallberg et al. 1991)
Chillocothe	17,490±230	20,710±272	Youngest of a set of 11 new radiocarbon analyses from 3 sites (Biers Run 2, Dry Run, and North Fork; Table 1) representing an interstadial interior to the Owl Creek Moraine (Lowell, 1993)
Cincinnati	19,590±35	23,180±41	Average of 21 samples from 4 adjacent sites (Beckett Road, Dimmick Road, Sharonville, and Princeton High School; Table 1) all interior to the Hartwell Moraine (Lowell et al., 1990)
Oxford	20,770±59	24,540±68	Average of 10 ages on in-place stumps overrun, probably associated with the terminal position at Cincinnati. Terminal ice position unknown.
Cuba Gully	20,360±84	24,080±97	Average of 3 analyses interior to the Cuba Moraine (Dubois, 1996).
Todd Fork	23,200±86	27,290±96	Average of 6 analyses interior to the Cuba Moraine (Dubois, 1996).

ronments is a temperature rise from 0.5 to 1.0 °C (Grove, 1988), the resulting comparable retreat distances demonstrate that ablation at the margin overwhelmed a range of individual influx rates. Taken as a whole, the Little Ice Age shows several centuries of cooling that drove glacier margins to expanded positions. Superposed were shorter term ice-margin changes that built complex moraine sets just before the warming of the late nineteenth century forced the glacier margins into retreat. Since direct observation of the millennial long Little Ice Age provides insights about the relationship of temperature changes, and hence ablation, to glacier margin position, we believe that the Little Ice Age can serve as a temporal and process model for the last glacial maximum.

EXAMPLES FROM THE LATE WISCONSIN

Below we compare expansions of southern Laurentide Ice Sheet lobes with climate trends. Our primary comparison is temporal, not the size or position of the moraine. These sites, except Lake Gribben, all lie near but interior to the moraine limit of an expansion. Thus they provide an age when the ice-sheet was still expanding, but before moraines were built (Fig. 1). These sites lie in different lobes of the ice sheet and were chosen because a large number of dates were secured at each site, providing as firm a chronology as possible. The chronological resolution indicates that each event discussed below is comparable in duration to the Little Ice Age. The examples start with the Younger Dryas and progress back to the last glacial maximum. We find, within the resolution of time scale conversions, that organic sites were overrun and the maximum ice extent was reached after an extended period of cooling but before a warming event.

Expansion to Lake Gribben

Near the end of the last glacial cycle, the Lake Superior Lobe expanded to south of its present shoreline. Several authors have reported wood, ranging in age from 11,000 and 10,000 ^{14}C yr B.P., associated with a red till representing this event (Black, 1976; Clayton and Moran, 1982; Hack, 1965). Near, Marquette, Michigan, Hughes and Merry (1978) studied a buried forest bed exposed during the construction of a mine tailings pond near Lake Gribben. A 5- to 10-cm-thick acid paleosol indicates some interval of exposure during which a boreal forest developed. At this site prograding outwash sediments buried this forest as the Outer Marquette Moraine was being built (Drexler et al., 1983). Thus this age represents the end of glacial expansion. At least one younger ice-margin deposit lies interior to this position suggesting an active ice margin.

This forest was buried at the end of the Younger Dryas interval. Hughes and Merry (1978) reported several ages ranging from $9,545 \pm 225$ (Dal-340) to $10,330 \pm 300$ (W-3896) that were interpreted to average 9,900 ^{14}C yr B.P. Recently Lowell et al. (1997) obtained a new series of radiocarbon ages (Table 1) from the original samples Hughes and Merry recovered, resulting in a mean age of $10,030 \pm 20$ ^{14}C yr B.P. or $11,550 \pm 25$ cal yr B.P. (Table 2). This age lies at the end of a long interval of cooling that started about 12,860 and ended at 11,650 cal yr B.P. (Fig. 2) when the oxygen isotope record from Greenland indicates a rapid warming that started the Holocene. Thus the Superior Lobe of the Laurentide Ice Sheet was advancing to the Lake Gribben position during a cold interval and was shedding meltwater and had began to retreat just as the climate signal reversed.

Expansion to Two Creeks

Perhaps the best documented ice-margin expansion is the Two Rivers advance over the Two Creeks Forest Bed and adjacent sites in Wisconsin. Broecker and Farrand (1963) reported an age of $11,850 \pm 100$ ^{14}C yr B.P. or $13,840 \pm 125$ cal yr B.P (using the Bard et al. (1990) conversion) from organic remains; many subsequent radiocarbon analyses have confirmed this age. These analyses reflect the time when the forest drowned from rising lake levels dammed by an advancing ice margin. If the lake formed when the ice margin passed the Straits of Mackinac on the north side of Lake Michigan, the actual time the ice margin completed its expansion out to the Two Rivers Moraine would be later. The ice-core record shows a cooling interval that started about 14,510 cal yr B.P. or about 700 cal yr before the site was drowned.

Expansion to Des Moines

The Des Moines Lobe extended into central Iowa about 13,900 B.P. (Kemmis et al., 1981) to reach its farthest extent in the last glacial cycle. Refined chronology is available because a flood in 1984 overtopped the emergency spillway of the Saylorville Dam on the Des Moines River at a location about 15 km north of the Bemis Moraine (Hallberg et al., 1991). At the base of exposed sections lie thinly bedded silt loam to silt alluvium that yielded radiocarbon ages to at least $16,930 \pm 180$ ^{14}C yr B.P. (Beta-10525; Bettis et al., 1985) or $20,050 \pm 215$ cal yr B.P. Within these deposits boreal taxa suggest cooler, but not tundra conditions at the time of alluvium deposition. Above the alluvium are glaciofluvial gravels and diamictons of the Dows Formation (Bettis et al., 1985). In 1994, Bettis et al. (1985) recorded additional samples of trees at the alluvium-till contact near section O (Table 1). These yielded an average age of $13,790 \pm 23$ ^{14}C yr B.P. or $16,250 \pm 28$ cal yr B.P. (Table 2), which is slightly younger than Ruhe (1969) and Kemmis et al. (1981) report. The stratigraphy records the ice margin overrunning the site on its way to its subsequent terminal moraine position at Des Moines.

The Des Moines Lobe was expanding during a cold interval and probably reached its maximum extent at the peak of that interval. The oxygen-isotope values show a cooling trend that started about 19,170 cal yr B.P. and lasted over 2,500 years to reach maximum cold peaks at both 16,460 and 15,920 cal yr B.P. Since the Des Moines must have started its advance prior to $16,250 \pm 28$ it would have started its advance when the climate was cooled and completed its advance just before or as the major warming started (Fig. 1).

Expansion to Chillocothe

The remaining examples come from southern Ohio where the Scioto and Miami sublobes built numerous, closely spaced moraines which are of various ages and exhibit some crosscutting relationships (Dreimanis and Goldthwait, 1973). These moraines represent expansions from 17,500 to 23,150 ^{14}C yr B.P. during the last glacial maximum. Several sites in the central part of the Scioto Lobe constrain the age of the youngest of these expansions (Lowell, 1993). The most distinct ice margin (local name Owl Creek Moraine Complex) consists of at least eleven ridge crests, some of which exceed 30 m in height and stretch for at least 35 km to block a large valley containing Paint Creek. These ridges extend up the west side of the valley, where the drift limit is traced to the west and eventually into the Lattaville Moraine Complex of Quinn and Goldthwait (1985). The westward extensions onto bedrock topography indicate a steep ice-sheet profile.

Transported and in-place trees and organic material from several closely spaced sites just inside the Owl Creek Moraine Complex yield ages from 17,500 to at least 18,800 ^{14}C yr B.P.

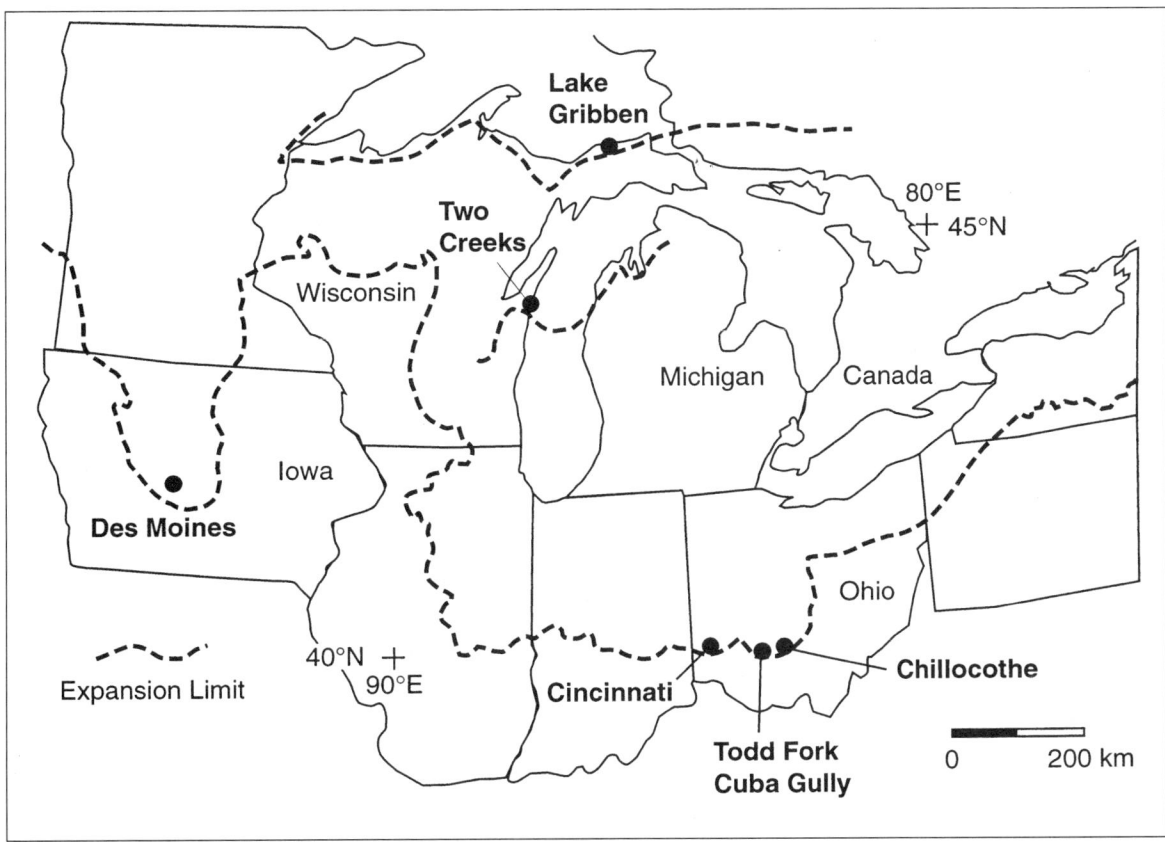

Figure 1. Index map showing outline of the southern lobes of the Laurentide Ice Sheet. Limits of expansions associated with the sites discussed in text shown. Not all margins are the same age. Many other margins between those shown here.

(Table 1). The transported samples probably originated in materials from different stratigraphic levels within an interstadial bed, therefore this range of ages likely represents the duration of an interstadial. We take the two youngest ages of this set (17,490 ± 230, UGa-6698; and 17,590 ± 280, UGa-6708) to bracket the glacial advance. Converted, this age is 20,710 ± 272 cal yr B.P. (Table 2). The ice-core record shows a cooling period that started about 23,300 and peaked at 21,310 cal yr B.P. (Fig. 2). In other words, the accumulation of organic material started just after a climatic cooling began and the Owl Creek Moraine Complex formed just after the peak of cooling.

Expansion to Cincinnati

The Hartwell Moraine marks the outer margin of the Laurentide Ice Sheet in the Miami Sublobe. This moraine is dated using several organic-rich sites within 5 km of the moraine. For example, at Sharonville, a basal till unit below ablation gravels has been thrust and stacked over a forest bed, indicating that a glacier moved over this horizon at least twice (Savage and Lowell, 1992).

Five sites (Table 1) indicate that the ice margin reached its terminal position about 19,590 ± 35 ^{14}C yr B.P. or 23,180 ± 41 cal yr B.P. This is the average of all sites and provides a refinement to Lowell et al. (1990), who reported an age of 19,700 ^{14}C yr B.P. from one site. One of the adjacent alluvium sequences (Creek Road, Table 1) provides basal ages of 22,730 ± 460 (ISGS-2837) and 22,800 ± 330 (ISGS-2829), which rule out the presence of any ice margin at this location from 26,840 ± 370 until it was overrun at 23,180 ± 41 cal yr B.P. The ice core records an overall cooling from 27,380 to a peak at 23,980 cal yr B.P. Or the same interval as the alluvium sequence was capped by till. Moreover, a glacial expansion was underway at 20,770 ^{14}C yr B.P. or 24,540 ± 68 cal yr B.P. in the Miami Sublobe. Several in-place stumps up-flowline at Oxford were killed at this time (Table 1). Thus Oxford was overrun during this cooling and the Hartwell Moraine was built just after this long cooling interval.

Expansion to Cuba

The Cuba Moraine complex of the Scioto Sublobe contains evidence for at least two expansions of the Laurentide Ice Sheet at its southern limit (Dubois, 1996). Sites reported in this and the next section all lie within 5 km of the limit of late Wisconsin drift. Ages gathered interior to the Cuba Moraine indicate that organic material was accumulating until at least 20,870 ± 130 ^{14}C yr B.P. (ISGS-3235) with the glacier margin overrunning trees as it reached the Cuba Gully site about 20,360 ± 84 ^{14}C yr B.P. or 24,080 ± 97 cal yr B.P. (Table 2). This advance, the second to

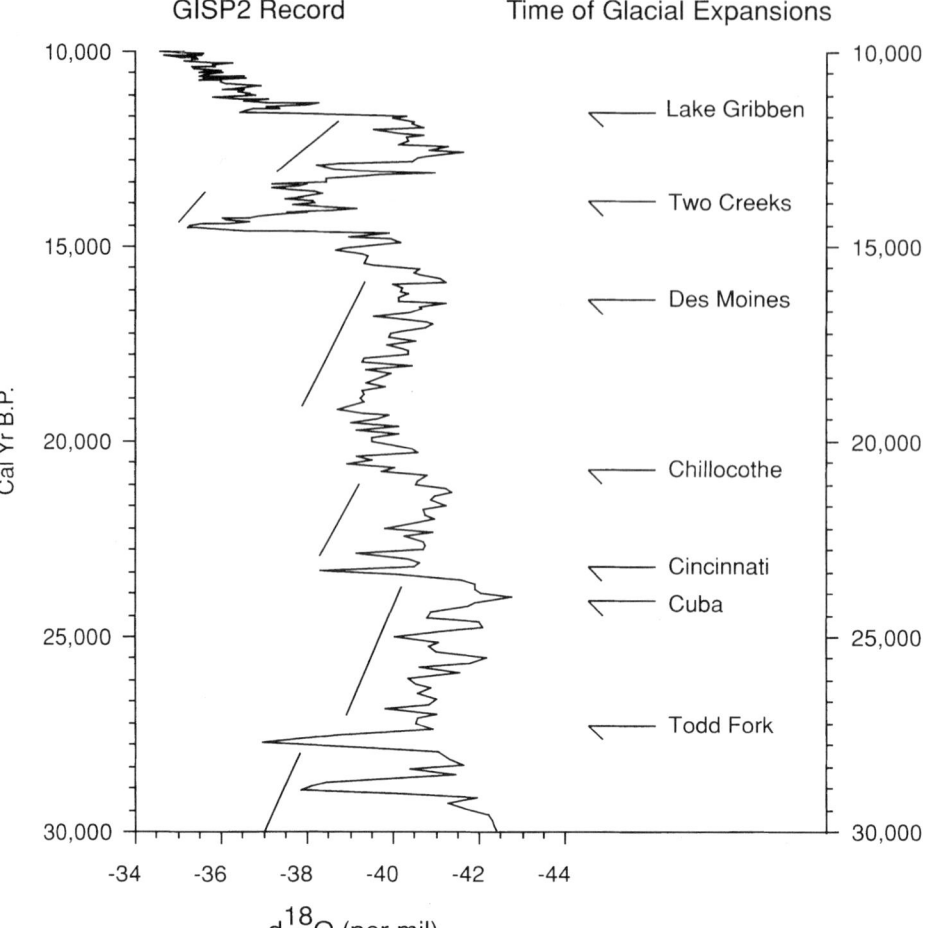

Figure 2. Comparison of the GISP2 ice-core record with the timing of glacial expansions along the southern margin of the Laurentide Ice Sheet. All of these expansions reflect the time the ice margin reached its terminal position and sloping lines along the oxygen isotope curve reflect the duration of the cooling associated with each expansion.

reach the Cuba Moraine, moved onto a very thin organic sequence. Parts of the Cuba Moraine were thus built at the peak of the cooling cycle that lasted from 27,380 to 23,980 cal yr B.P. We attribute this expansion to the same cooling interval that drove the adjacent Miami Sublobe forward. These examples show that one cooling interval can produce expansions that peak at slightly different times, much as different maximum expansions were recorded during the Little Ice Age (Grove, 1988).

Expansion to Todd Fork

Another site within the Cuba Moraine complex records the first expansion of the Laurentide Ice Sheet to its maximum extent. The Todd Fork site lies interior to the outermost Cuba Moraine and probably records overrunning as the glacier margin extended to less than 2 km from its maximum expansion (Dubois, 1996). Trees and other organic remains on deeply weathered drift yielded a mean age of 23,200 ± 86 ^{14}C yr B.P. or 27,290 ± 96 cal yr B.P. following the Bard et al. (1990) conversion (Table 1). The ice-core record in this time interval indicates a complex climate signal that includes relatively short duration warm intervals. The Todd Fork site was overrun at the end of the long cooling interval associated with the H3 event of Bond et al. (1997). As in the Cincinnati example, the moraine formed at this site in the warm portion of a climate cycle.

Summary of comparisons

The one-to-one matching of the major climate cycles recorded in the ice core with dated glacial expansions of the southern side of the Laurentide Ice Sheet suggests a linkage. This set of examples is intended to establish an overall pattern, not to argue that there is a perfect correspondence between the ice-core record and glacier margins. The inherent uncertainties in present dating techniques prevent this. However, we argue that consistent, repetitive occurrence of the glacial expansions starting during long cooling intervals and ending with moraine building consistent with the peak of every cooling interval implies a casual mechanism. We take this overall pattern to be representative, as we have no reason to assume that each of the reported ice sheet expansions were driven by different processes, and explore its implications.

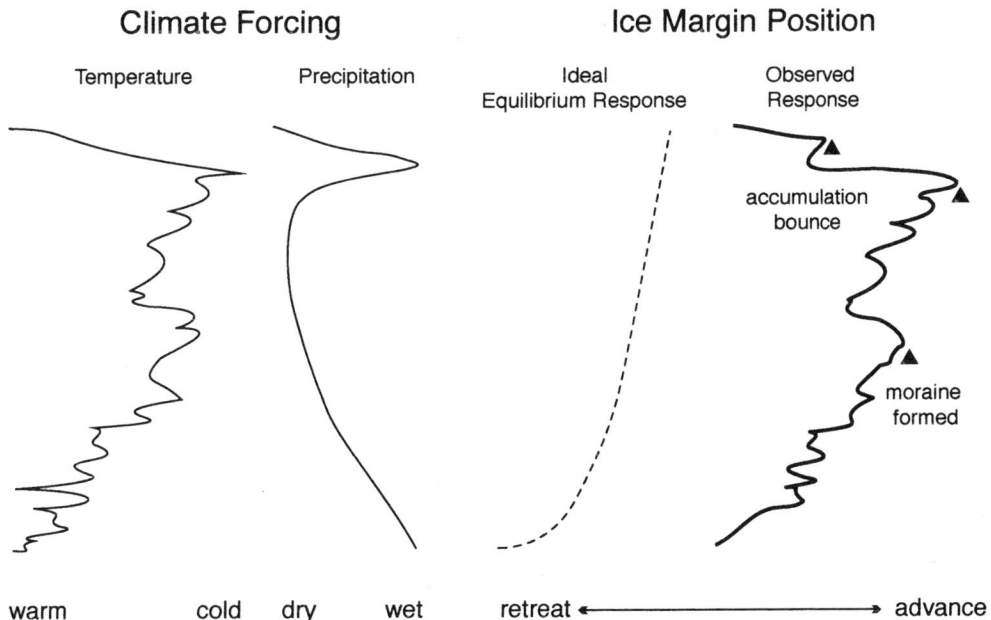

Figure 3. Schematic relationship of climate forcing and ice-margin position. Ablation effects are instantaneous whereas accumulation effects are laged proportional to the length of the flow line. The dashed line in the ice-margin position trace shows equilibrium position that would be achieved if the climate signal did not reverse itself. The bounce is a ice-margin readvance driven by the arrival of the accumulation flux and may vary it's expression in temporal position or magnitude.

DISCUSSION AND IMPLICATIONS

Why does it work?

Glacier systems are a good example of a system in dynamic equilibrium. Two of the primary controls on glacier dynamics, accumulation and ablation, are climate driven but have different response times. The effects of mass balance surpluses must be transmitted from the accumulation areas to the ablation areas before any adjustments of the glacier margin can take place. This lag time varies approximately with the size of the glacier so the lag may reach several thousand years for ice sheets. But during the last glacial cycle, climate was only static for about 1,000 years (Grootes et al., 1993). Existing models suggest that changes in the basal conditions also require more than 6,000 years to be effective (MacAyeal, 1993).

However, the climate impact on the ablation system is far more rapid (Fig. 3). One or several colder than normal summers will reduce the amount of ablation at the ice margin and, in response, the ice margin will advance. An extended cooling interval will act like a carrot on a stick—it will tease the ice margin out farther and farther. As the ice margin approaches an equilibrium position, its margin advance rate will slow. Other factors (e.g., topography, subglacial sediment type, existing moraines) combine to determine the spatial limit of this advance, but the start of low ablation intervals will dictate when the advance begins. The situation is considerably different when climate warms up. High summer melting or a longer ablation season will force the ice margin backward and the winter cooling cannot make up the lost ground. This response is rapid: one warm summer and the glacier is in retreat. How fast that retreat takes place depends on the balance between ablation rate, ice-volume present below the equilibrium line, and mass flux into the ablation area.

Ablation rate is only one factor in determining the position and type of any resulting moraine. However, we feel that is the primary one to predict when ice margins advance and retreat and hence the age of the moraine. It might be argued that the age of many moraines does not correspond with any climate signal; that is a difficult point to support or refute until sufficient age control for each moraine is established. The examples reported here show that the best dated terrestrial moraines in the Great Lakes Sector of the Laurentide Ice Sheet correspond to climate oscillations.

Possible tests

We propose three tests to determine the sensitivity of ice-margin position to ablation.

First, we advocate detailed comparisons of the ice-margin position with local to regional climate conditions. For example, high-resolution pollen cores could be used at sites adjacent to locations where the ice-margin chronology can be established. The sediments must represent stratigraphic intervals sufficiently older than the margin advance to firmly establish the general climate trend. Unfortunately, changes in ablation may be sensitive to temperature changes less than 1 °C, and those changes can be difficult to extract from the pollen records.

Second, we might ask how long it takes to form moraines. In the view held here they are rapid events, decades to centuries. Evidence from the Little Ice Age supports this view. The age of individual moraine ridges found within a moraine belt by Denton and Karlén (1973) could not be resolved for late Wisconsin moraines. Indeed the Little Ice Age seems to be about the duration between major expansions of the ice sheet (Mickelson et al., 1983; Bond, 1997). Since decade stratigraphic information will be difficult to recover from the geologic record, modeling experiments may substitute. If sediment flux rate studies show that volumes of existing moraines volumes require longer than one climate swing (millennium) to accumulate, it implies the ice margin must be stationary for that long. Since this duration is longer than the climate changes as recorded in the Greenland ice core, the margin can not be responding to climate changes.

A third test consists of modeling experiments designed to identify the sensitivity of the ice margin to small changes in ablation rates. Observed mass-balance data and modeling experiments from Greenland indicate that warming temperatures of 1 °K increase ablation 0.5 m water equivalent/year (Zuo and Oerlemans, 1996). Although this would be specific to the profile of the Greenland Ice Sheet, it probably represents a lower limit for the ice margins around the southern margin of the Laurentide Ice Sheet. A temperature change acting across the extensive surface area presented by these very low profile lobes would inflict a higher change in ablation than estimated for Greenland. Can a 1 °K temperature change force the Laurentide lobes?

Implications

The above implies:
- The timing of ice sheet advances corresponds to cold or cooling intervals when ablation is low.
- The age of many ice marginal landforms correspond to switches from cold intervals to warmer climate conditions and corresponding higher ablation.
- Ablation processes and effects must be more extensively considered in ice-sheet models especially those attempting to reconstruct the time evolution of ice sheets. This is a departure from an approach that considers the physical properties within the ice sheet to be the primary factors in ice-sheet dynamics.

ACKNOWLEDGMENTS

Lee Clayton started us thinking about this problem when he suggested that the Lake Gribben beds were too young to associated with the Younger Dryas event. Support from the NSF grant ERA-9205703 provided many of the radiocarbon analysis reported here. Art Bettis pointed us to and helped sample the Saylorville site. Patrick Colgan and Mark Johnson asked critical questions, pointed out weak arguments, and thus greatly improved the manuscript.

REFERENCES CITED

Bard, E., Hamelin, B., Fairbanks, R. G., and Zindler, A., 1990, Calibration of the ^{14}C timescale over the past 30,000 years using mass spectrometric U-Th ages from Barbados corals: Nature, v. 345, p. 405–410.

Bettis, E. A. I., Kemmis, T. J., Witzke, B. J., Howes, M. R., Quade, D. J., Littke, J. P., Hallberg, G. R., Baker, R. G., and Frest, T. J., 1985, After the Great Flood: Exposures in the Emergency Spillway, Saylorville Dam: Geological Society of Iowa, p. 2-1–2-42.

Black, R. F., 1976. Quaternary geology of Wisconsin and contiguous Upper Michigan, in Mahaney, W. C., ed., Quaternary stratigraphy of North America: Stroudsburg, Pennsylvania, Dowden, Hutchinson & Ross, Inc., p. 93–117.

Bond, G., Broecker, W., Johnsen, S., McManus, J., Labeyrie, L., Jouzel, J., and Bonani, G., 1993, Correlations between climate records from North Atlantic sediments and Greenland ice: Nature, v. 365, p. 143–147.

Bond, G., Showers, W., Cheseby, M., Lotti, R., Almasi, P., deMenocal, P., Priore, P., Cullen, H., Hajdas, I., and Bonani, G., 1997, A pervasive millennial-scale cycle in North Atlantic Holocene and glacial climates: Science, v. 278, p. 1257–1266.

Broecker, W. S., and Farrand, W. R., 1963, Radiocarbon age of the Two Creeks Forest Bed, Wisconsin: Geological Society of America Bulletin, v. 74, p. 795–802.

Clark, P. U., Licciardi, J. M., MacAyeal, D. R., and Jenson, J. W., 1996, Numerical reconstruction of a soft-bedded Laurentide Ice Sheet during the last glacial maximum: Geology, v. 24, no. 8, p. 679–682.

Clayton, L., and Moran, S. R., 1982, Chronology of late Wisconsinan glaciation in Middle North America: Quaternary Science Reviews, v. 1, p. 55–82.

Denton, G. H., and Karlén, W., 1973, Holocene climatic variations—their pattern and possible cause: Quaternary Research, v. 3, p. 155–205.

Denton, G. H., and Karlén, W., 1977, Holocene glacial and tree-line variations in the White River Valley and Skolai Pass, Alaska and Yukon Territory: Quaternary Research, v. 7, p. 63–111.

Dreimanis, A., and Goldthwait, R. P., 1973, Wisconsin glaciation in the Huron, Erie, and Ontario Lobes: Geological Society of America Memoir 136, p. 71–106.

Drexler, C. W., Farrand, W. R., and Hughes, J. D., 1983, Correlation of glacial lakes in the Superior Basin with eastward discharge events from Lake Agassiz: Geological Association of Canada, v. 26, p. 309–332.

Dubois, M., 1996, The Late Wisconsin Cuba Moraine: It's age and implications for ice sheet behavior [M.S. thesis]: University of Cincinnati, 53 p.

Ekberg, M. P., Lowell, T. V., and Stuckenrath, R., 1993, Late Wisconsin glacial advance and retreat patterns in southwestern Ohio, USA: Boreas, v. 22, p. 189–204.

Grootes, P. M., Stuiver, M., White, W. C., Johnsen, S., and Jouzel, J., 1993, Comparison of oxygen isotope records from the GISP2 and GRIP Greenland ice cores: Nature, v. 366, p. 552–554.

Grove, J. M., 1988, The Little Ice Age: London, Methuen, 498 p.

Hack, J. T., 1965, Postglacial drainage evolution and stream geometry in the Ontonagon area, Michigan: Washington, D.C., United States Government Printing Office, Geological Survey Professional Paper 504-B, Shorter Contributions to General Geology, p. B1–B40.

Hallberg, G. R., and 10 others, 1991, Quaternary geologic map of the Des Moines 4° × 6° Quadrangle, United States: U.S. Geological Survey, scale 1:1,000,000.

Hughes, L., and Merry, W. J., 1978, Marquette buried forest 9,850 years old: American Association for the Advancement of Science, Abstract for Annual Meeting, no. 12–14 February 1978.

Johnsen, S. J., Clausen, A. B., and Dansgaard, W., 1992, Irregular glacial interstadials recorded in a new Greenland ice core: Nature, v. 359, p. 311–313.

Kemmis, T. J., Hallberg, G. R., and Lutenegger, A. J., 1981, Iowa Geological Survey Guidebook: Depositional environments of glacial sediments and landforms on the Des Moines Lobe, Iowa: Iowa Geological Survey, 139 p.

Lowell, T. V., 1993, The late Wisconsin boundary of the Scioto Sublobe near Chillicothe, Ohio: Geological Society of America, Abstracts with Program,

v. 25, no. 2, p. 34.

Lowell, T. V., and Brockman, C. S., 1994, Quaternary sediment sequences in the Miami Lobe and environs; Midwest Friends of the Pleistocene Annual Meeting: Cincinnati, University of Cincinnati, 68 p.

Lowell, T. V., Savage, K. M., Brockman, C. S., and Stuckenrath, R., 1990, Radiocarbon analysis from Cincinnati, Ohio and their implications for glacial stratigraphic interpretations: Quaternary Research, v. 34, p. 1–11.

Lowell, T. V., Heusser, C. J., Andersen, B. G., Moreno, P. I., Hauser, A., Heusser, L. E., Schluchter, C., Marchant, D. R., and Denton, G. H., 1995, Interhemispheric correlation of late Pleistocene glacial events: Science, v. 269, p. 1541–1549.

Lowell, T. V., Larson, G., Dubois, M., and Denton, G., H, 1997, The role of rapid climate oscillations in determining ice margin position: examples from Ohio and Michigan: Abstracts with Programs, Geological Society of America, v. 29, no. 4, p. 32.

MacAyeal, D. R., 1993, Binge/purge oscillations of the Laurentide Ice Sheet as a cause of the North Atlantic's Heinrich events: Paleoceanography, v. 8, no. 6, p. 775–784.

Messerli, B., Messerli, P., Pfister, C., and Zumbühl, H. J., 1978, Fluctuations of climate and glaciers in the Bernese Oberland, Switzerland, and their geoecological significance, 1600 to 1975: Arctic and Alpine Research, v. 10, p. 247–260.

Mickelson, D. M., Clayton, L., Fullerton, D. S., and Borns, H. W., Jr., 1983, The late Wisconsin glacial record of the Laurentide Ice Sheet in the United States, *in* Porter, S. C., ed., Late Quaternary environments of the United States: Volume 1: Minneapolis, University of Minnesota Press, p. 3–37.

Mickelson, D. M., Clayton, L., Baker, R. W., Mode, W. H., and Schneider, A. F., 1984, Pleistocene stratigraphic units of Wisconsin: Wisconsin Geological and Natural Survey Miscellaneous Paper 84-1, 107 p.

Oerlemans, J., and van der Veen, C. J., 1984, Ice sheets and climate: Boston, Dordrecht, 217 p.

Quinn, M. J., and Goldthwait, R. P., 1985, Glacial geology of Ross County, Ohio: State of Ohio Geological Survey, Report of Investigations 127, 42 p.

Ruhe, R. V., 1969, Quaternary landscapes in Iowa: Ames, The Iowa State University Press, 255 p.

Savage, K. M., and Lowell, T. V., 1992, Dynamics of the marginal late Wisconsin Miami Sublobe, Cincinnati, Ohio: Ohio Journal of Science, v. 92, p. 107–118.

Stuiver, M., and Reimer, P. J., 1993, Extended ^{14}C data base and revised CALIB 3.0 ^{14}C age calibration program: Radiocarbon, v. 35, p. 215–230.

Ward, G. K., and Wilson, S. R., 1978, Procedures for comparing and combining radiocarbon age determinations: a critique: Archaemometry, v. 20, no. 1, p. 19–31.

Zuo, Z., and Oerlemans, J., 1996, Modelling albedo and specific balance of the Greenland Ice Sheet: calculations for the Sömfjord transect: Journal of Glaciology, v. 42, no. 141, p. 305–317.

MANUSCRIPT ACCEPTED BY THE SOCIETY OCTOBER 8, 1998